Microscopy Methods for Virus Research

Microscopy Methods for Virus Research

Guest Editor

Krishanu Ray

MDPI

Basel • Beijing • Wuhan • Barcelona • Belgrade • Novi Sad • Cluj • Manchester

Guest Editor
Krishanu Ray
Institute of Human Virology
University of Maryland
School of Medicine
Baltimore
USA

Editorial Office
MDPI AG
Grosspeteranlage 5
4052 Basel, Switzerland

This is a reprint of the Special Issue, published open access by the journal *Viruses* (ISSN 1999-4915), freely accessible at: https://www.mdpi.com/journal/viruses/special_issues/N3899663PM.

For citation purposes, cite each article independently as indicated on the article page online and as indicated below:

Lastname, A.A.; Lastname, B.B. Article Title. *Journal Name* **Year**, *Volume Number*, Page Range.

ISBN 978-3-7258-8163-5 (Hbk)
ISBN 978-3-7258-8164-2 (PDF)
https://doi.org/10.3390/books978-3-7258-8164-2

Contents

About the Editor

Krishanu Ray

Krishanu Ray is an Associate Professor at the Institute of Human Virology at the University of Maryland School of Medicine. His research interests center on developing advanced fluorescence, nanotechnology, and imaging-based approaches for biomedical applications. Currently, his laboratory focuses on applying advanced imaging and spectroscopic methods to study HIV-1 and SARS-CoV-2 proteins and virions, and on developing novel assays to quantify the binding dynamics of broadly neutralizing antibodies in complex biological fluids.

viruses

MDPI

Editorial

Special Issue: Microscopy Methods for Virus Research

Krishanu Ray [1,2]

[1] Division of Vaccine Research, Institute of Human Virology, University of Maryland School of Medicine, 725 West Lombard Street, Baltimore, MD 21201, USA; kray@som.umaryland.edu
[2] Department of Biochemistry and Molecular Biology, University of Maryland School of Medicine, 725 West Lombard Street, Baltimore, MD 21201, USA

Significant advances in microscopy methods for biomedical applications have been made in the past two decades, and these methods are now being actively pursued for virus research. Implementing quantitative advanced microscopic methods, including optical and electron microscopy imaging, can provide important information regarding the mechanism and dynamics of viral infection and identify targets for therapeutic intervention. These methods are opening new avenues for research in microbiology, particularly with respect to viruses. Furthermore, improvements in microscopy methods have allowed researchers to directly examine single viruses in their natural forms, generating new insights into their structure, function, and mechanism of action over time and/or in three dimensions. Specifically, recent progress in several super-resolved fluorescence microscopy, single molecule, and in vivo optical imaging methods, as well as novel fluorescence tags and electron microscopy methods including cryoEM and cryoET, opens the possibility of tracking viruses in host–pathogen interactions and evaluating dynamics at the molecular level. This is fundamental to providing unprecedented information for the development of interventions and therapeutic strategies. This Special Issue of *Viruses* brings together seven original research articles, two reviews and one CryoVirus data descriptor, highlighting the progress of imaging in the broad area of virus research including HIV-1 and SARS-CoV-2.

The contributions are diverse in scope: one review discussed RNA viral infections with multiscale microscopic methods, while the other review provided a snapshot of cryoEM for bridging the structural gap for the underrepresented viruses. One research article reported a flow cytometry-based application towards quantification of a viral infection. A solution-based quantitative fluorescence correlation spectroscopy approach conducted in-depth analyses on the differences between the antigenic properties of the soluble SARS-CoV-2 spike proteins versus the virion-associated envelope spikes. This study underscores the importance of studying viruses in their native, virion-associated states. A report on the multiplex microscopy assay evaluated the functioning of therapeutic and serum antibodies against emerging pathogens. Another article on quantitative fluorescence microscopy reported on the interactions between hantavirus nucleoprotein and glycoproteins. Additionally, the imaging of retroviral RNA genome heterodimers was achieved by applying the fluorescence complementation method. The application of super-resolution microscopy showed the responses of nuclear pores to mechanical stress. Finally, the use of expansion microscopy provided a new lens for studying HIV-1–cell interactions.

Briefly, the contributions in this Special Issue underscore that the future of virology is inherently multi-modal. Moreover, these articles point out that the integration of high-resolution imaging and quantitative analysis is not just an evolution of methods, but an important step forward in our ability to visualize and combat viral threats.

Conflicts of Interest: The author declares no conflict of interest.

https://doi.org/10.3390/v18050551

viruses

MDPI

Article

Super-Resolution Imaging of Nuclear Pore Responses to Mechanical Stress and Energy Depletion

Dariana Torres-Rivera [1], Sobhan Haghparast [2], Bernd Rieger [2] and Gregory B. Melikyan [1,3,*]

1 Department of Pediatrics, Emory University, Atlanta, GA 30322, USA; dariana.torres.rivera@emory.edu
2 Faculty of Applied Sciences, Delft University of Technology, 2628 Delft, The Netherlands; s.haghparast@tudelft.nl (S.H.); b.rieger@tudelft.nl (B.R.)
3 Children's Healthcare of Atlanta, Atlanta, GA 30329, USA
* Correspondence: gmeliki@emory.edu

Abstract

HIV-1 entry into host cells culminates in integration of the reverse transcribed double-stranded viral DNA into host genes. Several lines of evidence suggest that intact, or nearly intact, HIV-1 cores—large, ~60 nm-wide structures—pass through the nuclear pore complex (NPC), and that this passage is associated with pore remodeling. Cryo-electron tomography studies support the dynamic nature of NPCs and their regulation by cytoskeleton and ATP-dependent processes. To explore NPC remodeling, we used super-resolution Stochastic Optical Reconstruction Microscopy (STORM) of U2OS cells endogenously expressing nucleoporin 96 tagged with SNAP. Single-molecule localization imaging and computational averaging resolved 8-fold symmetric nuclear pores with an average radius of ~51 nm. Depletion of cellular ATP using sodium azide or antimycin A, previously reported to reduce the size of yeast NPCs, did not significantly alter the nuclear pore radius in U2OS cells. Similarly, stressing the nuclear envelope by hypotonic or hypertonic conditions failed to induce detectable expansion or contraction of NPCs. These results indicate that the NPCs in U2OS cells do not respond to ATP depletion nor mechanical stresses on changes in pore morphology that can be resolved by STORM. Since these cells are infectable by HIV-1, we surmise that direct multivalent interactions between HIV-1 capsid and phenylalanine-glycine nucleoporins lining the pore's interior drive the core penetration into the nucleus and the associated changes in the pore structure.

Keywords: nuclear pore complex; single-molecule localization microscopy; human immunodeficiency virus; osmotic stress; ATP depletion

1. Introduction

The nuclear pore complex (NPC) is a ~120 MDa multiprotein complex that spans the nuclear envelope (NE) and controls the transport of multiple proteins in and out of the nucleus [1–4]. The NPC is composed of nucleoporins (NUPs), which are classified as scaffold NUPs and phenylalanine-glycine-NUPs (FG-NUPs). The scaffold NUPs are structural proteins that form the pore and are directly attached to the NE [2–5]. The Y-complex is a part of the scaffold and consists of NUP96, NUP107, and NUP133, among other NUPs [2,5]. The FG-NUPs are flexible NUPs that contain FG repeats, which generates a liquid-phase separation inside the inner ring of the pore that serves as a selective filter for transport of host factors [2,6,7].

The NPC has been previously considered a static structure that allowed the passage of molecules in/out of the nucleus in a size-dependent manner. Small molecules permeate the pore through passive diffusion, while larger molecules require active transport to enter in the nucleus, which is mediated by transport factors like Importin β [2,3,8,9]. Passive diffusion through the NPC is generally limited for molecules larger than 60 kDa [2,3]. Furthermore, CryoEM imaging of isolated nuclei suggested that the diameter of the central pore was ~40 nm [5]. However, recent evidence using Cryo-FIB milling and imaging native, *in cellulo* pore structures has revealed that NPCs of mammalian cells have larger inner pore diameter, close to 64 nm [10–12]. In addition, Zimmerli et al. [13] have shown that yeast NPCs are dynamic, and that ATP is required to maintain an expanded state of the pore. In addition, this study and a recent work using expansion fluorescence microscopy have also found that hyperosmotic shock reduces the NPC size [13,14], suggesting that the nuclear envelope tension can modulate the pore size. It is worth noting that the pore size can be heterogeneous within the same cell, and that NPCs located at the bottom and top of the NE have been reported to have different diameters [14]. Neuronal cell differentiation and *Xenopus* oocyte development can also promote changes in the NPC architecture [15–17]. All these findings support the notion that the NPC is a dynamic structure that may, in principle, allow the passage of very large macromolecular complexes.

Perhaps one of the largest cargoes that can pass through the NPC is the cone-shaped HIV-1 capsid core, which measures ~60 nm at the widest end [10,18]. The presence of cone-shaped HIV-1 cores inside the nucleus of infected cells supports the ability of intact or nearly intact capsid cores to penetrate the NPCs [10,18–23]. Interestingly, a fraction of NPCs appeared deformed or cracked in the vicinity of HIV-1 capsids inside the nucleus [18,19,24]. These studies support a flexible model for the NPC that can facilitate the passage of very large macromolecular complexes, like the HIV-1 capsid core.

Based on the above findings, we hypothesize that the nuclear pore may respond to mechanical stimuli and/or require ATP to sustain an enlarged state. To explore the NPC dynamics, we used U2OS cells endogenously expressing NUP96-SNAP and performed Single Molecule Localization (SML) STORM microscopy to visualize the NPC architecture. We first confirmed that U2OS NUP96-SNAP cells are readily infectable by VSV-G pseudotyped HIV-1, and that the nuclear import of viral pre-integration complexes is not delayed compared to parental U2OS cells. Depletion of ATP with sodium azide or Antimycin A did not significantly affect the NPC radius. Similarly, mechanically stressing the nuclear envelope by osmotic swelling or by hyperosmotic shock did not cause a detectable increase or decrease of the NPC diameter, respectively. Taken together, nuclear pore size in U2OS cells appears to be tolerant to both ATP depletion and mechanical stresses. Comparing our data with previous studies reporting changes in NPC morphology [10,13,19,25,26], we surmise that the limited spatial resolution of SML microscopy (around 13 nm) and NPC heterogeneity may mask the potential effects of NPC stretching and energy depletion on their size. Additionally, differences in cell types or their physiological state can contribute to NPC's responses to mechanical stimuli.

2. Methods

2.1. Cells and Reagents

U2OS was obtained from Philip Santangelo (Georgia Institute of Technology, Atlanta, GA, USA), HEK293T/17 cells from ATCC (#CRL11268, Manassas, VA, USA), and TZM-bl cells from the BEI Resources (NIAID, NIH; HRP-8129, Manassas, VA, USA). U2OS NUP96-SNAP clone #33, in which SNAP was knocked-in at the C-terminus of NUP96 using CRISPR/Cas9 technology, was obtained from Cell Lines Services (CLS GmbH #300444, Eppelheim, Germany; see Thevathasan et al., 2019 [27]). All cells were maintained in Dulbecco's modified

https://doi.org/10.3390/v18020167

Eagle's medium (DMEM, Corning, New York, NY, USA, #10-013-CV), supplemented with 10% Fetal Bovine Serum (FBS, Hyclone, #SV30014.03, Logan, UT, USA) and 100 U/mL of Penicillin/Streptomycin (P/S, Gemini Bio-products #400-109, West Sacramento, CA, USA) at 37 °C, 5% CO_2. HEPES-based live cell imaging buffer was obtained from Invitrogen (#A14291DJ, Waltham, MA, USA). Sorbitol (PHR1006) and reagents used for ATP depletion, 2-deoxy-D-glucose (2-dDG, #D8375-1G), antimycin A (Ant-A, #A8674), and sodium azide (NaN_3, #S8032) were obtained from Sigma (St. Louis, MO, USA).

2.2. Production and Characterization of HIV-1 Pseudoviruses

To produce luciferase reporter viruses, HIV-1 pseudoviruses were made by transfecting HEK293T/17 cells with pVSV-G (0.5 µg, a gift from Mamuka Kvaratskhelia, U. Denver) and pNL4-3 R-E Luciferase (2 µg, from Anna Cereseto, University of Trento, Trento, Italy). For producing fluorescent viruses, cells were transfected with pVSV-G (0.5 µg), NL4-3 eGFP (2 µg, from Chris Aiken, Vanderbilt, Nashville, TN, USA), and Integrase-superfolder GFP (pVpr-PC-IN-sfGFP, 0.8 µg; from Anna Cereseto). Transfection was done using JetPrime reagent (PolyPlus transfection, #114-15), according to the manufacturer's instructions. Briefly, plasmids were mixed at 1:2 DNA:JetPrime ratio in 200 µL of JETPrime Buffer and added to cells. Transfection medium was changed to complete DMEM medium without Phenol Red (Gibco, ThermoFisher, #31053-028, Baltimore, MD, USA) at 6 h after transfection, and viruses were collected after 48 h. Cell supernatant was collected and filtered with 0.45 µm of polyethersulfone Nalgene syringe filters (ThermoFisher, #725-2545, Waltham, MA, USA), aliquoted and stored at −80 °C. Viruses were characterized by measuring reverse transcriptase activity (RT) using qPCR [28]. This method measures RT activity as a reference for mature virions.

For PF74 time-of-addition assay, 1×10^4 cells/well of TZM-bl, U2OS, and U2OS NUP96-SNAP cells were seeded on a 96-well black clear-bottom plate (Corning #3603) and cultured overnight. NL4-3 Luciferase-expressing pseudovirus in complete DMEM medium were added to cells (0.8 RTU/mL) and spinoculated at 1550× g for 30 min, 4 °C. Medium was changed to complete DMEM, and cells were incubated at 37 °C. Two µM of PF74 (PF-3450074, Sigma #SML0835, St. Louis, MO, USA) were added at indicated times post-infection (with a 2 h interval) up to 24 h.p.i. Luciferase activity was measured at 48 h.p.i. by mixing 1:1 Luciferase reagent Bright Glow (Promega #2620, Wisconsin, IL, USA) with 1X Glo Lysis Buffer (Promega #E2661, Madison, WI, USA). Luciferase signal was measured on a PerkinElmer TopCount NXT (Shelton, CT, USA) luminescence plate reader.

To measure nuclear import of fluorescently labeled viruses, TZM-bl, U2OS, and U2OS NUP96-SNAP cells were seeded on collagen-coated 8-well coverslips (Nunc Brand #177402, ThermoFisher, Waltham, MA, USA) at 4×10^4 cells/well and cultured overnight. Cells were then infected with NL4-3eGFP INsfGFP viruses (0.65 RTU/mL) in complete DMEM by spinoculation at 1550× g for 30 min, 4 °C. Medium was changed with fresh complete DMEM, and cells were incubated at 37 °C for 4 h. Cells were fixed, immunostained for Lamin with Rabbit Anti-Lamin B1 (AbCam #16048, Cambridge, UK) and secondary AlexaFluor 647 Goat anti-rabbit IgG (H + L) (ThermoFisher #A21245), and imaged on a Zeiss LSM880 confocal microscope (Carl Zeiss Group, Oberkochen, Germany).

2.3. ATP Depletion

For ATP depletion experiments, 50 µL of 5×10^5/mL U2OS NUP96-SNAP cells were seeded in 96-well black clear-bottom plates and cultured overnight. Next day, cells were washed with Phosphate Buffer Saline without calcium and magnesium ($PBS^{-/-}$, Corning #21-040-CV) and incubated with complete medium (DMEM, 10% FBS and 1% P/S, abbreviated (+)Gluc), glucose-free medium (Live cell imaging buffer, 10% FBS

https://doi.org/10.3390/v18020167

and 1% P/S, abbreviated (−)Gluc), or ATP depletion medium for different times. ATP depletion medium was made by supplementing (−)Gluc buffer with 10 mM of NaN_3 and 6 mM of 2-dDG, or with 10 µM of Antimycin A and 20 mM of 2-dDG. Cell's ATP levels were determined using the PerkinElmer ATPlite assay kit (#6016943), following manufacturer's instructions. Briefly, ATP standards were resuspended and serially diluted 1:10. Cells were then lysed with lysis solution from the kit and shaken for 5 min at 7000 rpm. After lysis, substrate solution was added to cells and shaken for an additional 5 min at 7000 rpm. Lysates were then incubated for 10 min in dark. The resulting luciferase activity was measured using the PerkinElmer TopCount NXT luminescence plate reader.

2.4. Osmotic Stretching and Shrinkage of Nuclear Envelope

For osmotic stretching of NE, U2OS NUP96-SNAP cells were seeded in 35 mm collagen-coated dishes (Corning # 430165) and cultured overnight. Cells were then transiently transfected with 0.2 µg of pSB-CMV-MCS-Puro-cPLA2-EGFP (Addgene #162568, Watertown, MA, USA) with 1:2 DNA:JETPrime reagent and mixed in 200 µL of JETPrime Buffer. After 6 h, transfection medium was replaced with complete DMEM medium supplemented with 3 µg/mL aphidicolin (Millipore-Sigma #178273-1MG, Burlington, VT, USA), to prevent cell division, and cultured overnight at 37 °C. To promote nuclear envelope swelling, cells were treated according to Shen et al. [29] and Enyedi et al. [30]. Briefly, cells were permeabilized for 5 min at 37 °C with 25 µg/mL of digitonin (#43065-0.1; Research Product International (RPI); Mount Prospect, IL, USA) in the presence of 5% Polyvinylpyrrolidone (PVP360, Sigma #9003-39-8) dissolved in base medium (BM; 123 mM of KCl, 12 mM of NaCl, 1.94 mM of $MgCl_2$, 0.28 mM of $CaCl_2$, 1 mM of EGTA pH 7.2 (Sigma #E3889-25G), 10 mM of HEPES (ThermoFisher Scientific #15630080)). Next, medium was changed to BM containing 5% or 0% PVP360 and incubated for 5 min or 15 min at 37 °C, respectively. Cells were then fixed with 4% Paraformaldehyde (PFA, Electron Microscopy Sciences #15710, Hatfield, PA, USA) for 10 min, and PFA activity was quenched with 20 mM Tris pH 8.0 (G-Biosciences #R002, St. Louis, MO, USA) for 5 min at room temperature. Samples were blocked with 20% FBS for 30 min and stained with 0.5 µM SNAP-Surface AF647 (New England Biolabs #S9136S, Ipswich, MA, USA) with 1 µM of dithiothreitol (DTT, American Bioanalytical #AB00490, Natick, MA, USA) for 1 h at room temperature.

For shrinkage of NE, U2OS NUP96-SNAP cells were seeded in collagen coated 8-well chamber coverslips with 3 µg/mL of aphidicolin incubated overnight at 37 °C. Cells were then cultured with 2 M of sorbitol (hyperosmotic condition) for 1 h at 37 °C. Cells were fixed with 2% PFA in the presence of 500 mM of sorbitol for 1 min, permeabilized with 0.1% digitonin also in the presence of 500 mM of sorbitol for 30 min and fixed again with 2% PFA for 10 min at room temperature. Fixation was quenched with 20 mM of Tris pH 8.0 for 5 min. Cells were then blocked with 20% FBS for 30 min and stained with 0.5 µM of SNAP-AF647 supplemented with 1 µM DTT.

2.5. Preparation of Samples for Imaging

For ATP depletion experiments, U2OS NUP96-SNAP cells (10^5 cells/well) were seeded on collagen-coated 8-well chamber coverslips (Nunc Brand #177402) with 3 µg/mL aphidicolin, to prevent cell division, and cultured overnight at 37 °C. Next day, cells were treated with (+)Gluc, (−)Gluc or ATP depletion medium (10 mM of NaN_3 with 6 mM 2-dDG or 10 mM Antimycin A with 20 mM 2-dDG) at indicated times. For Supplementary Figure S6, Antimycin A treated samples were co-treated with 2 µM nocodazole (Noco, Honeywell Fluka, Seelze, Germany), or left untreated, for 4 h. All samples were then pre-fixed with 2% PFA for 30 s, permeabilized with 0.1% digitonin for 30 min, and post-fixed with 2% PFA for

https://doi.org/10.3390/v18020167

10 min. PFA was quenched with 20 mM of Tris pH 8.0 for 5 min. Samples were blocked with 10–20% FBS for 30 min, stained with 0.5–1 μM SNAP-AF647 with 1 μM DTT for 1 h at room temperature, and washed with 10% FBS.

For STORM imaging, a STORM buffer was prepared based on Nikon's N-STORM buffer recommendations [31]. Briefly, Buffer A contained 10 mM of Tris pH 8.0 and 50 mM of NaCl, and Buffer B contained 50 mM of Tris pH 8.0, 10 mM of NaCl, and 10% glucose (D-Glucose anhydrous; Gibco #15023-021) in PBS+/+. GLOX solution was prepared by adding 14 mg of Glucose Oxidase (Sigma #G2133-50KU) and 50 μL of catalase (17 mg/mL; Sigma #C40-500MG) into 200 μL of Buffer A. Finally, 50–100 mM MEA (Cysteamine Hydrochloride; Sigma #M6500-25G) and 1% GLOX solution were mixed with Buffer B and added to samples. The wells were then sealed with parafilm and imaged immediately (see below). STORM buffer was replaced every 2 h.

2.6. Confocal and STORM Imaging

Confocal images were taken on a Zeiss LSM880 confocal microscope using a 63×, 1.4NA oil immersion objective. The image dimensions were 512 × 512 pixels, with zoom of 2 (pixel size of 0.132 μm). The pixel dwell time was 2.05 μs, and 2-line averaging was used to reduce background noise. Z-stacking was performed with 0.5 μm interval and a pinhole of 100 μm. Images were taken using the filter sets 488/561/647 nm. SNAP-AF647 was excited with 647 nm laser, and cPLA2-GFP was excited with 488 nm laser.

For STORM images, we used an Oxford Nanoimager (ONi, Oxford, England) SML microscope equipped with an Olympus 100×, 1.4NA oil immersion objective, and a 640 nm dichroic mirror. The excitation power density was ~4 kW/cm^2 for 488/561/647 nm lasers (0.24 kW/cm^2 for 405 nm UV laser). Before imaging, a 2-channel color mapping calibration was performed using 0.1 μm TetraSpeck microspheres (Invitrogen #T7279) emitting in blue, green, orange, and far red. A mapping calibration with less than 20 nm of Standard Deviation of Errors was accepted. After calibration, drift correction was tested by imaging coverslip-adhered TetraSpeck beads with 60 ms exposure time (16 frames per second) using 640 nm and 561 nm lasers, with 100 frames each and with a 45° (semiTIRF) illumination angle. Calibration/drift correction was confirmed by visual inspection of both colors colocalizing. After calibration, cells were imaged using 60 ms exposure time in a semiTIRF mode, and frames were acquired as follows: (1) 10–40 frames of 640 nm laser excitation with 1.7 mW of laser power; (2) pre-bleaching of SNAP-AF647 with 1500 frames at the highest power (~180 mW) of 640 nm laser; (3) blinking of SNAP-AF647 started at 120 mW of 640 nm laser and imaged for 5000 frames. Blinking was then enhanced with exposure of 0.2 mW of 405 nm laser in conjunction with 120 mW of 640 nm laser for an additional 5000 frames.

2.7. Image Analysis and Statistics

Confocal images were processed with Zeiss ZEN (black edition) version 8.0.0.273 software and ImageJ version 2.3.0. Normalized gray values of nuclear envelope fluorescence were obtained through line histogram plugin on ImageJ. The volume of nucleus was calculated using the HKMeans plugin on ICY software version 2.5.4.0 (https://icy.bioimageanalysis.org/, accessed on 24 January 2026). Half times for HIV-1 nuclear import ($T_{1/2}$) from PF74 addition were determined by fitting with exponential function using GraphPad Prism 10.3.1. Fluorescently labeled HIV-1 INsfGFP particles inside the nucleus were counted manually. Statistical analyses for ATP depletion luciferase assays, nuclear envelope volumes, and NPC radius were done using GraphPad. Brown-Forsythe and Welch ANOVA were used for all statistical analyses unless stated otherwise.

https://doi.org/10.3390/v18020167

For visualizing STORM images, drift correction, filtering, and exporting of localization coordinates, we used NimOS version 1.19.1 software and the cloud-based CODI (https://alto.codi.bio, accessed on 24 January 2026) program developed by ONi. Drift correction was performed using the built-in drift correction feature in NimOS. Localizations were filtered by removing localizations with less than 500 photons and with more than 5000 photons. In addition, events with localization precision worse than 20 nm were discarded. Averaging of NPCs were performed using a MATLAB 2024b script developed by the Rieger lab (Delft, The Netherlands) [32,33]. For this analysis, each NPC exhibiting symmetric circular appearance was manually picked from the field of view by cropping the point cloud with 180 × 180 nm boxes. For each dataset, all NPCs were initially centered by subtracting their mean localization coordinate from all coordinates and the cartesian coordinates were then transformed to polar coordinates ($(x,y) \rightarrow (\rho,\theta)$). To generate images of averaged NPCs, we applied the template-free registration-based particle averaging method developed by Wang et al. [33]. Briefly, individual NPC localizations were iteratively aligned using pairwise registration based on structural similarity. This process is corrected for translational and rotational variations across particles. The aligned NPCs were then combined to produce a high signal-to-noise composite image that emphasizes consistent structure of NPCs. The FRC curve was calculated, as previously established [34]. Briefly, the dataset was divided into two halves, and each half was registered, followed by the application of eight-fold symmetry to each particle. This ensures alignment accuracy and minimizes bias while enhancing the signal-to-noise ratio. After applying the rotational symmetry, the FRC curve was computed and plotted. By applying a resolution threshold to the FRC curve, the resolution of the reconstruction was determined, providing a quantitative assessment of the reconstructed structure.

The mean radius of the pores was computed as follows. First, the localizations were centered by subtracting their mean coordinates and by transforming the localizations from Cartesian to polar coordinates, as mentioned previously. The localizations were superimposed to create a 2D histogram of the NPC SMLs. The radial distance from center of the localizations were plotted, fitted by a Gaussian distribution and normalized to the number of localizations. Lastly, the mean μ and standard deviation σ were calculated from the Gaussian fit. The μ position of the Gaussian does not showcase the correct average of the pore size, so the peak position was corrected to first order by $-\frac{1}{2}\left(\frac{\sigma}{\mu}\right)^2$ [17,35]. A total of 100,000–400,000 SMLs from several hundreds to several thousands of NPCs were used to create the superimposed NPCs and to calculate the mean radius from the Gaussian fit after removing outliers (SMLs farther than 100 nm from center and closer than 10 nm from center).

3. Results

3.1. NUP96-SNAP Tag Does Not Affect the HIV-1 Nuclear Import Kinetics

We chose to visualize the NPC complexes by using tagged NUP96, 32 copies of which are present in the nuclear pore complex [36]. We employed the U2OS cell line in which the endogenous locus of NUP96 was tagged with SNAP using a CRISPR/Cas9 knock-in approach (described in Thevathasan et al. [27]). These cells were previously used to visualize the nuclear pore structure by super-resolution microscopy, yielding a pore radius of 53–56 nm [27,37]. We first tested whether the SNAP-tagged NUP96 can affect HIV-1 nuclear import and infectivity. Parental U2OS cells and U2OS NUP96-SNAP cells, as well as control HeLa-derived reporter TZM-bl cells, were infected with NL4-3 pseudoviruses encoding for luciferase. The kinetics of nuclear import was measured by adding 2 μM of PF74, a compound that binds to HIV-1 capsid [38,39], at different times post-infection (p.i.) and reading the resulting luciferase activity at 48 h.p.i. (Supplementary Figure S1A). Since 2 μM of PF74 are known to inhibit HIV-1 nuclear import but not reverse transcription [38,40–42], the luciferase signal as a function of time of PF74 addition reports the HIV-1 nuclear import kinetics. The overall efficiency of infection was evaluated at the 24 h time point, at which the signal reached a plateau (Supplementary Figure S1A). The overall efficiency of infection was ~2-fold lower in U2OS NUP96-SNAP compared to parental U2OS cells, and 4.5-fold lower than TZM-bl (Supplementary Figure S1A). However, the kinetics of nuclear import was not delayed by SNAP-tagging of NUP96. The estimated half times ($T_{1/2}$) of nuclear import in U2OS NUP96-SNAP and parental cells was 4.1 h and 5.1 h, respectively. These times were close to $T_{1/2}$ of 5.3 h for TZM-bl cells. To more directly evaluate the efficiency of nuclear import, we measured the number of fluorescently labeled HIV-1 cores inside the nucleus around the half-time of infection (4 h.p.i.) (Supplementary Figure S1B,C). Cells were infected with NL4-3 pseudoviruses labeled with integrase-superfolder GFP (INsfGFP) and counted the number of fluorescent spots per nucleus (Supplementary Figure S1C). Although U2OS NUP96-SNAP cells exhibited fewer HIV-1 cores per nucleus than parental cells, in agreement with their reduced infectivity (Supplementary Figure S1A), the difference did not reach statistical significance. These results suggest that the nuclear import of HIV-1 is not majorly reduced by tagging NUP96. In contrast, the number of nuclear cores in TZM-bl cells was significantly higher compared to both parental U2OS and U2OS NUP96-SNAP cells, highlighting cell type-dependent differences in the efficiency of HIV-1 infection.

3.2. ATP Depletion Does Not Cause Detectable Changes in NPC Size

Since the yeast NPCs have been reported to constrict upon depletion of cellular ATP [13], we asked whether depleting ATP in mammalian cells would have the same effect. Cells were depleted of ATP using two approaches: (1) incubation with 10 mM of sodium azide, in combination with 6 mM 2-deoxy-D-glucose (dubbed NaN_3), or (2) treatment with 10 μM of Antimycin A with 20 mM 2-deoxy-D-glucose (Ant-A). Our ATP depletion protocols were validated by measuring the ATP concentration in U2OS NUP96-SNAP cells (Figure 1). Cells incubated in HEPES-based imaging buffer without glucose ((−)Gluc) or in complete cell medium ((+)Gluc) were used as controls (see Section 2). Cellular ATP concentrations significantly dropped after NaN_3 and Ant-A treatment compared to (+)Gluc- or (−)Gluc-treated cells. ATP depletion with Ant-A was more efficient than with NaN_3. Interestingly, cells treated with (−)Gluc medium contained more ATP than cells treated in (+)Gluc, suggesting that glucose starvation activates alternative metabolic pathways to increase ATP production [43–45]. A dramatic depletion of cellular ATP using these protocols was reproducibly observed in multiple replicates, even after 10 min of incubation (Supplementary Figure S2). Also, the ATP levels remained the lowest after 4 h of incubation with Ant-A (Supplementary Figure S2B).

In single-molecule localization microscopy (SMLM), molecules transition between dark and emissive states, which are localized in each time frame and fitted with a Point Spread Function model. The fluorescent events are spatially sparsely distributed. Thus, the position of a molecule can be pinpointed with much higher accuracy than the diffraction limit. From the estimated positions of many thousands of localizations, a super resolution image can be reconstructed [46,47].

Using Stochastic Optical Reconstruction Microscopy (STORM), we next measured the NPC radii in control and ATP depleted cells (Figure 2). Since the average NPC diameter reported for the Y-complex, comprising NUP96, is ~107 nm [27,48], changes in the pore size exceeding 20% (~20 nm) should be resolvable by STORM. The NPC structure with NUP96 highlighted is shown in Figure 2A. Cells were depleted of ATP with NaN_3 or Ant-A for 30 min, fixed, and stained for NUP96-SNAP with the SNAP-reactive SNAP-AF647 dye.

 https://doi.org/10.3390/v18020167

Initial examination by confocal microscopy showed no apparent changes in NE morphology for cells in (+)Gluc vs. ATP depletion buffer (Figure 2B, top). STORM images revealed NPCs with different appearances with varied apparent spoke arrangements, including the expected 8-fold symmetry (Figure 2B, middle, *inset*). The varied pore appearance can be due to incomplete labeling of NUP96-SNAP, which is around 60% for the SNAP-AF647 dye [27], and/or may reflect the overall NPC heterogeneity [25,26]. To measure the NPC size, we manually selected NPCs, from 680 to 2470 NPCs, with a symmetric circular appearance. Next, the selected pores were centered by subtracting their center of mass, the cartesian coordinates were transformed into polar coordinates, and all SMLs were superimposed to form an ensemble of NPC localizations. We also applied a template-free registration-based composite to the averaged NPC pores to visualize the NPC symmetry (Figure 2B, bottom). The spatial resolution determined by Fourier Ring Correlation (FRC) analysis was about 13.6 nm for (+)Gluc-treated samples (Supplementary Figure S3), in excellent agreement with 13.3 nm reported by Thevathasan et al. [27] in SNAP-AF647-labeled NUP96-SNAP U2OS cells. The number of localizations per NPC, and, therefore, the NUP96-SNAP expression level and/or labeling efficiency were not affected by ATP depletion (Figure 2C,D). We also found that the nuclear volume did not change after depleting ATP (Figure 2E).

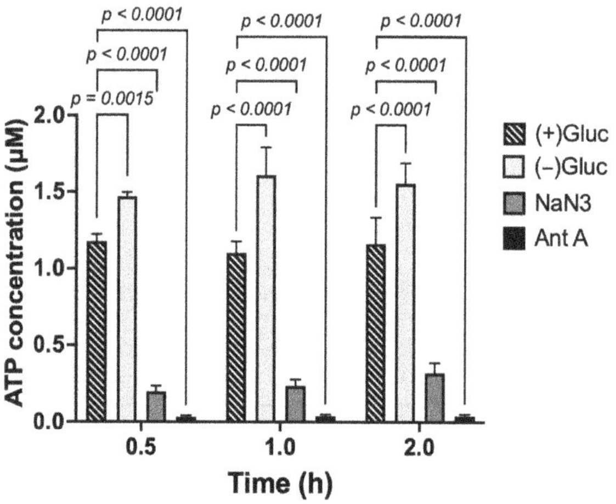

Figure 1. ATP depletion of U2OS NUP96-SNAP cells. Cells were incubated in complete DMEM with 10% FBS ((+)Gluc), in HEPES-based imaging buffer without glucose ((−)Gluc), or in ATP depletion buffer: 10 mM of sodium azide with 6 mM of 2-deoxy-D-glucose (NaN_3) or 10 μM of Antimycin A with 20 mM of 2-deoxy-D-glucose (Ant-A) at different times. The cellular ATP concentration (in μM) was measured in triplicate samples using a kit and plotted as mean ± SD. Statistical analysis was done with two-way ANOVA. (See Supplementary Figure S2 for an independent replicate).

We next calculated the mean radius of NPCs from the superimposed single-molecule localizations (Figure 3). After aligning the SMLs of the selected pores, the radial distance from center was measured by fitting the SML distributions with a Gaussian function, from which we calculated the mean (μ) and standard deviation (σ) (Figure 3A–C, top and bottom). The calculated mean NPC radius for (+)Gluc samples was 51 ± 12.9 nm, and for the ATP depleted samples, NaN_3 and Ant-A, it was 51 ± 12.8 nm and 51 ± 13.9 nm, respectively (Figure 3A–D). The difference between the NPC radius in ATP depleted and (+)Gluc samples was not significant (Figure 3D), and this result was consistent across three biological replicates (Figure 3E, and Supplementary Figures S4–S6).

 https://doi.org/10.3390/v18020167

We repeated the ATP depletion experiment by pre-incubating cells for 4 h with Ant-A and treating with 2 μM of the microtubule (MT) polymerization inhibitor Nocodazole (Noco, Supplementary Figure S6). Nocodazole was used since MTs are known to interact with the NPC [49], so their depolymerization could affect the NPC radius. We saw changes in nuclear envelope morphology in cells treated with Ant-A + Noco compared to (−)Gluc samples or cells treated with Ant-A alone (Supplementary Figure S6A). We then calculated the pore radius distributions for (−)Gluc and Ant-A + Noco cells. The mean pore radius did not significantly change between (−)Gluc (51 ± 13.4 nm) and Ant-A + Noco (51 ± 13.5 nm) (Supplementary Figure S6D,F). These results show that depleting ATP, alone or combined with MT depolymerization, does not cause changes in the NPC radius in U2OS cells that can be detected by single-molecule localization microscopy.

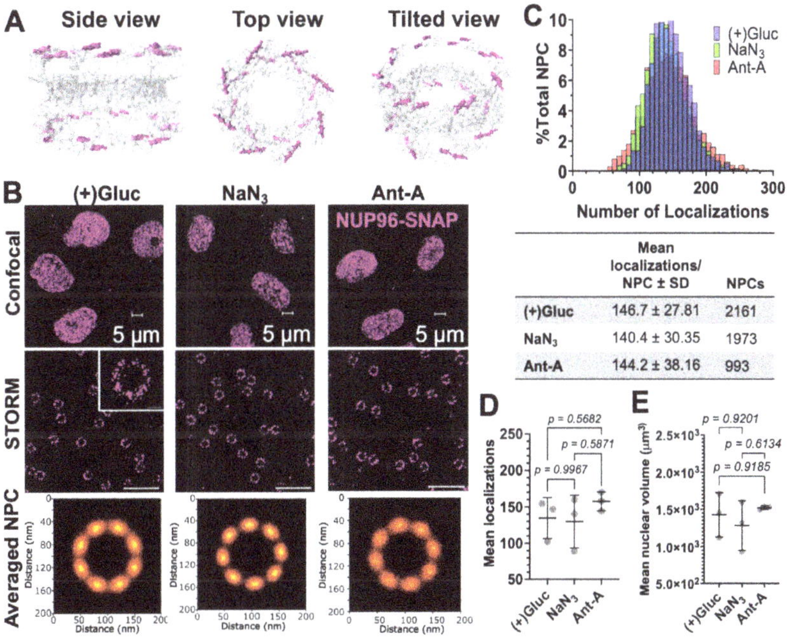

Figure 2. STORM imaging of NPCs on ATP depleted samples. (**A**) EM density (grey) images of the NPC with NUP96 highlighted (magenta) viewed from the side, top, and tilted. Structure was taken from the Protein Data Bank (PDB): 7TBJ. (**B**) *Top*: Confocal images of U2OS NUP96-SNAP cells incubated in complete medium ((+)Gluc) and ATP-depletion media, NaN_3 or Ant-A, for 30 min at 37 °C. Cells were then fixed with PFA, permeabilized, and NUP96-SNAP was stained with the SNAP-AF647 dye. Shown are representative images of confocal slices at the bottom of the nuclear envelope. *Middle*: STORM images (AF647 localizations) of NPCs at the bottom of the nuclear envelope. Scale bar: 0.5 μm. *Inset*: zoomed in of one NPC (scale bar 50 nm). *Bottom*: Images of averaged NPCs from STORM images. NPCs were selected manually, and averaging was performed using a MATLAB script. (**C**) Distribution of the number of SNAP-AF647 dye localizations per NPC obtained by STORM (panel (**B**)). Table shows mean localizations per NPC ± SD and the number of NPCs analyzed. (**D**) Mean number of localizations for each biological replicate. (See Supplementary Figures S4–S6 for independent replicates). Statistical analysis was done with Brown-Forsythe and Welch ANOVA. (**E**) Mean nuclear volumes (μm^3) for (+)Gluc, NaN_3, and Ant-A treated samples. Shown are mean ± SD from three biological replicates. Brown-Forsythe and Welch ANOVA were used for statistical analysis.

 https://doi.org/10.3390/v18020167

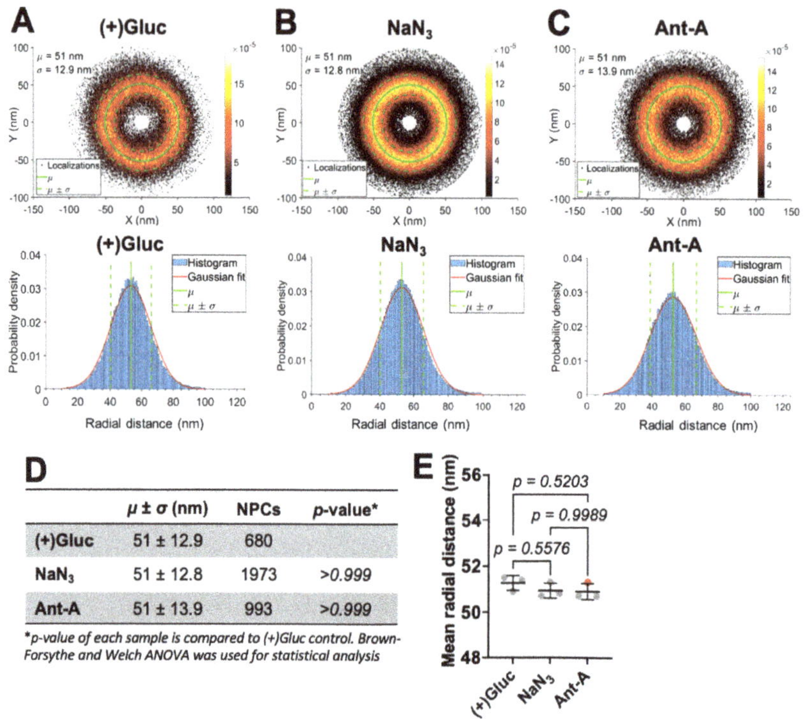

	μ ± σ (nm)	NPCs	p-value*
(+)Gluc	51 ± 12.9	680	
NaN₃	51 ± 12.8	1973	>0.999
Ant-A	51 ± 13.9	993	>0.999

**p-value of each sample is compared to (+)Gluc control. Brown-Forsythe and Welch ANOVA was used for statistical analysis*

Figure 3. ATP depletion does not change the NPC size. (**A–C**) *Top*: Scatter plots showing the localizations of super imposed NPCs analyzed in Figure 2 for (+)Gluc (**A**), NaN_3 (**B**), and Ant-A (**C**) samples. We aligned the centers of mass for each NPC SMLs, transformed localizations into polar coordinates, and then superimposed to obtain a density map of all localizations for the analyzed pores. Shown right is the pseudo color scale for the localization density. *Bottom*: SML radial distance histograms were fitted with a Gaussian function (red). The mean radius (μ) and standard deviation (σ) were calculated from the fitted Gaussian distribution (*Top* and *Bottom*, green lines). The correct average peak position of the localizations is found from the radial histogram peak by subtracting $0.5\sigma^2/\mu^2$ [17,35] and superimposed onto the combined NPC localizations (see Section 2; *Top*, green). (**D**) The table shows $\mu \pm \sigma$ calculated from A-C, number of NPCs analyzed for each condition, and *p*-values. (**E**) Mean SML radial distances from the center for three biological replicates. For one biological replicate (red symbol), the cells were treated with 2 μM of Nocodazole in addition to Ant-A ATP depletion (Supplementary Figure S6). Mean ± SD are plotted. (See Supplementary Figures S4–S6 for results and analyses of independent replicates).

3.3. NPC Size Is Resistant to Stretching and Shrinking of the Nuclear Envelope

Changes in nuclear membrane tension have been reported to alter the NPC diameter [13,14]. We sought to osmotically swell the nucleus to determine whether stretching of the nuclear membrane would expand the NPCs. To monitor the NE tension, we transfected U2OS NUP96-SNAP cells with the membrane tension sensor, cytosolic phospholipase A_2 fused to GFP ($cPLA_2$-GFP). This sensor binds to stretched nuclear membranes in a calcium-dependent manner [29,30]. We then permeabilized the cells with digitonin in the presence or absence of 5% Polyvinylpyrrolidone (PVP360), a polymer that prevents cell swelling by maintaining osmotic balance. $cPLA_2$-GFP is evenly distributed across the nucleus in non-swelled nuclear membranes of permeabilized cells in the presence of 5% PVP360 (Figure 4A). However, in cells with stretched nuclear membranes (0% PVP360), $cPLA_2$-GFP localizes primarily to the nuclear envelope

 https://doi.org/10.3390/v18020167

(Figure 4B), thus confirming the generation of tension in the NE. To independently verify swelling of the nucleus, we also measured the nuclear volume (Figure 4C). Cells incubated with 5% PVP360 had significantly smaller nuclear volume ($2.4 \pm 0.9 \times 10^3$ μm^3) than those incubated with 0% PVP360 ($3.1 \pm 1.5 \times 10^3$ μm^3), confirming nuclear membrane stretching.

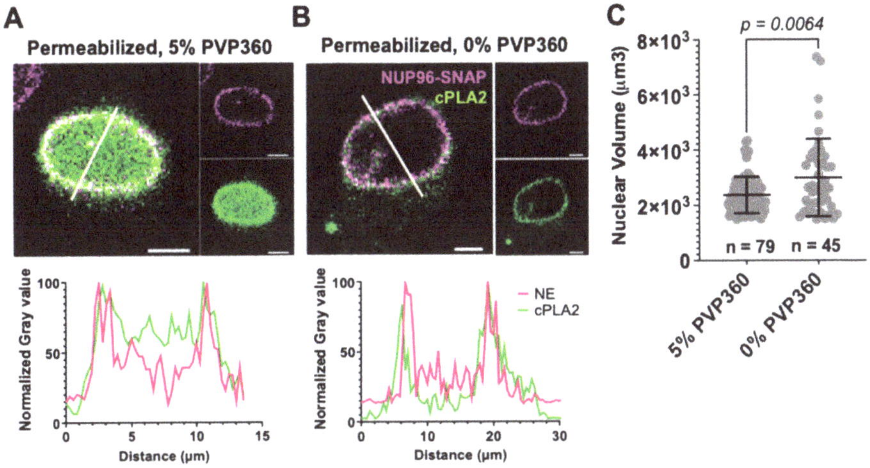

Figure 4. Osmotic swelling of nuclear membrane. (**A**,**B**) U2OS NUP96-SNAP cells were transfected with the membrane tension sensor, cPLA2-GFP (green). Twenty-four hours after transfection, cells were permeabilized for 5 min with 25 μg/mL digitonin in base medium in the presence of 5% Polyvinylpyrrolidone (PVP360) to prevent cell swelling. Scale bar 5 μm. (**A**,**B**). Permeabilization medium was removed, and cells were further incubated with 5% PVP360 for 5 min (**A**) or in 0% PVP360 for 15 min (**B**). Pre-permeabilized samples were fixed and stained with SNAP-AF647 dye (magenta). *Left*: Representative images of a middle section of the nuclear envelope (NE). *Right*: Images of individual cPLA2 and SNAP channels. Below each image are line histograms of the normalized intensities of cPLA2 (green) and SNAP (NE, magenta), corresponding to lines drawn across the nuclei on the confocal images. (**C**) Distributions of nuclear volumes (in μm^3) of stretched (0% PVP360) and non-stretched (5% PVP360) nuclei. Shown are means ± SD, n = Number of nuclei. Statistical analysis was done using Welch's t-test.

STORM imaging of stretched and unstretched nuclear membranes yielded the mean of NPC radius for permeabilized cells in the presence or absence of 5% PVP360 of 51 ± 13.3 nm and 51 ± 12.9 nm, respectively. In control experiments, the mean radius of samples fixed before permeabilization was about 51 ± 12.9 nm, all in excellent agreement with Figure 3 (Figure 5B–E). We found no significant change in the mean radial distance between each condition across three biological replicates (Figure 5F and Supplementary Figures S7 and S8). In summary, our results suggest that osmotically stressing the nuclear membrane does not cause detectable changes in the pore radius.

In addition to osmotically swelling the nucleus, we also tested the effect of shrinkage through exposure to hypertonic buffer [13,14]. We incubated U2OS NUP96-SNAP cells in 0 M or 2 M of sorbitol for 1 h, fixed and stained for NUP96-SNAP. As expected, the nuclear volume was significantly reduced in cells treated with 2 M of sorbitol compared to the control (Supplementary Figure S9A,B). Interestingly, the difference in the overall mean pore radius was highly significant (Supplementary Figure S9C–E). However, the difference was ~0.3 nm and, in addition, falls within the standard deviation, which suggests that the

https://doi.org/10.3390/v18020167

significant difference might be due to high sample size and does not necessarily reflect significant changes in the pore radius.

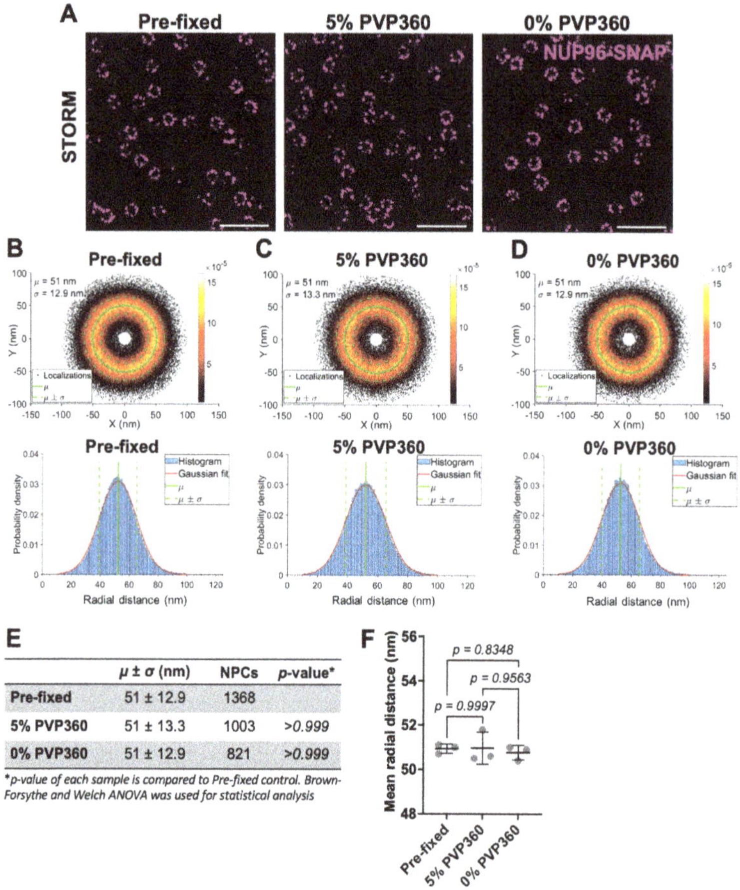

Figure 5. Stretching the nuclear envelope does not change the NPC radius. (**A**) STORM images of U2OS NUP96-SNAP cells either fixed before permeabilization (Pre-fixed) or pre-permeabilized with 25 μg/mL of digitonin and incubated under swelling (0% PVP360) or not swelling (5% PVP360) conditions, as in Figure 4. Cells fixed prior to permeabilization were used as a control. NUP96-SNAP was stained with the SNAP-AF647 dye. Shown are representative images at the bottom of the nuclear envelope. Scale bar: 0.5 μm. (**B–D**) Same as in Figure 3. *Top*: Scatter plots showing the radial distributions of single-molecule localization densities of superimposed NPCs selected from (**A**), aligned at their center of mass. Shown right is the pseudo color scale for the localization density. *Bottom*: SML radial distance histograms were fitted with a Gaussian function (red). The μ and σ were calculated from the fitted Gaussian distribution (*Top* and *Bottom*, green lines). (**E**) The table shows $\mu \pm \sigma$ calculated on (**B–D**), number of NPCs analyzed for each condition and p-values. (**F**) Mean SML radial distances from the center for three biological replicates. Mean ± SD are plotted. (See Supplementary Figures S7 and S8 for analysis of independent replicates).

4. Discussion

The nuclear pores have been used as reference for SML imaging by using antibodies, direct tags like self-labeling SNAP or Halo tags that covalently react with organic dyes, or by

 https://doi.org/10.3390/v18020167

transient interactions with labeled DNA strands to perform DNA-PAINT imaging [27,37,50]. DNA-PAINT affords better spatial resolution while yielding the NPC radius close to that measured by direct STORM of NUP96-SNAP using SNAP-AF647 dye [27,37]. Our STORM imaging of U2OS NUP96-SNAP cells yielded a mean NPC radius of ~51 nm. The spatial STORM resolution of AlexaFluor647-labeled NUP96-SNAP pores estimated by FRC, SML alignment, and Gaussian fitting was ~13 nm, which is consistent to the previously established spatial resolution of ~13.3 nm by the same group [27]. Within this limit, we did not detect significant changes in the mean NPC radius after ATP depletion or upon stretching/shrinking the nuclear envelope through osmotic stresses. In some cases, difference in the mean radius would appear significant, but the actual change in the radius was less than 1 nm, which is more than 10 times less than previous studies showing changes in NPC radius and would rather suggest sample-to-sample variation and/or a result of a very large SML sample size.

Our findings contrast with the cryo-ET study reporting contraction of yeast NPCs in the absence of ATP and under hypertonic conditions [13]. This discrepancy may be related to the differences in yeast and mammalian NPC composition. For example, *Saccharomyces cerevisiae* has 16 total copies of the Y-complex scaffold per NPC, compared to 32 copies in the human NPC [3,48,51,52]. Similarly, *Schizosaccharomyces pombe* has an asymmetric Y-complex distribution across the NPC, with more subunits in the nucleoplasmic ring (NR) than in the cytoplasmic ring (CR) [13,48]. ATP depletion in *S. pombe* resulted in a marked reduction of pore's diameter by ~20 nm at the CR and inner ring (IR), while the NR diameter was reduced by only ~8 nm [13]. This result suggests that the double Y-complex at the NR in *S. pombe* cells may provide more rigidity compared to CR and IR. In addition, the Y-complex of *S. cerevisiae* lacks two NUPs, which are present in the human Y-complex [3,48,51,52]. These structural differences may allow the yeast NPC to respond more readily to energy depletion and mechanical stimuli than the human nuclear pore. Note that depletion of the members of Y-complex, NUP96 or NUP133, resulted in a loss of NR and CR spokes and increased the IR diameter in mammalian cells [11,15]. This result supports the role of Y-complex in regulating the rigidity and symmetry of the NPC, while the IR might be more flexible and prone to change its diameter than the CR and NR [11,15]. Future studies using endogenously tagged NUPs located in the central channel, such as NUP54, may enable detection of NPC remodeling by STORM. STORM Imaging of NUP54 will likely yield a smaller apparent pore radius, thereby providing a positive control for our ability to resolve changes in the NPC size.

In addition to energy depletion, changes in NE tension have also been shown to impact the NPC size and permeability [13,14,53,54]. Hypertonic conditions appear to shrink the nucleus and reduce the NPC diameter in yeast and mammalian cells [13,14]. Conversely, stretching the NE increases the NPC size, as determined by cryo-ET, and induces nuclear translocation of the mechanosensing YAP transcription factor [53,54]. However, such mechanical stress-induced changes in pore diameter were not detectable by our STORM imaging. One reason for the apparent discrepancy with our data may be the inferior spatial resolution of STORM imaging compared to cryo-ET and TEM used by others. However, these studies showed changes of about 20 nm in NPC diameter, which falls within the resolution limit of STORM. In addition, expansion microscopy [14], which achieves a similar spatial resolution to that of STORM, detected NPC shrinking in human cells by hypertonic shock. We also cannot rule out possible paraformaldehyde fixation artifacts masking the pore remodeling, but such prefixation has also been used for expansion microscopy [14]. In addition, labeling efficiency of SNAP with SNAP-AF647 dye and SNAP-tagging of NUP96 might affect the resolution limit or the NPC structure and dynamics, respectively. While the reasons for these apparent discrepancies are presently unclear, detection of morphological changes in NPC could be affected by the cell type, method used for sample preparation, and/or imaging modality.

Our results indicate a surprising resistance of NPCs of human U2OS cells to mechanical stimuli and the lack of reliance on active ATP energy-dependent processes for maintaining the pore architecture. Since cryo-ET studies have revealed a marked reduction in pore diameter in isolated nuclei vs. intact cells [5,10,11,24], our future STORM experiments will compare the NPC diameters in intact U2OS cells and isolated nuclei. In addition, we will employ a DNA-PAINT approach to achieve a higher photon budget and better localization precision [37]. Once our ability to detect NPC remodeling is validated, we will employ 2-color STORM to assess the effects of HIV-1 docking on the shape and diameter of nuclear pores.

Supplementary Materials: The following supporting information can be downloaded at: https://www.mdpi.com/article/10.3390/v18020167/s1. Figure S1: HIV-1 infection of U2OS endogenously expressing NUP96-SNAP. Figure S2: ATP depletion in U2OS NUP96-SNAP cells, second biological replicate. Figure S3: Spatial resolution of (+)Gluc treated cells. Figure S4: Effects of ATP depletion on the NPC size distribution (second replicate). Figure S5: Effect of ATP depletion with NaN_3 on the NPC size, third replicate. Figure S6: Effects of ATP depletion with Antimycin A and microtubule depolymerization with Nocodazole on NPC size. Figure S7: Effects on nuclear envelope swelling on NPC size (second biological replicate). Figure S8: Effects on nuclear envelope swelling on NPC size (third biological replicate). Figure S9: Shrinkage of the nuclear envelope with hypertonic medium does not change the NPC radius.

Author Contributions: Conceptualization, D.T.-R. and G.B.M.; methodology, D.T.-R. and G.B.M.; software, S.H. and B.R.; formal analysis, D.T.-R. and S.H.; investigation, D.T.-R. and S.H.; resources, G.B.M.; writing—original draft preparation, D.T.-R. and G.B.M.; writing—review and editing, D.T.-R., S.H., B.R. and G.B.M.; visualization, D.T.-R. and B.R.; supervision, G.B.M. and B.R.; project administration, G.B.M.; funding acquisition, G.B.M. and B.R. All authors have read and agreed to the published version of the manuscript.

Funding: This research was funded by NIAID/NIH R01 AI148382 grant to G.B.M. and U54 AI170855 Center grant, and by Dutch Research Council (NWO), VICI grant no. 17046 to B.R.

Data Availability Statement: Original STORM data related to this research are available from the corresponding author upon request.

Acknowledgments: The authors wish to thank Yen-Cheng Chen for his help with training on an ONi STORM system, and Mariana Marin and Gokul Raghunath for helpful discussions and comments on the manuscript.

Conflicts of Interest: The authors declare no conflict of interest.

References

1. Beck, M.; Hurt, E. The nuclear pore complex: Understanding its function through structural insight. *Nat. Rev. Mol. Cell Biol.* **2016**, *18*, 73–89. [CrossRef]
2. Knockenhauer, K.E.; Schwartz, T.U. The Nuclear Pore Complex as a Flexible and Dynamic Gate. *Cell* **2016**, *164*, 1162–1171. [CrossRef] [PubMed]
3. Lin, D.H.; Hoelz, A. The Structure of the Nuclear Pore Complex (An Update). *Annu. Rev. Biochem.* **2019**, *88*, 725–783. [CrossRef] [PubMed]
4. Paci, G.; Caria, J.; Lemke, E.A. Cargo transport through the nuclear pore complex at a glance. *J. Cell Sci.* **2021**, *134*. [CrossRef]
5. Bui, K.H.; von Appen, A.; DiGuilio, A.L.; Ori, A.; Sparks, L.; Mackmull, M.-T.; Bock, T.; Hagen, W.; Andrés-Pons, A.; Glavy, J.S.; et al. Integrated Structural Analysis of the Human Nuclear Pore Complex Scaffold. *Cell* **2013**, *155*, 1233–1243. [CrossRef] [PubMed]

 https://doi.org/10.3390/v18020167

6. Celetti, G.; Paci, G.; Caria, J.; VanDelinder, V.; Bachand, G.; Lemke, E.A. The liquid state of FG-nucleoporins mimics permeability barrier properties of nuclear pore complexes. *J. Cell Biol.* **2019**, *219*, e201907157. [CrossRef]
7. Frey, S.; Richter, R.P.; Görlich, D. FG-Rich Repeats of Nuclear Pore Proteins Form a Three-Dimensional Meshwork with Hydrogel-Like Properties. *Science* **2006**, *314*, 815–817. [CrossRef]
8. Panté, N.; Kann, M. Nuclear Pore Complex Is Able to Transport Macromolecules with Diameters of ~39 nm. *Mol. Biol. Cell* **2002**, *13*, 425–434. [CrossRef]
9. Paci, G.; Zheng, T.; Caria, J.; Zilman, A.; A Lemke, E. Molecular determinants of large cargo transport into the nucleus. *eLife* **2020**, *9*, e55963. [CrossRef]
10. Zila, V.; Margiotta, E.; Turoňová, B.; Müller, T.G.; Zimmerli, C.E.; Mattei, S.; Allegretti, M.; Börner, K.; Rada, J.; Müller, B.; et al. Cone-shaped HIV-1 capsids are transported through intact nuclear pores. *Cell* **2021**, *184*, 1032–1046.e18. [CrossRef]
11. Schuller, A.P.; Wojtynek, M.; Mankus, D.; Tatli, M.; Kronenberg-Tenga, R.; Regmi, S.G.; Dip, P.V.; Lytton-Jean, A.K.R.; Brignole, E.J.; Dasso, M.; et al. The cellular environment shapes the nuclear pore complex architecture. *Nature* **2021**, *598*, 667–671. [CrossRef]
12. Mahamid, J.; Pfeffer, S.; Schaffer, M.; Villa, E.; Danev, R.; Cuellar, L.K.; Förster, F.; Hyman, A.A.; Plitzko, J.M.; Baumeister, W. Visualizing the molecular sociology at the HeLa cell nuclear periphery. *Science* **2016**, *351*, 969–972. [CrossRef]
13. Zimmerli, C.E.; Allegretti, M.; Rantos, V.; Goetz, S.K.; Obarska-Kosinska, A.; Zagoriy, I.; Halavatyi, A.; Hummer, G.; Mahamid, J.; Kosinski, J.; et al. Nuclear pores dilate and constrict in cellulo. *Science* **2021**, *374*, eabd9776. [CrossRef]
14. Morgan, K.J.; Carley, E.; Coyne, A.N.; Rothstein, J.D.; Lusk, C.P.; King, M.C. Visualizing nuclear pore complex plasticity with pan-expansion microscopy. *J. Cell Biol.* **2025**, *224*, e202409120. [CrossRef]
15. Taniguchi, R.; Orniacki, C.; Kreysing, J.P.; Zila, V.; Zimmerli, C.E.; Böhm, S.; Turoňová, B.; Kräusslich, H.-G.; Doye, V.; Beck, M. Nuclear pores safeguard the integrity of the nuclear envelope. *Nat. Cell Biol.* **2025**, *27*, 762–775. [CrossRef] [PubMed]
16. Sellés, J.; Penrad-Mobayed, M.; Guillaume, C.; Fuger, A.; Auvray, L.; Faklaris, O.; Montel, F. Nuclear pore complex plasticity during developmental process as revealed by super-resolution microscopy. *Sci. Rep.* **2017**, *7*, 14732. [CrossRef]
17. Löschberger, A.; van de Linde, S.; Dabauvalle, M.-C.; Rieger, B.; Heilemann, M.; Krohne, G.; Sauer, M. Super-resolution imaging visualizes the eightfold symmetry of gp210 proteins around the nuclear pore complex and resolves the central channel with nanometer resolution. *J. Cell Sci.* **2012**, *125*, 570–575. [CrossRef] [PubMed]
18. Hou, Z.; Shen, Y.; Fronik, S.; Shen, J.; Shi, J.; Xu, J.; Chen, L.; Hardenbrook, N.; Engelman, A.N.; Aiken, C.; et al. HIV-1 nuclear import is selective and depends on both capsid elasticity and nuclear pore adaptability. *Nat. Microbiol.* **2025**, *10*, 1868–1885. [CrossRef]
19. Kreysing, J.P.; Heidari, M.; Zila, V.; Cruz-León, S.; Obarska-Kosinska, A.; Laketa, V.; Rohleder, L.; Welsch, S.; Köfinger, J.; Turoňová, B.; et al. Passage of the HIV capsid cracks the nuclear pore. *Cell* **2025**, *188*, 930–943.e21. [CrossRef]
20. Hou, Z.; Fronik, S.; Shen, Y.; Chen, L.; Thompson, C.; Neumann, S.; Zhang, P. Direct visualization of HIV-1 core nuclear import and its interplay with the nuclear pore. *EMBO Rep.* **2025**, *26*, 5133–5153. [CrossRef]
21. Müller, T.G.; Zila, V.; Peters, K.; Schifferdecker, S.; Stanic, M.; Lucic, B.; Laketa, V.; Lusic, M.; Müller, B.; Kräusslich, H.-G. HIV-1 uncoating by release of viral cDNA from capsid-like structures in the nucleus of infected cells. *eLife* **2021**, *10*, e64776. [CrossRef] [PubMed]
22. Ay, S.; Burlaud-Gaillard, J.; Gazi, A.; Tatirovsky, Y.; Cuche, C.; Diana, J.-S.; Scoca, V.; Di Santo, J.P.; Roingeard, P.; Mammano, F.; et al. In vivo HIV-1 nuclear condensates safeguard against cGAS and license reverse transcription. *EMBO J.* **2024**, *44*, 166–199. [CrossRef]
23. Müller, T.G.; Klaus, S.; Zila, V.; Lucic, B.; Penzo, C.; Nopper, S.L.; Golani, G.; Anders-Össwein, M.; Sonntag-Buck, V.; Heuser, A.-M.; et al. Lenacapavir-induced capsid damage uncovers HIV-1 genomes emanating from nuclear speckles. *EMBO J.* **2025**, *45*, 449–470. [CrossRef]
24. Hou, Z.; Shen, Y.; Fronik, S.; Shen, J.; Shi, J.; Xu, J.; Chen, L.; Hardenbrook, N.; Thompson, C.; Neumann, S.; et al. Correlative In Situ Cryo-ET Reveals Cellular and Viral Remodeling Associated with Selective HIV-1 Core Nuclear Import. *bioRxiv* **2025**. [CrossRef]
25. Sabinina, V.J.; Hossain, M.J.; Hériché, J.-K.; Hoess, P.; Nijmeijer, B.; Mosalaganti, S.; Kueblbeck, M.; Callegari, A.; Szymborska, A.; Beck, M.; et al. Three-dimensional superresolution fluorescence microscopy maps the variable molecular architecture of the nuclear pore complex. *Mol. Biol. Cell* **2021**, *32*, 1523–1533. [CrossRef]
26. Ye, X.; Guan, M.; Guo, Y.; Liu, X.; Wang, K.; Chen, T.; Zhao, S.; Chen, L. Live-cell super-resolution imaging unconventional dynamics and assemblies of nuclear pore complexes. *Biophys. Rep.* **2023**, *9*, 206–214. [CrossRef] [PubMed]
27. Thevathasan, J.V.; Kahnwald, M.; Cieśliński, K.; Hoess, P.; Peneti, S.K.; Reitberger, M.; Heid, D.; Kasuba, K.C.; Hoerner, S.J.; Li, Y.; et al. Nuclear pores as versatile reference standards for quantitative superresolution microscopy. *Nat. Methods* **2019**, *16*, 1045–1053. [CrossRef]

28. Pizzato, M.; Erlwein, O.; Bonsall, D.; Kaye, S.; Muir, D.; McClure, M.O. A one-step SYBR Green I-based product-enhanced reverse transcriptase assay for the quantitation of retroviruses in cell culture supernatants. *J. Virol. Methods* **2009**, *156*, 1–7. [CrossRef]
29. Shen, Z.; Belcheva, K.T.; Jelcic, M.; Hui, K.L.; Katikaneni, A.; Niethammer, P. A synergy between mechanosensitive calcium- and membrane-binding mediates tension-sensing by C2-like domains. *Proc. Natl. Acad. Sci. USA* **2021**, *119*, e2112390119. [CrossRef] [PubMed]
30. Enyedi, B.; Jelcic, M.; Niethammer, P. The Cell Nucleus Serves as a Mechanotransducer of Tissue Damage-Induced Inflammation. *Cell* **2016**, *165*, 1160–1170. [CrossRef]
31. *Super Resolution Microscope N-STORM Protocol*; N-STORM PROTOCOL 2MJ-SCIH-4; Nikon Corporation: Tokyo, Japan, 2013.
32. Haghparast, S.; Stallinga, S.; Rieger, B. Detecting continuous structural heterogeneity in single-molecule localization microscopy data. *Sci. Rep.* **2023**, *13*, 19800. [CrossRef]
33. Wang, W.; Heydarian, H.; Huijben, T.A.P.M.; Stallinga, S.; Rieger, B. Joint registration of multiple point clouds for fast particle fusion in localization microscopy. *Bioinformatics* **2022**, *38*, 3281–3287. [CrossRef]
34. Nieuwenhuizen, R.P.J.; A Lidke, K.; Bates, M.; Puig, D.L.; Grünwald, D.; Stallinga, S.; Rieger, B. Measuring image resolution in optical nanoscopy. *Nat. Methods* **2013**, *10*, 557–562. [CrossRef]
35. van Vliet, L.J.; Verbeek, P.W. Edge localization by MoG filters: Multiple-of-Gaussians. *Pattern Recognit. Lett.* **1994**, *15*, 485–496. [CrossRef]
36. von Appen, A.; Kosinski, J.; Sparks, L.; Ori, A.; DiGuilio, A.L.; Vollmer, B.; Mackmull, M.-T.; Banterle, N.; Parca, L.; Kastritis, P.; et al. In situ structural analysis of the human nuclear pore complex. *Nature* **2015**, *526*, 140–143. [CrossRef] [PubMed]
37. Schlichthaerle, T.; Strauss, M.T.; Schueder, F.; Auer, A.; Nijmeijer, B.; Kueblbeck, M.; Sabinina, V.J.; Thevathasan, J.V.; Ries, J.; Ellenberg, J.; et al. Direct Visualization of Single Nuclear Pore Complex Proteins Using Genetically-Encoded Probes for DNA-PAINT. *Angew. Chem. Int. Ed. Engl.* **2019**, *58*, 13004–13008. [CrossRef] [PubMed]
38. Rankovic, S.; Ramalho, R.; Aiken, C.; Rousso, I. PF74 Reinforces the HIV-1 Capsid To Impair Reverse Transcription-Induced Uncoating. *J. Virol.* **2018**, *92*, e00845-18. [CrossRef]
39. Bhattacharya, A.; Alam, S.L.; Fricke, T.; Zadrozny, K.; Sedzicki, J.; Taylor, A.B.; Demeler, B.; Pornillos, O.; Ganser-Pornillos, B.K.; Diaz-Griffero, F.; et al. Structural basis of HIV-1 capsid recognition by PF74 and CPSF6. *Proc. Natl. Acad. Sci. USA* **2014**, *111*, 18625–18630. [CrossRef] [PubMed]
40. Bönisch, I.Z.; Dirix, L.; Lemmens, V.; Borrenberghs, D.; De Wit, F.; Vernaillen, F.; Rocha, S.; Christ, F.; Hendrix, J.; Hofkens, J.; et al. Capsid-Labelled HIV To Investigate the Role of Capsid during Nuclear Import and Integration. *J. Virol.* **2020**, *94*, e01024-19. [CrossRef]
41. Hulme, A.E.; Kelley, Z.; Foley, D.; Hope, T.J. Complementary Assays Reveal a Low Level of CA Associated with Viral Complexes in the Nuclei of HIV-1-Infected Cells. *J. Virol.* **2015**, *89*, 5350–5361. [CrossRef]
42. Balasubramaniam, M.; Zhou, J.; Addai, A.; Martinez, P.; Pandhare, J.; Aiken, C.; Dash, C. PF74 Inhibits HIV-1 Integration by Altering the Composition of the Preintegration Complex. *J. Virol.* **2019**, *93*, e01741-18. [CrossRef]
43. Robinson, G.L.; Dinsdale, D.; MacFarlane, M.; Cain, K. Switching from aerobic glycolysis to oxidative phosphorylation modulates the sensitivity of mantle cell lymphoma cells to TRAIL. *Oncogene* **2012**, *31*, 4996–5006. [CrossRef]
44. Song, S.B.; Hwang, E.S. A Rise in ATP, ROS, and Mitochondrial Content upon Glucose Withdrawal Correlates with a Dysregulated Mitochondria Turnover Mediated by the Activation of the Protein Deacetylase SIRT1. *Cells* **2018**, *8*, 11. [CrossRef]
45. Zhang, C.-S.; Hardie, D.G.; Lin, S.-C. Glucose Starvation Blocks Translation at Multiple Levels. *Cell Metab.* **2020**, *31*, 217–218. [CrossRef]
46. Wu, Y.-L.; Tschanz, A.; Krupnik, L.; Ries, J. Quantitative Data Analysis in Single-Molecule Localization Microscopy. *Trends Cell Biol.* **2020**, *30*, 837–851. [CrossRef]
47. Bates, M.; Jones, S.A.; Zhuang, X. Stochastic Optical Reconstruction Microscopy (STORM): A Method for Superresolution Fluorescence Imaging. *Cold Spring Harb. Protoc.* **2013**, *2013*, 498–520. [CrossRef]
48. Holzer, G.; Antonin, W. Breaking the Y. *PLoS Genet.* **2019**, *15*, e1008109. [CrossRef] [PubMed]
49. Goldberg, M.W. Nuclear pore complex tethers to the cytoskeleton. *Semin. Cell Dev. Biol.* **2017**, *68*, 52–58. [CrossRef]
50. Szymborska, A.; de Marco, A.; Daigle, N.; Cordes, V.C.; Briggs, J.A.G.; Ellenberg, J. Nuclear Pore Scaffold Structure Analyzed by Super-Resolution Microscopy and Particle Averaging. *Science* **2013**, *341*, 655–658. [CrossRef] [PubMed]
51. Rajoo, S.; Vallotton, P.; Onischenko, E.; Weis, K. Stoichiometry and compositional plasticity of the yeast nuclear pore complex revealed by quantitative fluorescence microscopy. *Proc. Natl. Acad. Sci. USA* **2018**, *115*, E3969–E3977. [CrossRef]
52. Allegretti, M.; Zimmerli, C.E.; Rantos, V.; Wilfling, F.; Ronchi, P.; Fung, H.K.H.; Lee, C.-W.; Hagen, W.; Turoňová, B.; Karius, K.; et al. In-cell architecture of the nuclear pore and snapshots of its turnover. *Nature* **2020**, *586*, 796–800. [CrossRef] [PubMed]

53. Elosegui-Artola, A.; Andreu, I.; Beedle, A.E.M.; Lezamiz, A.; Uroz, M.; Kosmalska, A.J.; Oria, R.; Kechagia, J.Z.; Rico-Lastres, P.; Le Roux, A.-L.; et al. Force Triggers YAP Nuclear Entry by Regulating Transport across Nuclear Pores. *Cell* **2017**, *171*, 1397–1410.e14. [CrossRef] [PubMed]
54. Meng, X.; Zhu, Y.; Tan, H.; Daraqel, B.; Ming, Y.; Li, X.; Yang, G.; He, X.; Song, J.; Zheng, L. The cytoskeleton dynamics-dependent LINC complex in periodontal ligament stem cells transmits mechanical stress to the nuclear envelope and promotes YAP nuclear translocation. *Stem Cell Res. Ther.* **2024**, *15*, 284. [CrossRef] [PubMed]

https://doi.org/10.3390/v18020167

viruses

MDPI

Article

Imaging Retroviral RNA Genome Heterodimers Using Bimolecular Fluorescence Complementation (BiFC)

Eunice C. Chen [1], Rebecca K. Maldonado [2,3] and Leslie J. Parent [2,3,*]

[1] Department of Medicine, Division of Pulmonary, Allergy, and Critical Care Medicine, Penn State College of Medicine, Hershey, PA 17033, USA; exc281@psu.edu

[2] Department of Molecular & Precision Medicine, Penn State College of Medicine, Hershey, PA 17033, USA; rjk297@psu.edu

[3] Department of Medicine, Division of Infectious Diseases and Epidemiology, Penn State College of Medicine, Hershey, PA 17033, USA

* Correspondence: lparent@psu.edu

Abstract: Retroviruses are single-stranded RNA viruses that package two copies of their positively stranded RNA genomes as a non-covalent dimer into newly formed virions. This process is evolutionarily conserved, and disruption of genome dimerization results in production of non-infectious virus particles. Genome dimers can be packaged as homodimers, containing two identical RNAs, or heterodimers, consisting of two genetically distinct copies. Genome dimerization generates genetic diversity, and different retroviruses have preferences for the type of genome dimers packaged into virions. We developed a novel imaging approach to specifically label and detect retroviral genome heterodimers in cells using a modified bimolecular fluorescence complementation (BiFC) technique. This method utilizes viral genomes encoding two different RNA stem-loop cassettes that each specifically binds to an RNA-binding protein conjugated to a split fluorophore. When two genetically different genomes are within close proximity, the fluorophore halves come together to reconstitute fluorescence. These BiFC-labeled RNA dimers can be visualized and tracked in living cells and interact with retroviral Gag proteins. This method has the advantage of low background fluorescence and can be applied to the study of dimeric or double-stranded RNAs of viruses and other organisms.

Keywords: RNA genome dimer; retroviruses; RSV; RNA-RNA intermolecular interactions; bimolecular fluorescence complementation; confocal microscopy; fluorescence imaging

1. Introduction

Retroviruses are significant pathogens of humans and animals, causing a variety of cancers and immunodeficiency syndromes. Their single-stranded, positive-sense RNA genomes are packaged as non-covalently linked dimers into virions that are released from the plasma membrane of infected cells. Genome dimerization is conserved amongst all *Orthoretrovirinae*, including Rous sarcoma virus (RSV), human immunodeficiency virus (HIV), and murine leukemia virus (MLV). Genome dimerization plays several significant roles in the virus replication cycle, including regulating the translation of viral proteins, facilitating strand transfer reactions during reverse transcription, and generating diversity through recombination of genetically distinct RNAs, all of which are advantageous for the virus [1–14]. Disruption of genome dimerization or mutation of the dimer initiation site near the 5′ end of the genome results in production of noninfectious virus particles.

The characteristics of retrovirus genome dimerization have been explored in vivo over the years, specifically to examine whether retroviruses package genome heterodimers (two genetically distinct copies of the genome dimerize and are co-packaged) or genome homodimers (two genomes identical in sequence dimerize and undergo packaging) [13,15–22]. Moloney MLV genomes dimerize co-transcriptionally in the nucleus, resulting in the preferential packaging of genome homodimers [16–19]. In lentiviruses, such as HIV type 1, genome dimerization occurs in a random fashion and there is no preference for packaging genome heterodimers or homodimers [9,13,20,22]. In the *Alpharetrovirus* RSV, genome heterodimers have been visualized in the nucleus, cytoplasm, and at the plasma membrane, indicating that dimers may arise during or shortly after transcription [15]. Both heterodimers and homodimers were detected in RSV particles, although homodimers were preferentially packaged.

The above in vivo studies utilized single-molecule RNA labeling methods to visualize retroviral RNA genomes within cells and single virions. These methods include the use of RNA aptamer and stem-loop systems as well as single-molecule RNA fluorescence in situ hybridization (smFISH). In the RNA aptamer and stem-loop system, a cassette of RNA stem-loops is inserted into the RNA of interest, and these loops are bound with high affinity by a protein that has been fused to a fluorophore (reviewed in [23–27]). The original stem-loop system used is the MS2 coat protein (CP) system, derived from the MS2 bacteriophage [28–33], in which 24 copies of the MS2 stem-loop inserted into the RNA are bound by the fluorophore-labeled MS2 CP (MCP). The MS2-MCP complex can be visualized by microscopy throughout all cellular compartments, including the nucleus, as long as the MCP contains a nuclear localization signal [28]. RNA stem-loops such as these do not naturally occur in mammalian cells, so the fluorophore-tagged MCP will not label other mammalian RNAs [25]. However, there can be diffuse background fluorescence signal present when MCPs not bound to their cognate RNAs accumulate in the nucleus. Other stem-loop aptamer systems that have been developed include the BglG binding protein (Bgl), based on the *Escherichia coli* BglG protein [20], and the PP7 CP (PCP), derived from the PP7 bacteriophage [31,34]. Similarly to the MS2-MCP system, the fluorophore-tagged Bgl and PP7 proteins contain a nuclear localization signal to allow visualization of RNA-protein complexes within the nucleus and dispersed throughout the cell.

In this study, we used two different stem-loop aptamer systems combined with bimolecular fluorescence complementation (BiFC) to specifically visualize retroviral genome heterodimers within cells. We constructed plasmids to express the MCP fused to the N-terminal half of the Venus fluorophore (MS2-VN) paired with constructs expressing the C-terminal half of the Venus fluorophore fused to either Bgl binding protein (Bgl-VC) or PCP (PP7-VC). When two retroviral genomes containing distinct RNA aptamers bound to VN or VC are in close proximity, the Venus fluorophore will be reconstituted to generate fluorescence visible by microscopy [35–39]. Normally, fluorescently tagged MS2, Bgl, and PP7 proteins have diffuse signals in the cell unless bound to their respective RNA aptamers. With BiFC, there is less diffuse background signal because there will only be fluorescence when the proteins bind their target RNAs and the RNAs dimerize, reforming the Venus fluorophore. Using this approach, we demonstrated that BiFC-labeled genome dimers are visible as functional fluorescently labeled punctate foci that traffic through the cell and interact with viral Gag proteins. The BIFC labeling system for RNA labeling can be used in both fixed-cell imaging and live-cell imaging with minimal background fluorescence.

2. Materials and Methods

2.1. Plasmid Construction

The replication-competent RSV provirus construct (RC.V8) [40] was used as the backbone to generate a construct containing a cassette of either 24 MS2 loops (RC. MS2) [28], 18 BglG loops (RC. Bgl) [20], or 24 PP7 loops (RC. PP7) [31] inserted between *gag* and *pol* to specifically label unspliced viral RNA. pCR4-24XMS2SL-stable was provided by Dr. Robert Singer (Albert Einstein Medical College, Bronx, NY, USA) (Addgene plasmid # 31865; http://n2t.net/addgene:31865; RRID:Addgene_31865) [28], as was pCR4-24XPP7SL (Addgene plasmid # 31864; http://n2t.net/addgene:31864; RRID:Addgene_31864) [31]. The BglG loops were removed from GagCeFP-BglSL, a gift from Dr. Wei-Shau Hu (National Cancer Institute at Frederick, Frederick, MD, USA) [20]. These constructs were then modified to insert the *cfp* gene after the *gag* gene (RC.GagCFP.MS2, RC.GagCFP.Bgl, and RC.GagCFP.PP7) to encode an internally expressed GagCFP as previously described [15,41]. The MS2-YFP construct was a gift from Dr. Robert Singer, the Bgl-mCherry was a gift from Dr. Wei-Shau Hu. phage-ubc-nls-ha-tdPCP-gfp was provided by Dr. Robert Singer (Addgene plasmid # 40650; http://n2t.net/addgene:40650; RRID:Addgene_40650). The PP7 coat protein was moved into the Bgl-mCherry backbone and the fluorophore was swapped out. The Venus fluorophore was split according to a previously published strategy [42]. The N-terminal 173 amino acids of Venus were fused to either BglG binding protein (Bgl-VN) or PP7 CP (PP7-VN) and the C-terminal 155 amino acids of Venus were fused to MCP (MS2-VC). pDEST-Sun1-mCherry was a gift from Dr. Jan Karlseder (Salk Institute for Biological Studies, La Jolla, CA, USA) [43].

2.2. Cell Culture and Transfection

Experiments were performed in QT6 cells [44], a chemically transformed quail fibroblast cell line maintained in F10 medium supplemented with 10% fetal calf serum, 9% tryptose phostphate, 1% chicken serum, penicillin and streptomycin, and amphotericin B and cultured at 38.5 °C with 5% CO_2. Cells were transfected with the calcium phosphate method [45]. For fixed-cell microscopy, cells were grown on glass coverslips. Twenty-four hours post-transfection, coverslips were fixed in 3.7% formaldehyde/PBS at room temperature, washed in PBS, 4′,6-diamidino-2-phenylindole (DAPI, MilliporeSigma, Rockville, MD, USA) stained, and mounted onto glass slides with Prolong diamond antifade (ThermoFisher Scientific, Waltham, MA, USA). For live-cell microscopy, cells were grown on glass-bottomed cell culture dishes (Mattek) prior to transfection, then washed once with 1x PBS, and phenol red-free imaging media was added prior to imaging (ThermoFisher Scientific, Waltham, MA, USA). For instances in which nuclei were labeled with DRAQ5 (ThermoFisher Scientific, Waltham, MA, USA), cells were incubated with primary growth medium containing DRAQ5 for 1 h at 37 °C.

2.3. Microscopy, Image Acquisition, and Image Analysis

Microscope images were obtained on a SP8 confocal microscope (Leica Microsystems, Wetzlar, Germany) using a 63x oil-immersion objective, and UV, white light, and argon lasers. Z-stacks were obtained with thickness of 0.3 μm through the width of the nucleus (as indicated by DAPI staining). Live-cell imaging was performed with the same microscope using a heated live-cell stage with a 63x water-immersion objective.

Images presented in Figures 2B,C and 3A were subjected to Huygens Professional 24.10 deconvolution (Scientific Volume Imaging, Hilversum, Netherlands) using the classical maximum likelihood estimation (CMLE) algorithm. Further image analysis and processing was completed using Imaris 10.2.0 (Bitplane an Oxford Instruments Company, Belfast, United Kingdom). Z-stacks were used to create cross-sections of the x, y; z, y; and x, z dimensions of the cell. Co-localization was assessed via signal-based analysis using Imaris and displayed as a white co-localization channel. The Imaris spot function was used to identify Gag, RNA dimers, and/or co-localized foci and to perform particle tracking. Histograms of all images were adjusted and images that were not deconvolved were subjected to a Gaussian filter.

3. Results

3.1. Visualization of Genome Heterodimers Within the Cell

To detect intracellular genomic RNA heterodimers, we co-transfected two genetically distinct retroviral genomes containing MS2 and Bgl RNA stem-loops or MS2 and PP7 RNA stem-loops with their respective binding proteins fused to a split Venus fluorophore (VN or VC). A schematic diagram of the viral constructs is shown in Figure 1A. The RNA stem-loops are located between *gag* and *pol* and therefore will only label unspliced viral RNAs, which are packaged into virions as the viral genome. Fluorescence is visible when the genomes are in close proximity, allowing the Venus fluorophore to reform (Figure 1B). To check for possible nonspecific background fluorescence, we co-expressed either PP7-VN or Bgl-VN with MS2-VC without their respective RNA stem-loops, and BiFC signal was not produced (Supplemental Figure S1). RSV genome constructs containing MS2 loops and PP7 loops with their respective split fluorophore-fused binding proteins were co-transfected into quail fibroblast QT6 cells, fixed with paraformaldehyde, and then imaged by confocal microscopy. Fluorescent Venus foci representing BIFC-labeled genome heterodimers were visible in the nucleus, cytoplasm, and along the plasma membrane (Figure 2A). A cross-section was generated from a Z-stack with cross-hairs (white arrows) marking a genome heterodimer, which can be seen in the nucleus (DAPI, white outline) in three dimensions. Genome heterodimers were also observed in the cytoplasm (magenta arrow) and appeared to be located along the plasma membrane (cyan arrow). Each fluorescent focus represented one or more genome heterodimers, composed of two genetically distinct retroviral RNA genomes, consistent with our previous studies showing that RSV genome heterodimers were present in all three locations in the cell [15]. When the BIFC-labeled MS2 (RC.GagCFP.MS2-24x + MS2-VC) and PP7 (RC.GagCFP.PP7-24x + PP7-VN) genome heterodimers (green) were visualized with Gag-CFP (red), colocalization (white) of genome heterodimers with Gag-CFP was seen within the nucleus (DAPI, white outline), as shown in three dimensions (Figure 2B). Similar results were obtained when MS2 (RC.GagCFP.MS2-24x + MS2-VC) and Bgl (RC.GagCFP.Bgl-18x + Bgl-VN) heterodimers (green) were expressed with Gag-CFP (red), and co-localized foci (white) were observed in the nucleus, cytoplasm, and at the plasma membrane (Figure 2C). Together these results demonstrated that both BiFC RNA stem-loop systems (MS2 + PP7 and MS2 + Bgl) were effective at labeling genome heterodimers in fixed cells.

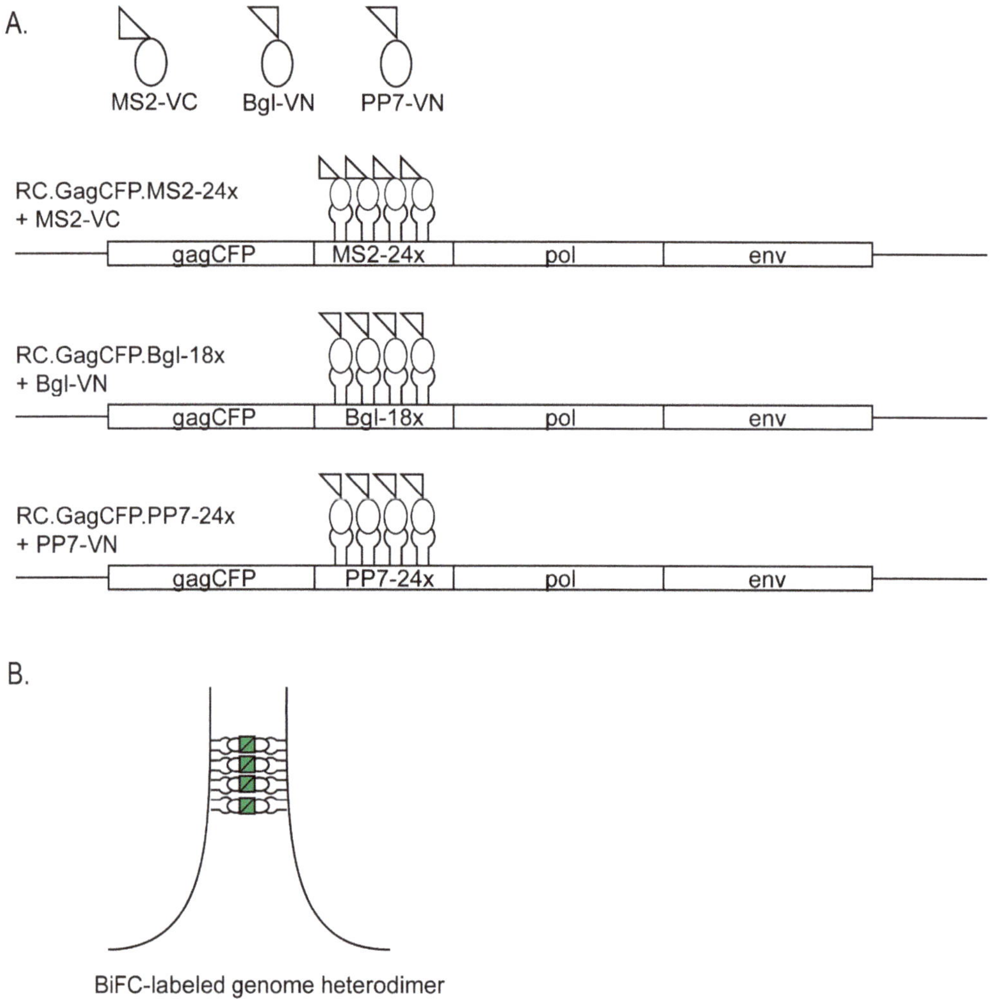

Figure 1. Overview of the bimolecular fluorescence complementation (BiFC) method used to visualize RSV genome dimers. (**A**) Schematic of the constructs used in this study. The MS2 CP was fused to the C-terminus of Venus to create MS2-VC, while Bgl binding protein or PP7 CP were fused to the Venus N-terminus (Bgl-VN and PP7-VN). Three RC.V8 constructs were generated to express GagCFP in *cis* and contained either 24 MS2 stem-loops (RC.GagCFP.MS2-24x), 18 Bgl stem-loops (RC.GagCFP.Bgl-18x), or 24 PP7 stem-loops (RC.GagCFP.PP7-24x). The split-fluorophore tagged proteins specifically bind their respective stem-loops. (**B**) When RC.GagCFP.MS2-24x bound to MS2-VC dimerized with either RC.GagCFP.Bgl-18x bound to Bgl-VN or RC.GagCFP.PP7-24x bound to PP7-VN, the Venus fluorophore will be reconstituted allowing for fluorescence to occur. In the absence of RNA binding, the coat proteins do not fluoresce (see Supplementary Figure S1).

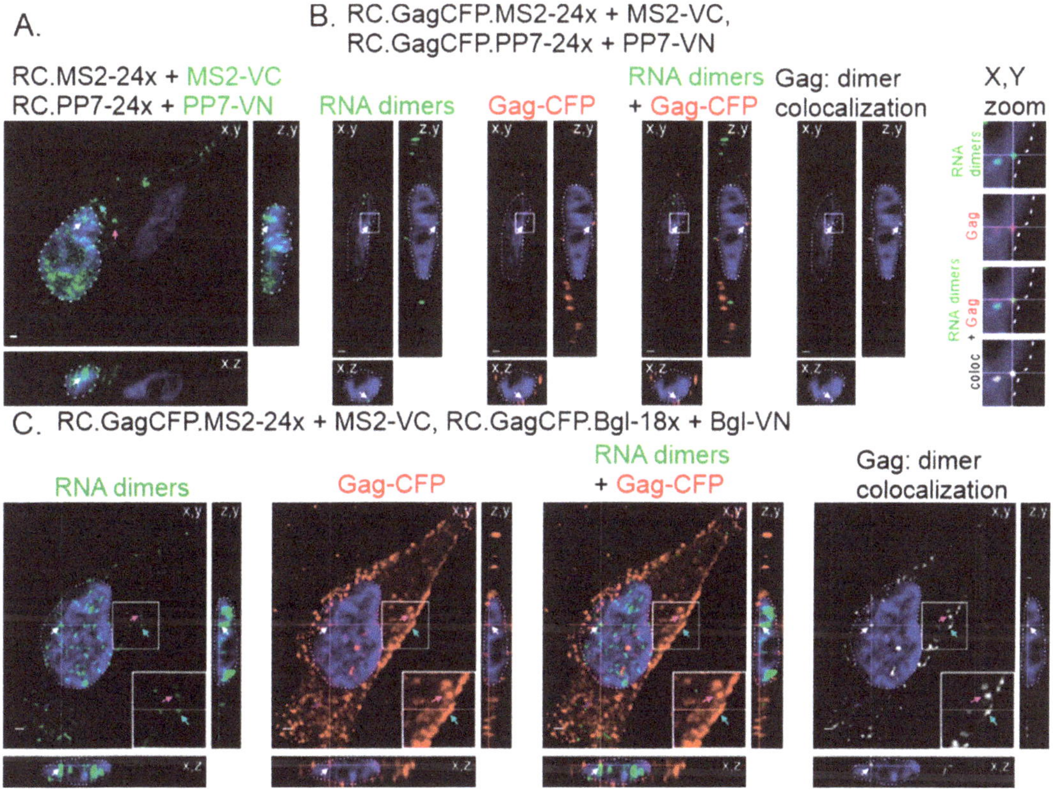

Figure 2. Labeling RSV genome heterodimers in fixed cells. (**A**) QT6 cells were transfected with RC.MS2-24x, MS2-VC, RC.PP7-24x, and PP7-VN. The RC.MS2-24x and RC.PP7-24x constructs did not express an internal Gag-CFP fusion protein. Z-stacks were collected via confocal microscopy and used to generate a cross-section. RNA genome heterodimers appeared as green foci throughout the cell. The white crosshairs and arrow indicate an RNA genome heterodimer in the nucleus (DAPI, dashed white outline). Foci were also observed in the cytoplasm (magenta arrow) and at the plasma membrane (cyan arrow). (**B**) Z-stacks of cells expressing RC.GagCFP.MS2-24x + MS2-VC and RC.GagCFP.PP7-24x + PP7-VN were used to generate cross-sections. An RNA genome heterodimer (green, left panel) was seen in the nucleus (DAPI, dashed white outline) as indicated by the white cross-hairs and arrow. The RC.V8 genome constructs express Gag fused to CFP (red). A Gag-genome heterodimer complex was observed in the nucleus (center panel, white arrow). A colocalization channel was generated and revealed Gag-genome heterodimer complexes (white) in the nucleus, cytoplasm and the plasma membrane (right panel). The focus of interest is outlined in the white square and included in the X,Y zoom in (right). (**C**) Cells co-expressing RC.GagCFP.MS2-24x + MS2-VC and RC.GagCFP.Bgl-18x + Bgl-VN (left panel) contained RNA genome heterodimers (green) in the nucleus (white arrow), cytoplasm (magenta arrow), and plasma membrane (green arrow). When Gag-CFP was visualized (red), Gag-genome heterodimer complexes were visualized in the nucleus, cytoplasm, and plasma membrane (white, magenta, and green arrows, respectively; center panel). Cytoplasmic and plasma membrane complexes are included in the inset. A colocalization channel was generated (white, right panel) revealing Gag-genome heterodimer complexes in the nucleus (white arrow), cytoplasm (magenta arrow), and at the plasma membrane (cyan arrow). Nuclear localization was confirmed by three-dimensional cross-sections of Z-stacks in the x,y; x,z; and z,y planes. Scale bar = 1 µm.

3.2. Live-Cell Imaging of BiFC-Labeled Genome Heterodimers in the Nucleus

Next, the BiFC system was used in living cells to visualize and characterize RSV genome heterodimers. QT6 cells were co-transfected with plasmids expressing RC.GagCFP.MS2-24x, RC.GagCFP.PP7-24x, MS2-VC, and PP7-VN. Cells were stained with DRAQ5 (blue) to visualize the nucleus (nuclear DNA) and live-cell microscopy was performed at a rate of approximately 3 frames sec^{-1}. An individual frame extracted from the imaging series shows a genome heterodimer in the nucleus (Figure 3A). In a subsequent experiment, Sun1-mCherry was co-transfected to mark the inner leaflet of the nuclear membrane (white dashed line) and QT6 cells were co-transfected with plasmids expressing RC.GagCFP.MS2-24x, RC.GagCFP.Bgl-18x, MS2-VC and Bgl-VN (Figure 3B). A genome heterodimer (green) and Gag (red) were visualized moving together inside the nucleus. Particle tracking was performed using Imaris software (indicated by red and green overlapping circles) at times 0, 18 and 54 secs. The visualization of a genome heterodimer: Gag complex within the nucleus suggests that genome dimerization in RSV can occur in the nucleus, as previously demonstrated [15], and dimers can be recognized by the nuclear Gag protein to form a viral ribonucleoprotein complex.

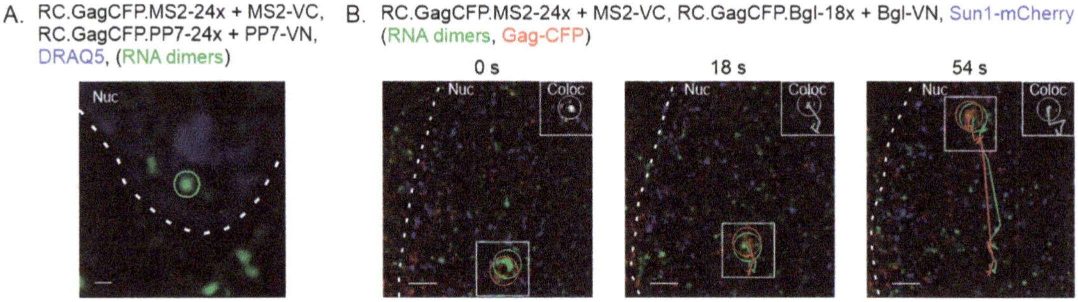

Figure 3. Live-cell imaging of nuclear RSV genome heterodimers. (**A**) QT6 cells were transfected with plasmids expressing RC.GagCFP.MS2-24x, RC.GagCFP.PP7-24x, PP7-VN, and MS2-VC. DRAQ5 was used to visualize the nucleus (blue). A genome heterodimer was tracked in the nucleus within a single confocal z-slice imaged approximately every 3 s. In a still image obtained at 24 s, a genome heterodimer (green focus) was visible within the nucleus (blue, dashed line). (**B**) QT6 cells were transfected with plasmids encoding RC.GagCFP.MS2-24x, RC.GagCFP.Bgl-18x, Bgl-VN, and MS2-VC. Sun1-mCherry was used to mark the inner leaflet of the nuclear membrane (blue). A complex containing RNA genome dimer (green) with Gag-CFP (red) colocalized in the nucleus and trafficked together. A colocalization channel (white) is shown in the upper right corner of the image, indicating that the red and green foci were overlapping. Scale bar = 0.5 µm.

3.3. BiFC-Labeled Genome Heterodimers Co-Localize with GagCFP in the Cytoplasm and at the Plasma Membrane

BiFC-labeled genome heterodimers were visualized co-localizing with Gag-CFP in the fixed cells shown in Figure 2. We sought to further characterize dynamic interactions of genome heterodimers and Gag using live-cell microscopy (Figure 4, Supplementary Videos S1–S3). QT6 cells were co-transfected with plasmids expressing retroviral genomes containing MS2 or Bgl RNA stem-loops and a Gag-CFP fusion protein (RC.GagCFP.MS2-24x and RC.GagCFP.Bgl-18x), the split fluorophore-tagged binding proteins (MS2-VC and Bgl-VN), and Sun1-mCherry to label the inner nuclear membrane. A single Z-plane was imaged approximately every 3 s. Colocalization between BiFC-labeled genome heterodimer foci (green) and Gag-CFP (red) was assessed via the generation of a colocalization channel (white). We then performed particle tracking of the heterodimers (green), Gag-CFP (red), and the colocalized signal (white). We focused on colocalized genome heterodimers and Gag-CFP foci that remained together throughout the duration of the movie (54 s), as demonstrated by the green (dimer), red (Gag), and white (colocalization) tracks in Figure 4. The Gag-genome heterodimer complex trafficked together over time in the cytoplasm, along the outer aspect of the nuclear membrane (labeled by Sun1 in blue and indicated by a white dashed line). The insets show an enlarged image of the tracked co-localized foci containing Gag and genomic RNA foci.

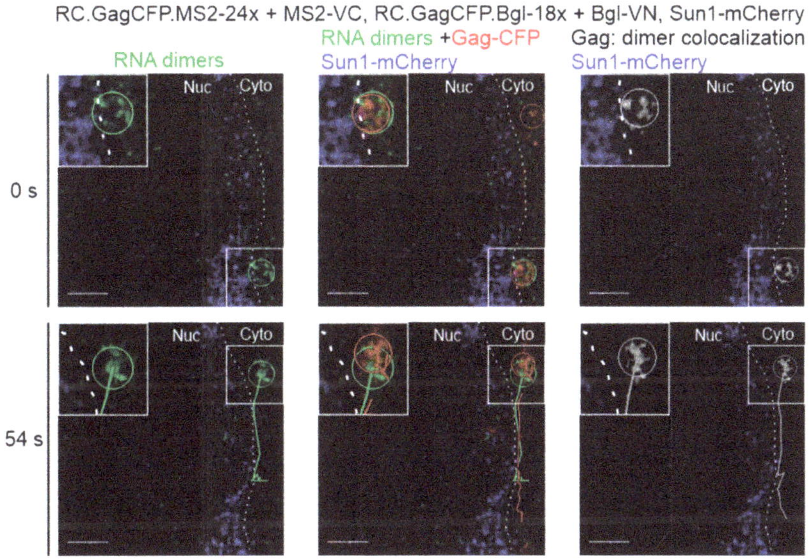

Figure 4. Live-cell imaging of RSV Gag-genome heterodimer complexes. QT6 cells were transfected with RC.GagCFPMS2-24x, MS2-VC, RC.GagCFP.Bgl-18x, and Bgl-VN. Sun1-mCherry was used to outline the inner nuclear membrane (blue). A single confocal Z-slice was imaged approximately every 3 s. RNA heterodimers appeared as green signal and form discrete foci. An RNA heterodimer (green line) and a Gag-CFP focus (red line) trafficked together in the cytoplasm (cyto) near the nuclear rim over time. A colocalization channel (white) was generated and tracked (white circle and line). Zoomed in images of the particles are included to show more detail (inset). Scale bar = 1 μm.

4. Discussion

We present a novel BiFC technique to selectively visualize retroviral genome heterodimers within fixed and living cells. We designed three-separate RSV constructs containing RNA stem-loop cassettes derived from MS2, Bgl, or PP7 that bind their cognate RNA binding proteins (MCP, Bgl, or PCP). The RNA binding proteins were fused to either the N-terminal or C-terminal half of the Venus fluorophore (MS2-VC, Bgl-VN, or PP7-VN). In this BIFC system, fluorescence was only visible when the genomes were within close proximity, allowing the Venus fluorophore to refold (Figure 1). The fluorescent foci visualized represent RSV genomic RNA heterodimers. When the genomic RNA constructs and their respective fluorophore-labeled binding proteins were visualized in fixed cells, BiFC-labeled genome heterodimers were seen within the nucleus, in the cytoplasm, and at the plasma membrane (Figure 2), recapitulating the intracellular pathway for RSV particle assembly, similar to our previous study [15]. When the binding proteins were expressed alone, without their respective RNA aptamer-containing RSV genomes, no BiFC fluorescence was visualized (Supplemental Figure S1). Using live-cell microscopy, we visualized BiFC-labeled genome heterodimers in complex with Gag proteins moving within the nucleus and in the cytoplasm (Figures 3 and 4, Supplementary videos S1–S3). Together, these findings demonstrate that BiFC-labeled genome heterodimers maintained the typical trafficking pathway and function of a genome dimer and were bound by the Gag protein. From these experiments, we cannot discern whether more than a single genome dimer was present in each of the Venus-labeled RNA puncta, and additional methods would be needed to attempt to estimate the number of RNA molecules present in each focus. This study provides evidence for the use of BiFC as a means to visualize retroviral genome heterodimers trafficking within a cell and in complex with the viral Gag protein. We believe this technique opens the door to further characterize genome heterodimer formation and dynamic movement during retroviral genome packaging and virion assembly. In addition, a similar BIFC approach could be adapted for the study of double-stranded viruses and cellular processes involving RNA multimers [46,47].

There are many benefits to using a BiFC-labeling system for the study of RNA dimers and multimers. One notable improvement over other detection systems is the decrease in background fluorescence from unbound fluorescently tagged binding proteins. In traditional RNA stem-loop and binding protein systems, the binding protein contains a nuclear localization signal and generates a low level of background fluorescence in the nucleus. However, in the BiFC-labeling system, fluorescence is only visible when the fluorophore halves are able to reform, resulting in less diffuse background fluorescence. Importantly, we showed that expression of the coat proteins alone without their respective RNA stem-loops did not result in visible fluorescence (Supplementary Figure S1). This result also suggests that the RNA binding proteins fused to Venus VN and VC do not produce pre-formed fluorescent signals prior to binding their cognate RNA stem-loops. Additionally, the MS2 and PP7 stem-loops have been shown to bind their binding proteins in a co-transcriptional fashion [28]. This idea is supported by the finding that fluorescently tagged BiFC genome heterodimer foci were visualized in nuclei, as demonstrated in an earlier study using a different detection system [15]. Therefore, it is more likely that the stem-loop containing genomes bind their respective binding proteins first and then dimerize when the RNAs move into close approximation, resulting in refolding of the Venus fluorophore.

Importantly, the BiFC-labeling system is flexible in terms of the RNA stem-loop cassettes utilized and compatibility with live-cell imaging. We could detect BiFC-labeled genome heterodimer formation using both the MS2/Bgl stem-loop and MS2/PP7 stem-loop systems, allowing us to visualize and characterize genome heterodimers as they functioned within living cells. These genome heterodimers were observed moving within different subcellular compartments (Figures 2–4), suggesting the inclusion of RNA stem-loops and binding of the MCP, BglG, and PCP did not interfere with trafficking of the viral RNA genome. Additionally, we visualized genome heterodimers that colocalized and trafficked with Gag-CFP in living cells (Figures 3 and 4, Supplementary Videos S1–S3), suggesting the inclusion of RNA stem-loops and binding of the VN and VC proteins did not preclude viral RNA:Gag protein interactions.

Of note, the PP7 binding protein does undergo nucleolar localization [48], although this quality does not appear to interfere with visualization of genome heterodimer in other parts of the cell. However, this point is important to keep in mind if the RNA of interest is predominantly localized to nucleoli, as there may be an unacceptable level of background fluorescence. Additionally, the BiFC interaction is irreversible once the fluorophore has reformed following interaction of the molecules of interest. This property can be helpful when studying transient interactions, as it allows for stabilization of brief binding events. However, because the interaction is locked in, there may be overrepresentation or skewing of downstream events from a transient interaction, especially when used in live-cell imaging. For example, in the case of dynamic interactions, once the binding between the two molecules occurs, this initial complex could be overrepresented and could prevent binding of new partners. It is important to keep these considerations in mind when employing BiFC-labeling to be certain that the characteristics of the system are applicable to the unique traits of the molecules of interest.

In recent years, there has been an increased interest in RNA-RNA interactions (RRI) in areas such as gene expression regulation, RNA metabolism, and RNA virus replication (reviewed in [49]). These interactions can be intermolecular (between two different RNA molecules) or intramolecular (between two distinct regions of the same RNA molecule). Numerous innovative biochemical and imaging techniques have been developed to allow researchers to detect and map RRIs (reviewed in [49]). Additionally, there have been many new methods to visualize single-RNA molecules using live-cell microscopy (reviewed in [27]). We present BiFC as an additional technique to further characterize known RRIs within fixed and living cells. As we have demonstrated, the strength of BiFC-labeling of RRIs is the ability to visualize subcellular localization and interaction with other host or viral proteins in both fixed and living cells with minimal background fluorescence.

Insertion of long RNA stem-loop cassettes can potentially affect structure, function, of transport of RNAs and should be considered when designing experiments to study retroviral genomes. Importantly, these aptamers have been used to study retroviral RNA trafficking and encapsidation in several previous publications, as reviewed recently [50]. However, we and others have shown that the addition of RNA stem-loops does not appear to interfere significantly with the trafficking and packaging of retroviral RNAs [13,15,20,41,51–53], and if placed in a nonessential region, infectivity can be maintained [41]. Importantly, the ability to visualize BiFC-tagged RNA dimers within living cells allows the further characterization of RNA-RNA intermolecular interactions in a novel way. Because "seeing is believing" [54] when it comes to RNA-RNA intermolecular interactions, this technique can be the starting point to developing new experiments to better understand their lifespan and function.

Supplementary Materials: The following supporting information can be downloaded at: https://www.mdpi.com/article/10.3390/v17081112/s1, Video S1: Trafficking of heterodimer in the cytoplasm; Video S2: Gag-RNA heterodimer co-trafficking in the cytoplasm, merged images of Gag and RNA channels from Video S1.; Video S3: Gag-heterodimer co-localization channel and particle tracking from Video S3; Video S4: A white colocalization channel was generated of RSV Gag-genome heterodimer complexes presented in Supplementary Videos 2 and 3; Figure S1: Expression of coat proteins without their respective RNAs.

Author Contributions: Conceptualization, E.C.C. and L.J.P.; methodology, E.C.C. and L.J.P.; formal analysis, E.C.C. and R.K.M.; investigation, E.C.C.; resources, L.J.P.; data curation, E.C.C. and R.K.M.; writing—original draft preparation, E.C.C., R.K.M., L.J.P.; writing—review and editing, E.C.C., R.K.M., L.J.P.; visualization, E.C.C. and L.J.P.; supervision, L.J.P.; project administration, L.J.P.; funding acquisition, L.J.P. All authors have read and agreed to the published version of the manuscript.

Funding: This work was supported by grants from the National Institutes of Health, R01 GM139392 (L.J.P.), R01 CA76534 (L.J.P.), and F30 CA214010 (E.C.C.), as well as support from Penn State College of Medicine including the Summer Bridge Award (L.J.P.). The content is solely the responsibility of the authors and does not necessarily represent the official views of the National Institutes of Health, Penn State University, or the Penn State College of Medicine.

Data Availability Statement: Data and reagents will be made available upon request. Correspondence and requests for materials can be addressed to Leslie J. Parent.

Acknowledgments: We thank Jordan Chang, Kevin M. Tuffy, and Breanna L. Rice (Penn State College of Medicine) for helpful discussions. We also thank Malgorzata Sudol for technical assistance. Microscopy images were generated and processed in the Penn State College of Medicine Advanced Light Microscopy Core (RRID: SCR_022526). The Advanced Light Microscopy Core services and instruments used in this project were funded, in part, by the Pennsylvania State University College of Medicine via the Office of the Vice Dean of Research and Graduate Students and the Pennsylvania Department of Health using Tobacco Settlement Funds (CURE). The Pennsylvania Department of Health specifically disclaims responsibility for any analyses, interpretations or conclusions.

Conflicts of Interest: The authors declare no conflicts of interest.

Abbreviations

The following abbreviations are used in this manuscript:

BiFC	Bimolecular fluorescence complementation
RSV	Rous sarcoma virus
HIV	Human immunodeficiency virus
MLV	Murine leukemia virus
smFISH	Single-molecule RNA fluorescence in situ hybridization
CP	Coat protein
MCP	MS2 coat protein
Bgl	BglG system
PCP	PP7 coat protein
RRI	RNA-RNA interactions

References

1. Berkhout, B.; van Wamel, J.L. Role of the DIS hairpin in replication of human immunodeficiency virus type 1. *J. Virol.* **1996**, *70*, 6723–6732. [CrossRef]
2. Garbitt, R.A.; Albert, J.A.; Kessler, M.D.; Parent, L.J. trans-acting inhibition of genomic RNA dimerization by Rous sarcoma virus matrix mutants. *J. Virol.* **2001**, *75*, 260–268. [CrossRef]
3. Laughrea, M.; Jetté, L.; Mak, J.; Kleiman, L.; Liang, C.; Wainberg, M.A. Mutations in the kissing-loop hairpin of human immunodeficiency virus type 1 reduce viral infectivity as well as genomic RNA packaging and dimerization. *J. Virol.* **1997**, *71*, 3397–3406. [CrossRef]
4. Parent, L.J.; Cairns, T.M.; Albert, J.A.; Wilson, C.B.; Wills, J.W.; Craven, R.C. RNA dimerization defect in a Rous sarcoma virus matrix mutant. *J. Virol.* **2000**, *74*, 164–172. [CrossRef]
5. Sakuragi, J.; Ueda, S.; Iwamoto, A.; Shioda, T. Possible role of dimerization in human immunodeficiency virus type 1 genome RNA packaging. *J. Virol.* **2003**, *77*, 4060–4069. [CrossRef] [PubMed]
6. Houzet, L.; Paillart, J.C.; Smagulova, F.; Maurel, S.; Morichaud, Z.; Marquet, R.; Mougel, M. HIV controls the selective packaging of genomic, spliced viral and cellular RNAs into virions through different mechanisms. *Nucleic Acids Res.* **2007**, *35*, 2695–2704. [CrossRef] [PubMed]
7. Dubois, N.; Marquet, R.; Paillart, J.C.; Bernacchi, S. Retroviral RNA Dimerization: From Structure to Functions. *Front. Microbiol.* **2018**, *9*, 527. [CrossRef]
8. Johnson, S.F.; Telesnitsky, A. Retroviral RNA dimerization and packaging: The what, how, when, where, and why. *PLoS Pathog.* **2010**, *6*, e1001007. [CrossRef]
9. Moore, M.D.; Hu, W.S. HIV-1 RNA dimerization: It takes two to tango. *AIDS Rev.* **2009**, *11*, 91–102.
10. Paillart, J.C.; Shehu-Xhilaga, M.; Marquet, R.; Mak, J. Dimerization of retroviral RNA genomes: An inseparable pair. *Nat. Rev. Microbiol.* **2004**, *2*, 461–472. [CrossRef]
11. Hu, W.S.; Temin, H.M. Genetic consequences of packaging two RNA genomes in one retroviral particle: Pseudodiploidy and high rate of genetic recombination. *Proc. Natl. Acad. Sci. USA* **1990**, *87*, 1556–1560. [CrossRef]
12. Mikkelsen, J.G.; Pedersen, F.S. Genetic reassortment and patch repair by recombination in retroviruses. *J. Biomed. Sci.* **2000**, *7*, 77–99. [CrossRef]

13. Moore, M.D.; Nikolaitchik, O.A.; Chen, J.; Hammarskjold, M.L.; Rekosh, D.; Hu, W.S. Probing the HIV-1 genomic RNA trafficking pathway and dimerization by genetic recombination and single virion analyses. *PLoS Pathog.* **2009**, *5*, e1000627. [CrossRef]
14. Beemon, K.; Duesberg, P.; Vogt, P. Evidence for crossing-over between avian tumor viruses based on analysis of viral RNAs. *Proc. Natl. Acad. Sci. USA* **1974**, *71*, 4254–4258. [CrossRef] [PubMed]
15. Chen, E.C.; Maldonado, R.J.K.; Parent, L.J. Visualizing Rous Sarcoma Virus Genomic RNA Dimerization in the Nucleus, Cytoplasm, and at the Plasma Membrane. *Viruses* **2021**, *13*, 903. [CrossRef] [PubMed]
16. Flynn, J.A.; An, W.; King, S.R.; Telesnitsky, A. Nonrandom dimerization of murine leukemia virus genomic RNAs. *J. Virol.* **2004**, *78*, 12129–12139. [CrossRef]
17. Flynn, J.A.; Telesnitsky, A. Two distinct Moloney murine leukemia virus RNAs produced from a single locus dimerize at random. *Virology* **2006**, *344*, 391–400. [CrossRef] [PubMed]
18. Maurel, S.; Houzet, L.; Garcia, E.L.; Telesnitsky, A.; Mougel, M. Characterization of a natural heterodimer between MLV genomic RNA and the SD' retroelement generated by alternative splicing. *RNA* **2007**, *13*, 2266–2276. [CrossRef]
19. Kharytonchyk, S.A.; Kireyeva, A.I.; Osipovich, A.B.; Fomin, I.K. Evidence for preferential copackaging of Moloney murine leukemia virus genomic RNAs transcribed in the same chromosomal site. *Retrovirology* **2005**, *2*, 3. [CrossRef]
20. Chen, J.; Nikolaitchik, O.; Singh, J.; Wright, A.; Bencsics, C.E.; Coffin, J.M.; Ni, N.; Lockett, S.; Pathak, V.K.; Hu, W.S. High efficiency of HIV-1 genomic RNA packaging and heterozygote formation revealed by single virion analysis. *Proc. Natl. Acad. Sci. USA* **2009**, *106*, 13535–13540. [CrossRef]
21. Chen, J.; Rahman, S.A.; Nikolaitchik, O.A.; Grunwald, D.; Sardo, L.; Burdick, R.C.; Plisov, S.; Liang, E.; Tai, S.; Pathak, V.K.; et al. HIV-1 RNA genome dimerizes on the plasma membrane in the presence of Gag protein. *Proc. Natl. Acad. Sci. USA* **2016**, *113*, E201–E208. [CrossRef] [PubMed]
22. Ferrer, M.; Clerté, C.; Chamontin, C.; Basyuk, E.; Lainé, S.; Hottin, J.; Bertrand, E.; Margeat, E.; Mougel, M. Imaging HIV-1 RNA dimerization in cells by multicolor super-resolution and fluctuation microscopies. *Nucleic Acids Res.* **2016**, *44*, 7922–7934. [CrossRef] [PubMed]
23. Bann, D.V.; Parent, L.J. Application of live-cell RNA imaging techniques to the study of retroviral RNA trafficking. *Viruses* **2012**, *4*, 963–979. [CrossRef]
24. Urbanek, M.O.; Galka-Marciniak, P.; Olejniczak, M.; Krzyzosiak, W.J. RNA imaging in living cells—Methods and applications. *RNA Biol.* **2014**, *11*, 1083–1095. [CrossRef]
25. Tingey, M.; Schnell, S.J.; Yu, W.; Saredy, J.; Junod, S.; Patel, D.; Alkurdi, A.A.; Yang, W. Technologies Enabling Single-Molecule Super-Resolution Imaging of mRNA. *Cells* **2022**, *11*, 3079. [CrossRef] [PubMed]
26. Le, P.; Ahmed, N.; Yeo, G.W. Illuminating RNA biology through imaging. *Nat. Cell Biol.* **2022**, *24*, 815–824. [CrossRef]
27. Braselmann, E.; Rathbun, C.; Richards, E.M.; Palmer, A.E. Illuminating RNA Biology: Tools for Imaging RNA in Live Mammalian Cells. *Cell Chem. Biol.* **2020**, *27*, 891–903. [CrossRef]
28. Bertrand, E.; Chartrand, P.; Schaefer, M.; Shenoy, S.M.; Singer, R.H.; Long, R.M. Localization of ASH1 mRNA particles in living yeast. *Mol. Cell* **1998**, *2*, 437–445. [CrossRef]
29. Wan, Y.; Anastasakis, D.G.; Rodriguez, J.; Palangat, M.; Gudla, P.; Zaki, G.; Tandon, M.; Pegoraro, G.; Chow, C.C.; Hafner, M.; et al. Dynamic imaging of nascent RNA reveals general principles of transcription dynamics and stochastic splice site selection. *Cell* **2021**, *184*, 2878–2895.e2820. [CrossRef]
30. Coulon, A.; Ferguson, M.L.; de Turris, V.; Palangat, M.; Chow, C.C.; Larson, D.R. Kinetic competition during the transcription cycle results in stochastic RNA processing. *Elife* **2014**, *3*, e03939. [CrossRef]
31. Larson, D.R.; Zenklusen, D.; Wu, B.; Chao, J.A.; Singer, R.H. Real-time observation of transcription initiation and elongation on an endogenous yeast gene. *Science* **2011**, *332*, 475–478. [CrossRef] [PubMed]
32. Fusco, D.; Bertrand, E.; Singer, R.H. Imaging of single mRNAs in the cytoplasm of living cells. *Prog. Mol. Subcell. Biol.* **2004**, *35*, 135–150.
33. Fusco, D.; Accornero, N.; Lavoie, B.; Shenoy, S.M.; Blanchard, J.M.; Singer, R.H.; Bertrand, E. Single mRNA molecules demonstrate probabilistic movement in living mammalian cells. *Curr. Biol.* **2003**, *13*, 161–167. [CrossRef]
34. Chao, J.A.; Patskovsky, Y.; Almo, S.C.; Singer, R.H. Structural basis for the coevolution of a viral RNA-protein complex. *Nat. Struct. Mol. Biol.* **2008**, *15*, 103–105. [CrossRef]
35. Kudla, J.; Bock, R. Lighting the Way to Protein-Protein Interactions: Recommendations on Best Practices for Bimolecular Fluorescence Complementation Analyses. *Plant Cell* **2016**, *28*, 1002–1008. [CrossRef]
36. Lai, H.T.; Chiang, C.M. Bimolecular Fluorescence Complementation (BiFC) Assay for Direct Visualization of Protein-Protein Interaction. *Bio Protoc.* **2013**, *3*, e935. [CrossRef]
37. Kerppola, T.K. Bimolecular fluorescence complementation: Visualization of molecular interactions in living cells. *Methods Cell Biol.* **2008**, *85*, 431–470. [CrossRef]

38. Hu, C.D.; Grinberg, A.V.; Kerppola, T.K. Visualization of protein interactions in living cells using bimolecular fluorescence complementation (BiFC) analysis. *Curr. Protoc. Cell Biol.* **2006**, *29*, 21.3.1–21.3.21. [CrossRef]
39. Hu, C.D.; Chinenov, Y.; Kerppola, T.K. Visualization of interactions among bZIP and Rel family proteins in living cells using bimolecular fluorescence complementation. *Mol. Cell* **2002**, *9*, 789–798. [CrossRef]
40. Craven, R.C.; Leure-duPree, A.E.; Weldon, R.A., Jr.; Wills, J.W. Genetic analysis of the major homology region of the Rous sarcoma virus Gag protein. *J. Virol.* **1995**, *69*, 4213–4227. [CrossRef] [PubMed]
41. Maldonado, R.J.K.; Rice, B.; Chen, E.C.; Tuffy, K.M.; Chiari, E.F.; Fahrbach, K.M.; Hope, T.J.; Parent, L.J. Visualizing Association of the Retroviral Gag Protein with Unspliced Viral RNA in the Nucleus. *mBio* **2020**, *11*, e00524-20. [CrossRef] [PubMed]
42. Shyu, Y.J.; Liu, H.; Deng, X.; Hu, C.D. Identification of new fluorescent protein fragments for bimolecular fluorescence complementation analysis under physiological conditions. *Biotechniques* **2006**, *40*, 61–66. [CrossRef]
43. Crabbe, L.; Cesare, A.J.; Kasuboski, J.M.; Fitzpatrick, J.A.; Karlseder, J. Human telomeres are tethered to the nuclear envelope during postmitotic nuclear assembly. *Cell Rep.* **2012**, *2*, 1521–1529. [CrossRef]
44. Moscovici, C.; Moscovici, M.G.; Jimenez, H.; Lai, M.M.; Hayman, M.J.; Vogt, P.K. Continuous tissue culture cell lines derived from chemically induced tumors of Japanese quail. *Cell* **1977**, *11*, 95–103. [CrossRef]
45. Fujiwara, T.; Oda, K.; Yokota, S.; Takatsuki, A.; Ikehara, Y. Brefeldin A causes disassembly of the Golgi complex and accumulation of secretory proteins in the endoplasmic reticulum. *J. Biol. Chem.* **1988**, *263*, 18545–18552. [CrossRef] [PubMed]
46. Ying, Y.; Yang, Y.; Chen, A.K. Roles of RNA scaffolding in nanoscale Gag multimerization and selective protein sorting at HIV membranes. *Sci. Adv.* **2024**, *10*, eadk8297. [CrossRef]
47. Bevilacqua, P.C.; Williams, A.M.; Chou, H.L.; Assmann, S.M. RNA multimerization as an organizing force for liquid-liquid phase separation. *RNA* **2022**, *28*, 16–26. [CrossRef]
48. Ferguson, M.L.; Larson, D.R. Measuring transcription dynamics in living cells using fluctuation analysis. *Methods Mol. Biol.* **2013**, *1042*, 47–60. [CrossRef]
49. Ye, R.; Zhao, H.; Wang, X.; Xue, Y. Technological advancements in deciphering RNA-RNA interactions. *Mol. Cell* **2024**, *84*, 3722–3736. [CrossRef]
50. Guerrero, J.F.; Lesko, S.L.; Evans, E.L.; Sherer, N.M. Studying Retroviral Life Cycles Using Visible Viruses and Live Cell Imaging. *Annu. Rev. Virol.* **2024**, *11*, 125–146. [CrossRef] [PubMed]
51. Kemler, I.; Meehan, A.; Poeschla, E.M. Live-cell coimaging of the genomic RNAs and Gag proteins of two lentiviruses. *J. Virol.* **2010**, *84*, 6352–6366. [CrossRef] [PubMed]
52. Pocock, G.M.; Becker, J.T.; Swanson, C.M.; Ahlquist, P.; Sherer, N.M. HIV-1 and M-PMV RNA Nuclear Export Elements Program Viral Genomes for Distinct Cytoplasmic Trafficking Behaviors. *PLoS Pathog.* **2016**, *12*, e1005565. [CrossRef] [PubMed]
53. Jouvenet, N.; Simon, S.M.; Bieniasz, P.D. Imaging the interaction of HIV-1 genomes and Gag during assembly of individual viral particles. *Proc. Natl. Acad. Sci. USA* **2009**, *106*, 19114–19119. [CrossRef]
54. Carmo-Fonseca, M. RNA imaging: Seeing is believing. *RNA* **2015**, *21*, 580–581. [CrossRef] [PubMed]

viruses

MDPI

Article

Interactions Between Hantavirus Nucleoprotein and Glycoproteins: A Quantitative Fluorescence Microscopy Study

Amit Koikkarah Aji, Titas Mandal and Salvatore Chiantia *

University of Potsdam, Institute of Biochemistry and Biology, Karl-Liebknecht-Straße 24-25, 14476 Potsdam, Germany; amitkoikkarah93@gmail.com (A.K.A.); titas.mandal@uni-potsdam.de (T.M.)
* Correspondence: chiantia@uni-potsdam.de

Abstract: Orthohantaviruses are tri-segmented negative-sense RNA viruses that can cause severe pathologies in humans. Currently, limited information exists on the molecular interactions driving orthohantavirus assembly in infected cells. Specifically, it is not clear how its glycoproteins (i.e., Gn and Gc) interact with other viral or host molecules. In this study, we use one- and two-color Number and Brightness fluorescence microscopy approaches to quantitatively characterize the interactions between orthohantavirus glycoproteins and the nucleoprotein in transfected cells. Our results indicate that orthohantavirus nucleoprotein homo-interactions are strongly affected by the host environment. Furthermore, we report evidence of Gc–nucleoprotein interactions, based on (i) the high fluorescence cross-correlation between these two proteins and (ii) the increased Gc-Gc interactions observed in the presence of nucleoprotein. Finally, experiments on a Gc deletion mutant suggest that the observed protein–protein interactions are mediated by the cytoplasmic tail of Gc. In conclusion, this study provides new insights into the role of the interactions between orthohantavirus glycoproteins and nucleoprotein in the context of viral assembly.

Keywords: protein–protein interactions; fluorescence microscopy; multimerization; virus assembly; fluorescence fluctuations; fluorescent proteins

1. Introduction

Orthohantaviruses (HVs) (Family: Hantaviridae; Order: Bunyavirales) belong to an emerging group of zoonotic viruses that cause sporadic outbreaks around the world [1]. HVs are tri-segmented negative-sense RNA viruses: the S segment encodes the nucleocapsid protein, the M segment encodes the glycoproteins (GPs), and the L segment encodes the RNA-dependent RNA polymerase (RdRp) [2]. The interplay between HV GPs (named Gn and Gc) to form the outer envelope spike, consisting of Gn-Gc hetero-octamers, is the initial step of the viral assembly process [2–4]. However, limited information exists on the precise mechanisms regulating the molecular interactions involved in the subsequent stages of viral assembly. In this regard, a key event is the association between ribonucleoproteins (RNPs: comprised of viral RdRP, nucleoprotein-NP, and viral RNA) [5] and GP spikes (Figure 1). This interaction is critical for viral genome packaging and influences the progression of the viral assembly by bringing pre-assembled GP spikes into proximity, eventually leading to the formation of a new virion [2].

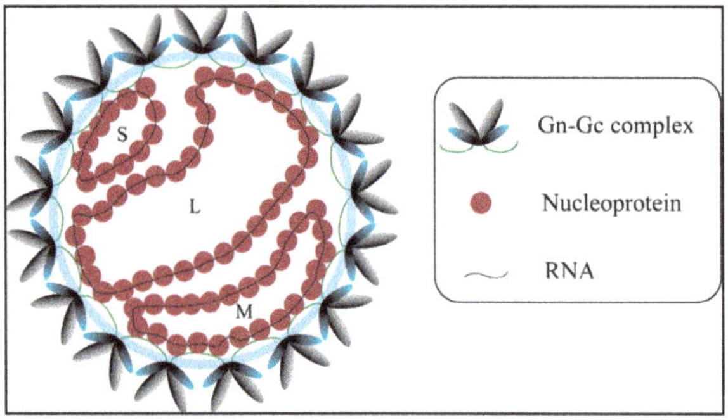

Figure 1. Hantavirus structure and structural proteins. The glycoproteins Gn and Gc are embedded in the lipid bilayer (light blue) and form hetero-complexes. Nucleoproteins bound to the viral RNA might interact with the cytoplasmic tails (green lines) of glycoproteins. The three RNA segments S, M, and L are labelled accordingly. Adapted from Hepojoki et al. [2].

Our current understanding of the interaction between GP spikes and RNPs in the context of virion assembly is described by the model proposed by Hepojoki et al. [2]. This states that the cytoplasmic tail (CT) of Gn initially associates with the RNP after GP spike formation. This interaction alters the orientation of the Gn CT, thus allowing Gc to bind to the RNP via its CT. The RNP-Gn complex further regulates viral transcription and eventually promotes viral assembly. Previous biochemical studies have attempted to validate this model by primarily identifying the interacting domains between the RNP complex and GP spikes. First, the regions within the CTs of Gn and Gc responsible for the binding to NP in multiple HV strains were mapped [6,7]. Further, it was reported that the zinc finger domains in the Gn CT and the middle domain of NP are needed for NP-Gn interactions [7]. Additionally, the NP residues and nucleotides from genomic RNA that facilitate the binding of RNP to GP CTs were also characterized [8]. Finally, the observation of colocalization of Hantaan virus NP and GPs in transfected cells indicates that NP promotes Gc partition into the Golgi apparatus and that NP-Gc interactions are essential for the stabilization of the GP spike complex [9].

Despite these significant advances, some key mechanistic aspects of the proposed assembly model have not yet been explored quantitatively beyond the GP spike formation step. In this context, the molecular details of the association between NP and individual GPs are particularly relevant. HVs lack a matrix protein that mediates the interaction between the GP spike and the RNP complex and organizes the assembly of the progeny virus. The direct NP-GP association is in fact hypothesized to act as a substitute for the lack of a matrix protein [10]. However, details regarding the affinity between NP and individual GPs and the role of other viral components in stabilizing the NP-GP complex remain unclear. This is primarily due to limited quantitative information on NP-GP interactions in HV assembly, particularly within living cells. Such insights are essential for accurately understanding how the viral components are packaged to form the final structure of the nascent virion.

Here, we present a quantitative characterization of NP-GP interactions using Puumala orthohantavirus (PUUV) constructs expressed in multiple living cell models. Recently, we showed significant colocalization between fluorescently labelled PUUV NP and GPs upon co-expression of NP, Gn, and Gc [11]. However, these findings could only qualitatively suggest the presence of NP-Gc and NP-Gn interactions. In this new study, we aim to quantify NP-NP and NP-GP interactions in human or rodent epithelial cell models.

2. Materials and Methods

In this work, we examine NP-GP association by assessing the variations in protein multimerization and quantifying protein–protein interactions, upon co-expression of fluorescently labeled PUUV NP and GPs (Gn and Gc) in living cells. This is achieved by applying Number and Brightness (N&B) and cross-correlation N&B (ccN&B) approaches, with single-cell resolution. These methods based on the statistical analysis of fluorescence fluctuations are particularly sensitive at very low protein expression levels (between μM and nM) and can assess protein behavior in a concentration-dependent manner [12]. We have previously applied such techniques to analyze the role of specific protein–protein interactions in viral assembly [4,13].

Cloning and generation of chimeric proteins: The constructs used for the transcription and translation of PUUV proteins originate from the PUUV Sotkamo/V-2969/81 strain. Plasmids encoding PUUV SP-FP-Gn and PUUV SP-FP-Gc, (SP: signal peptide sequence present at the N terminus of PUUV glycoprotein precursor (GPC); FP: fluorescent protein) were previously described [4,14]. In brief, PUUV SP was introduced at the N-terminus of FP-GP (GP: Glycoprotein) to ensure physiological membrane incorporation and localization of the GP constructs [14]. The plasmid encoding PUUV YFP-NP (YFP: yellow FP) has been previously described [11]. The construct mTurqoise2 (mTurq2)-NP was designed by substituting the mEYFP of the PUUV YFP-NP plasmid with the mTurq2. The constructs SP-mEGFP-GPΔCT (mEGFP: enhanced green FP) and SP-mCh2-GPΔCT were produced by polymerase chain reaction amplification of the GPΔCT sequence (GP devoid of the cytoplasmic tail) and the subsequent product inserted between the restriction sites AgeI and Bsp1407I within the PUUV SP-FP plasmid (details on the design of PUUV FP-GP are available [4]). GnΔCT and GcΔCT sequences were previously described [8,15]. All sequences were verified using the Sanger sequencing facility provided by LGC (LGC, Biosearch Technologies, Berlin, Germany).

All SP-FP-GP constructs, where the FP can be mEGFP/mEYFP/mCh2/mTurq2, are referred to as FP-GP. Any deviation from this nomenclature is explicitly highlighted in the text.

Cell Culture and transfection: NP-GP interactions were examined in this study using different human and rodent epithelial cell models, since the pathogenesis of HV infections is influenced by the identity of the host [16]. Chinese hamster ovary (CHO) and Baby hamster kidney (BHK-21) rodent cells were chosen since they were previously used for the characterization of virus-like particle formation [17], GP multimerization [4,14], NP multimerization [18], and production of antibodies against New World HV infection [19]. Human embryonic kidney (HEK-293T) and adenocarcinomic human alveolar basal epithelial (A549) cell models were used in studies regarding host regulation factors and NP expression [20,21], cellular immunity in infection [22,23], and lentiviral expression of NP [24].

Chinese hamster ovary cells CHO-K1 (ATCC, CCL-61™), Baby hamster kidney cells BHK-21 (ATCC, CCL-10™), adenocarcinomic human alveolar basal epithelial cells A549 (ATCC, CCL-185™), and human embryonic kidney epithelial cells (HEK) from the 293T line (CRL-3216™) were purchased from ATCC (Kielpin Lomianki, Poland), and maintained in Dulbecco's modified Eagle's medium containing 10% fetal bovine serum, 100 U/mL penicillin, 0.1 mg/mL streptomycin, and 4 mM L-glutamine at 37 °C and 5% CO_2. Cells were passaged every 3 to 5 days, no more than 15 times. All solutions, buffers, and media used for cell culture were purchased from PAN-Biotech (Aidenbach, Germany). Cells (3 to 6 × 105) were plated on glass-bottom 35 mm-diameter plates (CellVis, Mountain View, CA or MatTek Corp., Ashland, MA, USA) 48 h before experiments. Fusion protein expression plasmids were transfected into (70 to 90% confluent) CHO and HEK 293T cells using Turbofect (Thermo Fisher Scientific, Vilnius, Lithuania) according to the manufacturer's protocol, 20 to 24 h prior to experiments. BHK-21 and A549 cells were transfected with Lipofectamine 3000, according to the manufacturer's protocol (Thermo Fisher Scientific, Vilnius, Lithuania), 20 to 24 h prior to experiments.

Confocal microscopy imaging: 3 to 6 $\times$ 10^5 cells were plated onto 35 mm glass-bottom dishes (CellVis, Mountain View, CA or MatTek Corp., Ashland, MA, USA) 48 h prior to the experiment and transfected 16 to 24 h post-seeding. Confocal images were acquired on a Zeiss LSM780 system (Carl Zeiss, Oberkochen, Germany) using a 40$\times$ DIC M27 water immersion objective. To limit the out-of-focus light, a pinhole with size corresponding to one Airy unit (~39 μm) was used. Samples were excited with a 488 nm argon laser and a 561 nm diode laser. The fluorescence signal was collected by a Zeiss QUASAR multichannel GaAsP detector in photon-counting mode. The laser power was set so that the photon count rate and bleaching remained below 1 MHz and ca. 20%, respectively (typically ~3 μW for 488 nm, and ~5 μW for 561 nm). Measurements exhibiting stronger bleaching were discarded. Fluorescence was detected between 499 and 552 nm (mEGFP) and between 570 and 695 nm (mCh2), after passing through an MBS 488/561 dichroic mirror.

Number and Brightness (N&B): Number and Brightness analysis was performed as previously described [4]. Briefly, images of 128 $\times$ 128 pixels were acquired with pixel dimensions of 400 nm and a pixel dwell time between 25 and 50 μs. Image time-stacks of 100 scans were collected using the Zeiss Blu ZEN 3.8 software. The intensity time-stacks data were imported into MATLAB 2022 using the Bioformat [25] package and analyzed using a custom-written code (The MathWorks, Natick, MA, USA). The algorithm uses the equations from [26] for the specific case of photon-counting detectors to obtain the molecular brightness and number as a function of pixel position. Upon selecting the required region of interest (ROI), fluorophore bleaching and minor cell movements within a few pixels are suitably corrected. ROIs purposedly did not include any immobile structures, such as large aggregates or immobile protein filaments. Pixel intensities are then used to calculate the average brightness of the FP detected within the cellular ROI. This average brightness is utilized to evaluate the multimerization state of the FP upon comparison to a reference monomer brightness [27]. Thus, the oligomeric size of the protein complex is evaluated in terms of monomer units.

Cross Correlation Number and Brightness (ccN&B): The ccN&B analysis is based on a published workflow [13], modified suitably for the evaluation within the cytosolic region of the cell. Briefly, an image stack is acquired over time typically consisting of 100 frames. Images of 256 $\times$ (y< 256) pixels with a pixel size of 0.21 μm and a pixel dwell time of 12.6 μs were acquired, alternating two different excitation wavelengths. CZI-format image output files were imported into MATLAB using the Bioformat [25] package and analyzed using a self-written script implementing the framework described by Digman et al. [28] for the specific case of photon-counting detectors and two-color excitation. Pixels corresponding to a ROI are manually selected in an image map. Next, to account for lateral drift during the acquisition, frames are aligned to the first frame by maximizing the spatial correlation between sub-selections in consecutive frames, averaged over both channels, as a function of arbitrary translations. Additional steps regarding corrections for bleaching, minor cell movements, and specific detector response are applied as previously described [13].

The ccN&B analysis calculates the dimensionless parameter relative cross correlation (Relative CC). Relative CC is calculated as [13]:

$$\text{Relative CC} = max\ (N_{cc}/N_g, N_{cc}/N_r)$$

Here, N_g and N_r are the particle numbers detected in the image channel corresponding to the green FP and red FP, respectively. N_{cc} is obtained by dividing the product of the average intensities of the green and red FP channels by the covariance of the two channels.

Statistical Analysis: The statistical analysis in this work was performed using GraphPad Prism version 9.0.0 (GraphPad Software, LCC, San Diego, CA, USA). For pairwise comparisons of the derived parameters, the Kruskal–Wallis Test and modified Tukey's multiple comparisons test were used.

3. Results

3.1. NP-NP Interactions in Human Epithelial Cells Are Stronger than in Rodent Epithelial Cells

Hägele et al. [29] observed significant variations in NP localization and function between infected Vero E6 cells and primary human renal cells. This allowed them to conclude that the behavior of NP might depend on the specific cell type. Here, we aimed to determine whether different host environments (namely, human vs. rodent epithelial cell models) might influence homotypic NP-NP interactions in transfected cells.

NP multimerization in cells imaged via confocal microscopy was examined using N&B analysis, a quantitative approach based on the statistical fluctuations of the fluorescence signal within the probed confocal volume. This approach was previously used to monitor NP multimerization in CHO cells [11]. In this work, we extend the analysis to compare the average NP multimerization detected in human epithelial cells (HEK and A549) and rodent epithelial cell models (BHK-21) 16 to 20 hpt.

As previously shown by Davies et al. [30], NP initially appears homogenously distributed in the cytoplasm. This is followed by the formation of small mobile clusters, for all cell types tested (Figure 2A–D), ca. 4 hours post-transfection. At later time points (i.e. after ca. 12 to 16 hours), NP starts to coalesce, forming large immobile filaments [11], but such samples were not included in the current analysis.

According to N&B analysis, NP forms small oligomers, up to ca. decamers on average, for all cell models tested, within the entire concentration regime explored (Figure S1). More specifically, NP multimerization values in BHK cells increase strongly with protein concentration, as previously observed for CHO cells [11] (shown in Figure S1, as reference). On the other hand, multimerization values measured in A549 and HEK cells are higher (especially at lower concentrations) and display a less marked correlation with protein concentration (Figure S1). Since different cell models displayed variability in the expression levels, we compared the average NP multimerization in human and rodent cell models within a restricted concentration range (0.1 to 0.5 μM): within this concentration range, we were able to collect a similar amount of data points for all cell models and, therefore, a direct comparison of average values is reasonable. In these conditions, NP forms on average tetramers in rodent epithelial cells and roughly octamers in human epithelial cells (Figure 2E).

We further attempted a statistical analysis of the multimerization–concentration curves, for different cell models and across the whole available protein concentration range, by fitting an empirical model to the data [31], as shown in Figure S1. It is worth noting that such a fitting is not meant to deliver precise parameters, but rather to allow a statistical comparison of the different datasets. In agreement with the above-mentioned observations, NP multimerization behavior in rodent (CHO [11] and BHK-21) and human epithelial cell models (HEK-293T and A549) appears to be significantly different over the observed concentration range (Supplementary Table S1).

Altogether, these results indicate that the average NP-NP interactions observed in human epithelial cells are stronger than in rodent epithelial cell models, at least at the early stages of NP expression and assembly.

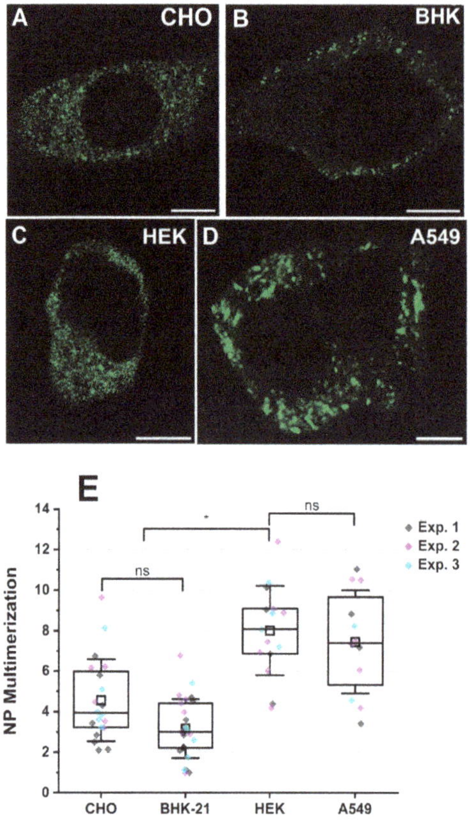

Figure 2. Average multimerization of fluorescently labelled PUUV NP is higher in human cell models compared to rodent epithelial cell models. YFP-NP was expressed in multiple epithelial cell models and observed 16 to 20 hpt. Panels (**A–D**) show representative confocal microscopy images of (**A**) CHO, (**B**) BHK-21, (**C**) HEK 293T, and (**D**) A549 cells expressing YFP-NP. Panel (**E**) shows a quantitative comparison of YFP-NP multimerization between different cell models. For this analysis, cells expressing YFP-NP within the concentration range 0.1–0.5 μM were chosen. Data points for NP multimerization in CHO cells are from Welke et al. [11] and incorporated as reference values. Each point in the boxplot represents the average multimerization in one cell. At least 12 cells were analyzed for each case, with three technical replicates denoted in different colors (see Supplementary Information for quantitative details). In all box plots, the horizontal line is the median, '□' marks the mean, with the first and third quartile as the boundaries and whiskers as standard deviation (SD). Statistical analysis was performed using the pairwise Kruskal–Wallis one-way ANOVA test (ns: not significant, * $p < 0.01$). Scale bars are 10 μm.

3.2. NP Interacts with Gc Within Few Small Intracellular Regions

We proceeded to compare the interactions between PUUV NP and each PUUV GP (Gn or Gc) upon co-expression in different human (HEK 293T and A549) and rodent epithelial (CHO and BHK-21) cell models. Welke et al. [11] previously indicated that NP, Gn, and Gc displayed significant spatial colocalization in transfected CHO cells. While these findings indicate that NP and GPs are found in close vicinity to each other, they do not directly imply an interaction between the proteins. Therefore, we performed two-color ccN&B analysis to investigate whether NP significantly interacts with any of the GPs. This method provides a parameter (i.e., relative CC) that indicates the relative amounts of co-diffusing hetero complexes.

Figure 3A–C and Figure 3F–H show representative confocal microscopy images of cells co-expressing YFP-NP and either mCh2-Gc or mCh2-Gn in CHO cells, respectively. The intracellular localization of Gn and Gc within the ER and/or perinuclear region is similar to what was previously observed [4].

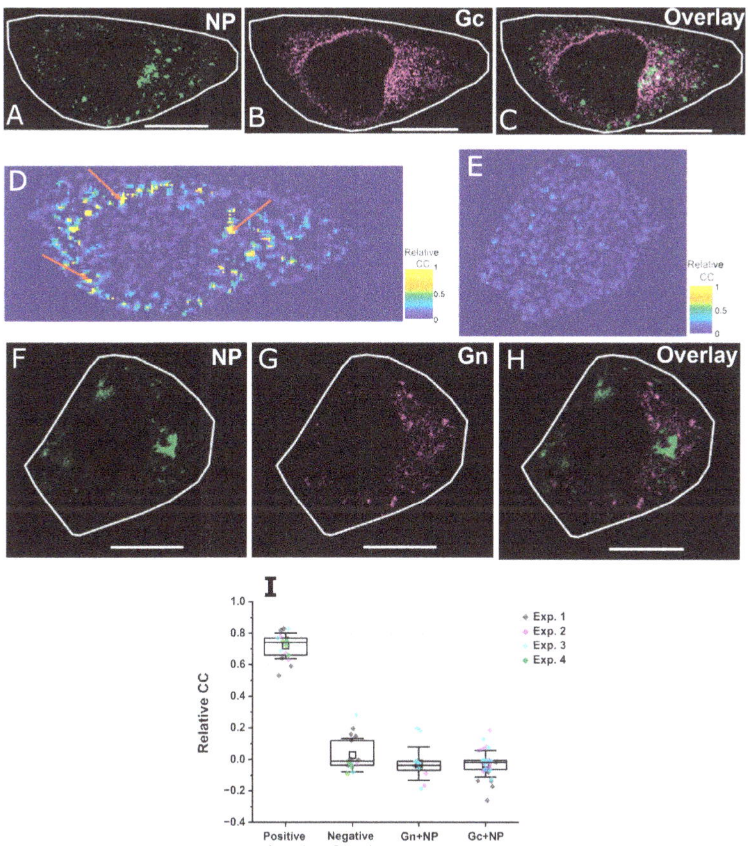

Figure 3. Single-cell analysis and spatial maps of the interactions between NP and GPs. Panels (**A–C**) show representative images of CHO cells co-expressing PUUV YFP-NP and PUUV mCh2-Gc, 20 to 24 hpt. Panel (**D**) shows the pixel-by-pixel relative CC map of the same CHO cell, within a ROI as indicated by a white line in panels (**A–C**). Red arrows in panel (**D**) denote localized regions characterized by high relative CC values. Panel (**E**) shows the pixel-by-pixel relative CC map of the same CHO cell, within a ROI as indicated by a white line in panels (**F–H**). Panels (**F–H**) show representative images of CHO cells co-expressing PUUV YFP-NP and PUUV mCh2-Gn, 20 to 24 hpt. Panel (**I**) shows a box plot of the relative CC values averaged over whole-cell ROIs and measured for different constructs expressed in CHO cells. Positive control is a tandem YFP-mCh2 construct that localizes in the cytoplasm. The negative control refers to cells co-expressing YFP-NP and cytosolic mCherry2. ROIs in cells with intensity values less than 1 MHz in the YFP and mCherry2 channel were chosen for evaluating the average relative CC within each cell. Each point in panel (**I**) represents the average relative CC value from one ROI in one cell. Number of points for each case >15, from three independent experiments denoted in individual colors (see Supplementary Information for quantitative details). In all box plots, the horizontal line is the median, '□' marks the mean, with first and third quartile as the boundaries and whiskers as SD. Scale bars are 5 μm.

To quantify hetero-interactions, ccN&B analysis was performed, thus obtaining relative CC maps, as shown in Figure 3D and E for NP-Gc and NP-Gn interactions, respectively. Additional examples of relative CC maps are shown in Figure S2I,J. Overall protein–protein interactions were first quantified by averaging the relative CC from each pixel in the selected ROIs (roughly corresponding to the whole intra-cellular region). As shown in Figure 3I for CHO cells, significant interactions were detected for the positive control (i.e., a tandem cytosolic construct of YFP and mCh2) but not for the negative control (consisting of independently co-expressed YFP-NP and cytosolic mCh2). Similar results were obtained for the controls in all cell models tested (Figure S3). The interactions between NP and either GP (i.e., Gn or Gc), averaged over the whole cell, were relatively weak for all cell models tested (CHO cells, see Figure 3I; HEK cells, see Figure S3A; and A549 cells, see Figure S3B) and not distinguishable from the negative control. Co-expression of fluorescently labelled NP and GP constructs in BHK-21 cells did not yield a GP expression high enough for quantifying relative CC and, therefore, this cell model was not used for subsequent analysis.

Next, we analyzed in greater detail the spatial variations of the relative CC parameter in cells expressing the different constructs. As expected, CHO cells expressing the tandem mEYFP-mCh2 construct (positive control) exhibited positive relative CC values homogenously distributed throughout the analyzed intracellular ROI (Figure S2E–H). Similarly, the analysis of CHO cells expressing YFP-NP and cytosolic mCherry2 (negative control) resulted in homogeneous low relative CC values throughout the cell (Figure S2A–D). Interestingly, similarly to the negative control, we observed relatively homogeneous intracellular distribution of low relative CC values for cells co-expressing Gn and NP (Figure 3E and Figure S2I). On the other hand, we consistently noticed localized regions characterized by high relative CC values (i.e., high local abundance of co-diffusing Gc-NP complexes) in CHO cells co-expressing Gc and NP (see red arrows in Figure 3D and Figure S2J). Such regions were mostly observed around the perinuclear region and, most importantly, did not display an evident correlation with the spatial distribution of either Gc or NP (see Figure 3A,B,D).

Altogether, these results demonstrate that PUUV GPs interact, on average, weakly with PUUV NP in intracellular regions, for all cell models tested. Nevertheless, PUUV Gc appears to interact significantly with PUUV NP within localized intracellular regions close to the perinuclear region.

3.3. *Apparent Gc Multimerization Increases in the Presence of NP*

The experiments described above suggest that NP might play a role in intermolecular interactions involving Gc, but not Gn. To further explore this interaction network, we measured GP-GP interactions (i.e. the multimerization state of each GP) in the presence of NP, using single-color N&B in different cell models. As for the ccN&B analysis, cells displaying a quasi-homogeneous distribution of NP within the cytoplasm (see Figure 2A–D) were chosen.

PUUV Gc multimerization in the presence of PUUV NP was analyzed by co-expressing YFP-NP and mCh2-Gc in CHO (Figure 3A–C), HEK (Figure S4A–C), and A549 (Figure S4D–F) cells and performing N&B measurements 24 hpt. Our previous work shows that the expression of Gc alone results in a mixture of monomers and dimers in the ER of different cell models [4,14]. Strikingly, in the presence of NP, Gc forms large multimers (ca. up to octamers) instead, for all the cell models tested, within the explored concentration range (Figure 4A). Thus, Gc-Gc interactions in the presence of NP are significantly stronger than those previously observed in the absence of NP (shown in black in Figure 4A as reference, [4]).

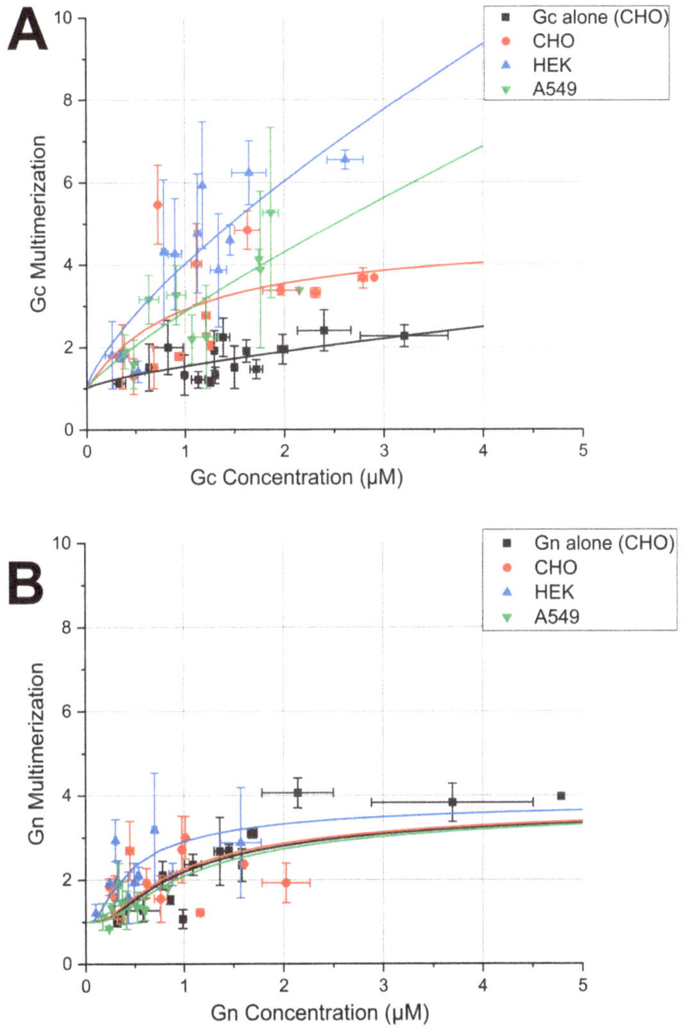

Figure 4. PUUV Gc (but not Gn) multimerization behavior is altered in the presence of NP. Panel (**A**) shows the concentration-dependent multimerization analysis of PUUV Gc in the presence of PUUV NP using N&B, in different epithelial cell models: CHO (red), A549 (green), and HEK 293T (dark green). PUUV Gc multimerization in the presence of PUUV NP is compared to that measured in the absence of NP in CHO cells (black, data from [4]). Each point in the graph represents the binned average multimerization values from two cells. The solid lines represent a fit to an empirical power growth model [31] in the form of $y = 1 + a \cdot x^k$. Fit results and statistical analysis are shown in Table S2. Panel (**B**) shows the concentration-dependent multimerization analysis of PUUV Gn in the presence of PUUV NP using N&B, in the same epithelial cell models as for panel A. PUUV Gn multimerization in the presence of PUUV NP is compared to that measured in the absence of NP in CHO cells (black, data from [4]). Each point in the graph represents the binned average multimerization values from two cells. The solid lines represent a fit to a monomer–tetramer equilibrium model [32]. Fit results and statistical analysis are shown in Table S3.

To compare different datasets, the multimerization curves were analyzed using a simple empirical growth model with two free parameters [32]. The resulting curves for all the examined cells expressing both Gc and NP lay considerably above the multimerization curve obtained in CHO cells expressing only Gc ($p < 0.01$), thereby confirming the initial qualitative observation. Only minor differences can be observed between the other curves, mostly due to the large data spread (see Table S2).

3.4. Gn Remains in a Monomer–Tetramer Equilibrium, Independent of the Presence of NP

We further examined the multimerization of Gn in the presence of NP. CHO (Figure 3E–G), HEK (Figure S5A–C), and A549 (Figure S5D–F) cells co-expressing fluorescently labelled Gn and NP were observed 20 to 24 hpt. Cells/ROIs in which NP formed homogeneously distributed small mobile oligomers were selected, thus excluding from the analysis large immobile NP structures. We have previously shown that Gn forms up to tetramers in the absence of other viral proteins [4]. The multimerization values obtained in this work (Figure 4B) in the presence of NP are compatible with similar multimerization behavior, for all the cell models tested, at least within the restricted concentration range that could be explored.

To quantitatively compare the different datasets, multimerization curves were fitted to an analytical model that describes a monomer to tetramer equilibrium through the association constant K_4 [32]. This parameter was not statistically distinguishable among the different samples ($0.2 < p < 0.8$) (see Table S3). This result indicates that Gn-Gn interactions are not strongly affected by the presence of NP.

3.5. Deletion of Gc-CT Domain Abolishes the Enhancement of Gc-Gc Interactions Induced by NP

Finally, the influence of the GP-CT domain on GP multimerization in the presence of NP was examined, since NP was proposed to interact with GPs via their CTs [6,7]. Truncated GPs without CT domains (GPΔCT) were initially expressed in cells in the presence and absence of NP (Figure 5A–D for GcΔCT, Figure S6A–D for GnΔCT). The intracellular distribution of the GPs does not appear to be strongly influenced by the presence of NP (Figure 5A,C and Figure S6). Also, the spatial organization of NP is similar to what was previously observed 24 hpt in CHO cells in the absence of other viral proteins [11].

We thus proceeded to analyze GPΔCT multimerization 20 to 24 hpt. Figure 5E shows that fluorescently labelled GcΔCT, in the absence of NP, displays an average multimerization of ~1.4 at the highest measured concentration. According to this value and assuming a monomer–dimer equilibrium, the protein appears to form mostly monomers (ca. 75 mol%), in agreement with previous measurements on non-fluorescent constructs in vitro [33]. In the presence of YFP-NP, the multimerization of GcΔCT did not change substantially (Figure 5E). Similar results were obtained with A549 cells (Figure S7). The concentration-dependent multimerization data points in Figure 5E were analyzed with an analytic model describing monomer–dimer equilibrium [32]. The derived association constant K_2 did not change significantly in the presence or absence of NP (Table S4).

The influence of Gn-CT on Gn-Gn interactions was similarly analyzed. Figure S6E shows that fluorescently labelled GnΔCT, in the absence of NP, displays an average multimerization value compatible with the presence of a large number of monomers in equilibrium with dimers, in agreement with the results obtained using non-fluorescent constructs [17]. In the presence of YFP-NP, the average multimerization of GnΔCT remained similar. Analogous results were obtained with A549 cells (Figure S8).

In summary, the increased formation of Gc-Gc complexes induced by NP appears to be mediated by the Gc-CT domain, for all the cell models tested. On the other hand, we did not observe any influence of NP on the multimerization of Gn, irrespective of the presence of its CT.

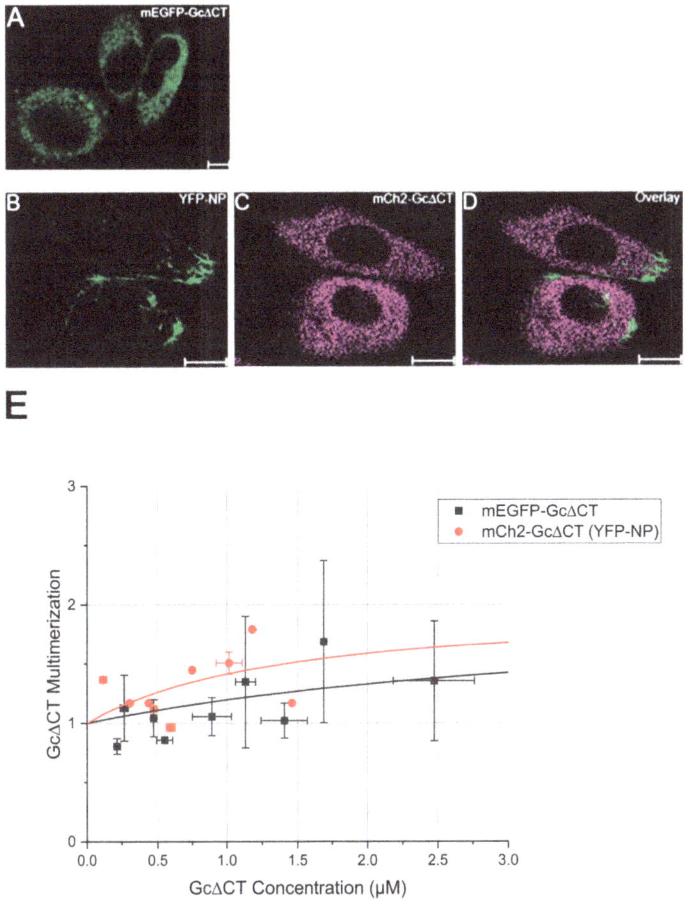

Figure 5. Deletion of GcΔCT inhibits the formation of large Gc assemblies in the presence of NP. Panel (**A**) shows a representative image of CHO cells expressing PUUV mEGFP-GcΔCT and observed 24 hpt. Panels (**B–D**) show typical confocal image of CHO cells co-expressing YFP-NP and mCh2-GcΔCT. Panel (**E**) shows the concentration-dependent multimerization analysis of PUUV GcΔCT in the absence (black) and presence (red) of PUUV NP, calculated using N&B analysis. Each point in the graph represents the binned average multimerization from two cells. The solid lines represent a fit to a monomer–dimer equilibrium model [32]. Fit results and statistical analysis are shown in Table S4. Scale bars are 10 μm.

4. Discussion

NP is a multifunctional HV protein that is capable of interacting with different host proteins and other viral components and mediates the association between the RNP and GPs [34]. While previous studies have focused on the protein regions possibly involved in the binding between NP and GPs, there is limited information regarding the molecular details of the actual interactions between NP and either Gc or Gn occurring in vivo. In this study, we address this important question by quantifying the interaction between NP and GPs upon co-expression, at the single-cell level. In order to pinpoint specific interactions between NP and Gn or Gc, we evaluated the behavior of these proteins in the absence of other viral components (e.g. viral RNA) in live cell models, through transient transfection.

HV infections display distinct cellular tropism [16,22,29]. Therefore, we initially assessed if a specific cellular environment could influence NP-NP binding, by comparing human and rodent epithelial cell models. We observed that average homotypic NP-NP interactions are significantly stronger in human epithelial cell models compared to rodent epithelial cell models, at a given time point post-transfection. This observation is in line with previous studies [29] that reported point-like NP structures in infected Vero E6 cells, smaller than the filamentous NP assemblies present instead in infected human renal cells. Also, it has been shown that cytoskeletal components, such as actin, vimentin, and microtubules, influence NP organization in a fashion that varies among different HV strains and cell models [29,35]. It is therefore possible that the modulation of NP-NP interactions described in this study may be determined by cell type-specific interactions with cytoskeletal components. Alternatively, post-translational modifications, including phosphorylation [36,37] or SUMOylation [38] may also contribute to the regulation of NP multimerization in a cell type-dependent manner. A third possibility is that NP multimerization is influenced by other host-specific factors [39], e.g., proteins binding to the NP multimerization interface. Independent from the specific molecular mechanism, it would be very interesting to investigate whether alterations in NP multimerization equilibrium are truly correlated to the susceptibility of a specific host cell to HV infection.

In the second part of our investigation, we observed weak overall NP-GP interactions, independent of the tested cell model, while NP still mostly formed small mobile clusters. Interestingly, within single cells, localized regions were detected in which NP showed an apparently strong interaction with Gc. We are currently investigating whether locally enhanced NP-Gc interactions might exhibit a spatial correlation with, e.g., cytoskeletal elements or other intracellular structures. Interestingly, such local interactions with NP were never observed for Gn. Previous studies have shown significant interactions between these two proteins via the Gn CT domain in infected cells using co-immunoprecipitation [6]. Nevertheless, it must be noted that these experiments were performed several days post-infection, i.e., at a time point in which NP forms large immobile structures [30] and in the presence of other viral components. This observation suggests that NP-Gn interactions might possibly be mediated by viral RNA [6,8]. Also, our results additionally show that the quantitative microscopy approach with single-cell resolution described here can provide more detailed information, compared to bulk methods that average over a large number of cells.

While the fluorescence fluctuation microscopy analysis used in this work (N&B) delivers quantitative information about inter-molecular interactions and has a high spatial resolution, it also requires the observed molecules to be mobile during the observation. Although we restricted our analysis to cells which did not show large immobile structures, it is known that NP might form a small fraction of larger very slow NP aggregates, as described by Welke et al. [11]. If the reported higher CC values (i.e. stronger interactions) do indeed originate from such almost immobile NP fractions, NP-GP interactions cannot be quantified precisely with this approach. For this reason, we complemented these measurements by separately observing GP multimerization in the presence of NP.

In the case of Gc, we observed a striking difference in homotypic interactions when NP is also present. While not a direct measurement of NP-Gc interaction, this is compatible with the above-mentioned results, under the hypothesis that the observed Gc multimer formation is indeed somehow mediated by interactions with NP (e.g., through the binding of several individual Gc molecules to a large NP multimer). Gc localization within NP clusters was also previously observed for other Bunyaviruses, such as Tomato spotted wilt virus [40,41].

Furthermore, our experiments on truncated Gc constructs indicate that NP-Gc interaction occurs via the Gc CT domain, as also suggested by previous studies [2,9]. This result strengthens the hypothesis that Gc CT can act analogously to a matrix protein [10], connecting the envelope of the virus with the enclosed genetic material. Such interactions might play a role in the large-scale assembly of the GP spikes during virus assembly and confer mechanical stability to the new virions.

Alternatively, it is also possible that the removal of the CT affects the stability and structure of Gc [14] and, thus, its interaction with NP. Taken together, our findings suggest that HV NP can form complexes with Gc (possibly including other molecules as well), also in the absence of Gn or RNPs.

On the other hand, the lack of direct interactions between Gn and NP as observed via ccN&B is corroborated by the fact that, in the observed cell models, Gn multimerization is independent of the presence of NP, with this GP consistently forming monomers to tetramers, as expected.

In conclusion, we have investigated here the interactions between HV NP and GPs using quantitative fluorescence microscopy approaches. According to N&B experiments, NP can interact with Gc in the absence of other viral components, but not with Gn. The observed interactions appear to be mediated by the CT of Gc. Notably, all the experiments performed in this work involve the transient expression of up to two viral proteins at the same time, with the purpose of specifically identifying direct interactions (i.e., not involving other viral proteins). As mentioned above, a more comprehensive picture of the interaction network can be obtained if additional viral components are systematically added to the studied system (e.g. RNA to further characterize NP-Gn interactions). Future studies will, for example, focus on quantifying interactions between fluorescently labelled NP, Gn, and Gc, also using third-order correlation functions [42]. Such experiments are complicated by the interference of fluorescent labels on inter-molecular interactions (e.g. between Gn and Gc) [4]. By using more advanced methodologies for labelling (see, e.g., ALFA tag [43], bioorthogonal click chemistry [44]), the approach described in this work could open new avenues for the precise characterization of binding between other HV proteins and further elucidate the steps involved in the HV virion formation.

Supplementary Materials: The following supporting information can be downloaded at: https://www.mdpi.com/article/10.3390/v17070940/s1. Figure S1: Homotypic NP-NP interactions differ between human and rodent epithelial cell models; Table S1: Statistical analysis of the fit results deriving from the comparison of NP multimerization between cell models; Figure S2: Intensity and relative CC maps of CHO cells expressing positive and negative control; Figure S3: Average interactions between NP and GPs co-expressed in HEK and A549 cells are weak; Figure S4: Co-expression of PUUV NP and PUUV Gc in HEK293T and A549 cells; Table S2: Statistical comparison of fit parameters describing Gc multimerization between cell models co-expressing NP and Gc; Figure S5: Co-expression of PUUV NP and PUUV Gn in HEK293T and A549 cells; Table S3: Statistical comparison of fit parameters describing Gn multimerization between cell models co-expressing NP and Gn; Table S4: Statistical comparison of fit parameter describing GcΔCT multimerization between cell models co-expressing NP and GcΔCT; Figure S6: PUUV GnΔCT remains mostly monomeric, independent of the presence of NP, upon expression in CHO cells; Figure S7: Lack of GcΔCT inhibits the formation of large Gc assemblies in the presence of NP, also in A549 cells; Figure S8: PUUV GnΔCT is mostly monomeric, independent of the presence of NP, upon expression in A549 cells; Table S5: Statistical parameters of the data shown in Figure 2 of the main manuscript; Table S6: Statistical parameters of the data shown in Figure 3 of the main manuscript. Numerical data and raw data (i.e., before binning) for all figures are provided in the form of two spreadsheets.

Author Contributions: Conceptualization, A.K.A. and S.C.; Methodology, A.K.A. and S.C.; Software, A.K.A. and S.C.; Validation, A.K.A. and S.C.; Formal Analysis, A.K.A.; Investigation, A.K.A.; Resources, S.C.; Data Curation, A.K.A.; Writing—Original Draft Preparation, A.K.A. and S.C.; Writing—Review and Editing, A.K.A., T.M. and S.C.; Visualization, A.K.A. and T.M.; Supervision, S.C.; Project Administration, S.C.; Funding Acquisition, S.C. All authors have read and agreed to the published version of the manuscript.

Funding: This work was supported by the German Research Foundation (Deutsche Forschungsgemeinschaft), grant 407961559 to S.C.

Institutional Review Board Statement: Not applicable.

Informed Consent Statement: Not applicable.

Data Availability Statement: The original contributions presented in this study are included in the Supplementary Material. Further inquiries can be directed to the corresponding author.

Acknowledgments: We thank Anja Thalhammer for helpful suggestions with the text.

Conflicts of Interest: The authors declare no conflicts of interest.

Abbreviations

The following abbreviations are used in this manuscript:

HVs	Orthohantaviruses
GPs	Glycoproteins
RdRp	RNA-dependent RNA polymerase
RNPs	Ribonucleoproteins
NP	Nucleoprotein
CT	Cytoplasmic tail
PUUV	Puumala orthohantavirus
N&B	Number and Brightness
ccN&B	Cross-correlation N&B
CHO	Chinese hamster ovary
HEK	Human embryonic kidney
A549	Adenocarcinomic human alveolar basal epithelial
SP	Signal peptide
GPC	Glycoprotein precursor
FP	Fluorescent protein
YFP	Yellow fluorescent protein
mTurq2	Monomeric Turqoise2
mEGFP	Monomeric enhanced green FP
ΔCT	devoid of cytoplasmic tail
SD	standard deviation

References

1. Kim, W.K.; Cho, S.; Lee, S.H.; No, J.S.; Lee, G.Y.; Park, K.; Lee, D.; Jeong, S.T.; Song, J.W. Genomic Epidemiology and Active Surveillance to Investigate Outbreaks of Hantaviruses. *Front. Cell Infect. Microbiol.* **2020**, *10*, 532388. [CrossRef] [PubMed]
2. Hepojoki, J.; Strandin, T.; Lankinen, H.; Vaheri, A. Hantavirus structure--molecular interactions behind the scene. *J. Gen. Virol.* **2012**, *93*, 1631–1644. [CrossRef]
3. Bignon, E.A.; Albornoz, A.; Guardado-Calvo, P.; Rey, F.A.; Tischler, N.D. Molecular organization and dynamics of the fusion protein Gc at the hantavirus surface. *Elife* **2019**, *8*, e46028. [CrossRef]
4. Petazzi, R.A.; Koikkarah, A.A.; Tischler, N.D.; Chiantia, S. Detection of Envelope Glycoprotein Assembly from Old-World Hantaviruses in the Golgi Apparatus of Living Cells. *J. Virol.* **2021**, *95*, 10–1128. [CrossRef] [PubMed]
5. Olal, D.; Daumke, O. Structure of the Hantavirus Nucleoprotein Provides Insights into the Mechanism of RNA Encapsidation. *Cell Rep.* **2016**, *14*, 2092–2099. [CrossRef] [PubMed]

6. Hepojoki, J.; Strandin, T.; Wang, H.; Vapalahti, O.; Vaheri, A.; Lankinen, H. Cytoplasmic tails of hantavirus glycoproteins interact with the nucleocapsid protein. *J. Gen. Virol.* **2010**, *91*, 2341–2350. [CrossRef]
7. Wang, H.; Alminaite, A.; Vaheri, A.; Plyusnin, A. Interaction between hantaviral nucleocapsid protein and the cytoplasmic tail of surface glycoprotein Gn. *Virus Res.* **2010**, *151*, 205–212. [CrossRef]
8. Strandin, T.; Hepojoki, J.; Wang, H.; Vaheri, A.; Lankinen, H. The cytoplasmic tail of hantavirus Gn glycoprotein interacts with RNA. *Virology* **2011**, *418*, 12–20. [CrossRef]
9. Shimizu, K.; Yoshimatsu, K.; Koma, T.; Yasuda, S.P.; Arikawa, J. Role of nucleocapsid protein of hantaviruses in intracellular traffic of viral glycoproteins. *Virus Res.* **2013**, *178*, 349–356. [CrossRef]
10. Strandin, T.; Hepojoki, J.; Vaheri, A. Cytoplasmic tails of bunyavirus Gn glycoproteins-Could they act as matrix protein surrogates? *Virology* **2013**, *437*, 73–80. [CrossRef]
11. Welke, R.W.; Sperber, H.S.; Bergmann, R.; Koikkarah, A.; Menke, L.; Sieben, C.; Kruger, D.H.; Chiantia, S.; Herrmann, A.; Schwarzer, R. Characterization of Hantavirus N Protein Intracellular Dynamics and Localization. *Viruses* **2022**, *14*, 457. [CrossRef]
12. Wohland, T.; Maiti, S.; Macháň, R. *An Introduction to Fluorescence Correlation Spectroscopy*; IOP Publishing Ltd.: Bristol, UK, 2020.
13. Petrich, A.; Dunsing, V.; Bobone, S.; Chiantia, S. Influenza A M2 recruits M1 to the plasma membrane: A fluorescence fluctuation microscopy study. *Biophys. J.* **2021**, *120*, 5478–5490. [CrossRef]
14. Sperber, H.S.; Welke, R.W.; Petazzi, R.A.; Bergmann, R.; Schade, M.; Shai, Y.; Chiantia, S.; Herrmann, A.; Schwarzer, R. Self-association and subcellular localization of Puumala hantavirus envelope proteins. *Sci. Rep.* **2019**, *9*, 707. [CrossRef] [PubMed]
15. Guardado-Calvo, P.; Bignon, E.A.; Stettner, E.; Jeffers, S.A.; Pérez-Vargas, J.; Pehau-Arnaudet, G.; Tortorici, M.A.; Jestin, J.-L.; England, P.; Tischler, N.D.; et al. Mechanistic Insight into Bunyavirus-Induced Membrane Fusion from Structure-Function Analyses of the Hantavirus Envelope Glycoprotein Gc. *PLoS Pathog.* **2016**, *12*, e1005813. [CrossRef] [PubMed]
16. Noack, D.; Goeijenbier, M.; Reusken, C.; Koopmans, M.P.G.; Rockx, B.H.G. Orthohantavirus Pathogenesis and Cell Tropism. *Front. Cell Infect. Microbiol.* **2020**, *10*, 399. [CrossRef] [PubMed]
17. Li, S.; Rissanen, I.; Zeltina, A.; Hepojoki, J.; Raghwani, J.; Harlos, K.; Pybus, O.G.; Huiskonen, J.T.; Bowden, T.A. A Molecular-Level Account of the Antigenic Hantaviral Surface. *Cell Rep.* **2016**, *16*, 278. [CrossRef]
18. Ravkov, E.V.; Compans, R.W. Hantavirus nucleocapsid protein is expressed as a membrane-associated protein in the perinuclear region. *J. Virol.* **2001**, *75*, 1808–1815. [CrossRef]
19. Engdahl, T.B.; Kuzmina, N.A.; Ronk, A.J.; Mire, C.E.; Hyde, M.A.; Kose, N.; Josleyn, M.D.; Sutton, R.E.; Mehta, A.; Wolters, R.M.; et al. Broad and potently neutralizing monoclonal antibodies isolated from human survivors of New World hantavirus infection. *Cell Rep.* **2021**, *36*, 109453. [CrossRef]
20. Wang, Z.; Ren, S.; Li, Q.; Royster, A.D.; Lin, L.; Liu, S.; Ganaie, S.S.; Qiu, J.; Mir, S.; Mir, M.A. Hantaviruses use the endogenous host factor P58IPK to combat the PKR antiviral response. *PLoS Pathog.* **2021**, *17*, e1010007. [CrossRef]
21. Vera-Otarola, J.; Castillo-Vargas, E.; Angulo, J.; Barriga, F.M.; Batlle, E.; Lopez-Lastra, M. The viral nucleocapsid protein and the human RNA-binding protein Mex3A promote translation of the Andes orthohantavirus small mRNA. *PLoS Pathog.* **2021**, *17*, e1009931. [CrossRef]
22. Gallo, G.; Kotlik, P.; Roingeard, P.; Monot, M.; Chevreux, G.; Ulrich, R.G.; Tordo, N.; Ermonval, M. Diverse susceptibilities and responses of human and rodent cells to orthohantavirus infection reveal different levels of cellular restriction. *PLoS Negl. Trop. Dis.* **2022**, *16*, e0010844. [CrossRef]
23. Sola-Riera, C.; Garcia, M.; Ljunggren, H.G.; Klingstrom, J. Hantavirus inhibits apoptosis by preventing mitochondrial membrane potential loss through up-regulation of the pro-survival factor BCL-2. *PLoS Pathog.* **2020**, *16*, e1008297. [CrossRef] [PubMed]
24. Muyangwa, M.; Garanina, E.E.; Martynova, E.V.; Rizvanov, A.A. Lentivirus Expression of Hantavirus Nucleocapsid Proteins. *BioNanoScience* **2016**, *6*, 403–406. [CrossRef]
25. Linkert, M.; Rueden, C.T.; Allan, C.; Burel, J.-M.; Moore, W.; Patterson, A.; Loranger, B.; Moore, J.; Neves, C.; MacDonald, D.; et al. Metadata matters: Access to image data in the real world. *J. Cell Biol.* **2010**, *189*, 777–782. [CrossRef] [PubMed]
26. Digman, M.A.; Dalal, R.; Horwitz, A.F.; Gratton, E. Mapping the number of molecules and brightness in the laser scanning microscope. *Biophys. J.* **2008**, *94*, 2320–2332. [CrossRef]
27. Dunsing, V.; Luckner, M.; Zühlke, B.; Petazzi, R.A.; Herrmann, A.; Chiantia, S. Optimal fluorescent protein tags for quantifying protein oligomerization in living cells. *Sci. Rep.* **2018**, *8*, 10634. [CrossRef]
28. Digman, M.A.; Wiseman, P.W.; Choi, C.; Horwitz, A.R.; Gratton, E. Stoichiometry of molecular complexes at adhesions in living cells. *Proc. Natl. Acad. Sci. USA* **2009**, *106*, 2170–2175. [CrossRef]
29. Hagele, S.; Muller, A.; Nusshag, C.; Reiser, J.; Zeier, M.; Krautkramer, E. Virus- and cell type-specific effects in orthohantavirus infection. *Virus Res.* **2019**, *260*, 102–113. [CrossRef]
30. Davies, K.A.; Chadwick, B.; Hewson, R.; Fontana, J.; Mankouri, J.; Barr, J.N. The RNA Replication Site of Tula Orthohantavirus Resides within a Remodelled Golgi Network. *Cells* **2020**, *9*, 1569. [CrossRef] [PubMed]

31. Merten, J.A.; Schultz, K.M.; Klug, C.S. Concentration-dependent oligomerization and oligomeric arrangement of LptA. *Protein Sci.* **2012**, *21*, 211–218. [CrossRef]
32. Paul, M.D.; Rainwater, R.; Zuo, Y.; Gu, L.; Hristova, K. Probing Membrane Protein Association Using Concentration-Dependent Number and Brightness. *Angew. Chem. Int. Ed. Engl.* **2021**, *60*, 6503–6508. [CrossRef] [PubMed]
33. Willensky, S.; Bar-Rogovsky, H.; Bignon, E.A.; Tischler, N.D.; Modis, Y.; Dessau, M. Crystal Structure of Glycoprotein C from a Hantavirus in the Post-fusion Conformation. *PLoS Pathog.* **2016**, *12*, e1005948. [CrossRef] [PubMed]
34. Reuter, M.; Kruger, D.H. The nucleocapsid protein of hantaviruses: Much more than a genome-wrapping protein. *Virus Genes.* **2018**, *54*, 5–16. [CrossRef]
35. Ramanathan, H.N.; Jonsson, C.B. New and Old World hantaviruses differentially utilize host cytoskeletal components during their life cycles. *Virology* **2008**, *374*, 138–150. [CrossRef] [PubMed]
36. Yoon, J.J.; Lee, Y.T.; Chu, H.; Son, S.Y.; Kim, M. Phosphorylation of the nucleocapsid protein of Hantaan virus by casein kinase II. *J. Microbiol.* **2015**, *53*, 343–347. [CrossRef]
37. Gorbunova, E.E.; Mackow, E.R. Binding of the Andes Virus Nucleocapsid Protein to RhoGDI Induces the Release and Activation of the Permeability Factor RhoA. *J. Virol.* **2021**, *95*, e0039621. [CrossRef]
38. Kaukinen, P.; Vaheri, A.; Plyusnin, A. Non-covalent interaction between nucleocapsid protein of Tula hantavirus and small ubiquitin-related modifier-1, SUMO-1. *Virus Res.* **2003**, *92*, 37–45. [CrossRef]
39. Lin, Y.; Zhu, Y.; Jing, L.; Lei, X.; Xie, Z. Regulation of viral replication by host restriction factors. *Front. Immunol.* **2025**, *16*, 1484119. [CrossRef]
40. Snippe, M.; Willem Borst, J.; Goldbach, R.; Kormelink, R. Tomato spotted wilt virus Gc and N proteins interact in vivo. *Virology* **2007**, *357*, 115–123. [CrossRef]
41. Ribeiro, D.; Borst, J.W.; Goldbach, R.; Kormelink, R. Tomato spotted wilt virus nucleocapsid protein interacts with both viral glycoproteins Gn and Gc in planta. *Virology* **2009**, *383*, 121–130. [CrossRef]
42. Dunsing, V.; Petrich, A.; Chiantia, S. Multicolor fluorescence fluctuation spectroscopy in living cells via spectral detection. *Elife* **2021**, *10*, e69687. [CrossRef] [PubMed]
43. Götzke, H.; Kilisch, M.; Martínez-Carranza, M.; Sograte-Idrissi, S.; Rajavel, A.; Schlichthaerle, T.; Engels, N.; Jungmann, R.; Stenmark, P.; Opazo, F.; et al. The ALFA-tag is a highly versatile tool for nanobody-based bioscience applications. *Nat. Commun.* **2019**, *10*, 4403. [CrossRef] [PubMed]
44. Arsic, A.; Hagemann, C.; Stajkovic, N.; Schubert, T.; Nikic-Spiegel, I. Minimal genetically encoded tags for fluorescent protein labeling in living neurons. *Nat. Commun.* **2022**, *13*, 314. [CrossRef] [PubMed]

viruses

MDPI

Article

Expanding Insights: Harnessing Expansion Microscopy for Super-Resolution Analysis of HIV-1–Cell Interactions

Annett Petrich [1,†], Gyu Min Hwang [1,†] Laetitia La Rocca [1], Mariam Hassan [1], Maria Anders-Össwein [1], Vera Sonntag-Buck [1], Anke-Mareil Heuser [1], Vibor Laketa [1,2], Barbara Müller [1], Hans-Georg Kräusslich [1,2] and Severina Klaus [1,*]

[1] Department of Infectious Diseases, Virology, Heidelberg University, 69120 Heidelberg, Germany
[2] German Center for Infection Research (DZIF), Partner Site Heidelberg, 69120 Heidelberg, Germany
* Correspondence: severina.klaus@med.uni-heidelberg.de
† These authors contributed equally to this work.

Abstract: Expansion microscopy has recently emerged as an alternative technique for achieving high-resolution imaging of biological structures. Improvements in resolution are achieved by physically expanding samples through embedding in a swellable hydrogel before microscopy. However, expansion microscopy has been rarely used in the field of virology. Here, we evaluate and characterize the ultrastructure expansion microscopy (U-ExM) protocol, which facilitates approximately four-fold sample expansion, enabling the visualization of different post-entry stages of the HIV-1 life cycle, focusing on nuclear events. Our findings demonstrate that U-ExM provides robust sample expansion and preservation across different cell types, including cell-culture-adapted and primary CD4+ T-cells as well as monocyte-derived macrophages, which are known HIV-1 reservoirs. Notably, cellular targets such as nuclear bodies and the chromatin landscape remain well preserved after expansion, allowing for detailed investigation of HIV-1–cell interactions at high resolution. Our data indicate that morphologically distinct HIV-1 capsid assemblies can be differentiated within the nuclei of infected cells and that U-ExM enables detection of targets that are masked in commonly used immunofluorescence protocols. In conclusion, we advocate for U-ExM as a valuable new tool for studying virus–host interactions with enhanced spatial resolution.

Keywords: expansion microscopy; ultrastructure expansion microscopy; super-resolution microscopy; HIV-1; HIV-1 post-entry; HIV-1 nuclear import; HIV-1 capsid; virus–host interaction

1. Introduction

Fluorescence microscopy has been a cornerstone of biological and medical research since its conception over a century ago. However, many structures of interest, particularly in the field of virology, cannot be resolved using conventional fluorescence microscopy due to the diffraction limit of light microscopy (i.e., ~250 nm laterally and 500 nm axially). For instance, the human immunodeficiency virus type 1 (HIV-1) has a size of ca. 100–150 nm, which limits investigation of subviral structures and virus–host interactions using fluorescence microscopy [1].

To overcome this resolution barrier, several advanced microscopy techniques have been developed [2]. These methods were readily embraced by the virus research community, leading to critical insights into, e.g., virus morphology, assembly, and release [3–5]. However, most super-resolution techniques rely on expensive equipment and specific fluorescent dyes, are technically challenging, and demand substantial expertise in instrument operation, sample handling, as well as data analysis [6]. As a result, their widespread application remains limited.

Complementing these existing approaches, an alternative category of resolution-enhancing techniques known as expansion microscopy (ExM) has recently gained prominence [7]. Unlike other approaches that aim to improve resolution through modifications of optical elements or single-molecule localization microscopy (SMLM), ExM operates on the principle of isotropically expanding the sample in all dimensions. This expansion increases the distance between epitopes, allowing the visualization of sub-diffraction targets using conventional imaging techniques [7–9]. Sample expansion is achieved by the attachment of gel-linkable anchors to biomolecules within fixed and crosslinked samples before embedding the samples in a polycharged hydrogel [9]. To ensure uniform expansion, the samples subsequently undergo homogenization, which can be accomplished through proteolytic digestion or a combination of heat and detergent treatment [9]. Finally, incubation of the hydrogel in deionized water leads to sample expansion due to the repulsion of the charges within the hydrogel [9].

Following an initial proof-of-concept study by Boyden and co-workers [7], which demonstrated an approximately four-fold sample expansion in brain tissue slices and mammalian cells, ExM was rapidly adapted and optimized for different applications, resulting in a wide range of protocols tailored to meet different research needs (a detailed review of ExM principles and techniques has been provided by Wen 2023 and more recently by Hümpfer 2024 [8,9]). While some ExM adaptations have focused on retaining specific biomolecules such as lipids or RNA, others have aimed to optimize sample preservation or to increase the expansion factor, resulting in protocols capable of achieving sample expansion of up to 20-fold for a variety of targets [10–12].

A critical aspect of ExM is the visualization of structures of interest within the gel. Two principal strategies have emerged for visualizing protein targets: one involves staining samples prior to gel embedding and expansion [13–15], while the other entails staining targets after expansion within the gel [16–18]. Pre-expansion staining protocols typically utilize proteolytic digestion for homogenization, which destroys most epitopes and renders them unrecognizable for antibody labeling [9]. To overcome this limitation, proteins and other targets are often labeled via immunostaining before the expansion procedure, as proteolytic digestion generally allows retention of fluorescence [9]. While indirect antibody staining provides an easy and accessible method for localization of targets, it also introduces a linkage error of ca. 30 nm due to the size of the two antibodies necessary for visualization. To reduce this linkage error, the use of directly conjugated antibodies, which are commercially available, is a viable alternative. These antibodies can reduce the linkage error to around 15 nm compared to traditional indirect staining methods. Employing directly conjugated Fab fragments or nanobodies may further enhance localization precision, achieving a linkage error of approximately 3 nm. However, as with direct immunofluorescence with conventional antibodies, these approaches show a lack of signal amplification. In pre-expansion staining protocols, antibodies are subsequently expanded alongside targets, resulting in relative linkage errors akin to those observed in conventional microscopy [17]. Additionally, gel-embedding and proteolytic digestion can diminish the brightness of fluorophores [9].

In contrast, post-expansion labeling approaches utilize a gentler homogenization process via heat denaturation in the presence of detergent, which more effectively preserves sample epitopes compared to proteolytic digestion but may impair the fluorescence of dyes introduced before expansion [16–18]. This approach results in improved labeling efficiency due to the enhanced antibody access to target epitopes after expansion [17,19]. In addition, as fluorophores are generally introduced after expansion, fluorescence is not affected by the sample preparation. Post-expansion labeling may also reveal previously masked epitopes that become accessible during the expansion process [19]. Expanded samples can be imaged using various microscopy techniques, facilitating super-resolution imaging with classical fluorescence microscopy. By further combining post-expansion labeling strategies with advanced super-resolution techniques such as Airyscan or stimulated emission depletion (STED) microscopy [20,21], it is possible to increase the spatial resolution even further, approaching molecular resolution (<10 nm) and effectively bridging the gap between light and electron microscopy [17,19].

Despite the potential of ExM, it has not yet been extensively applied in virology, and, in particular, not in the field of HIV-1 research. In this study, we present an ExM strategy tailored for the analysis of virus-infected cells with a focus on subviral replication complexes in the nucleus. The nucleus serves as a key compartment for the replication of many viruses, which rely on the nuclear machinery for viral genome replication and gene expression [22,23]. Among the viruses that depend on the nuclear environment for replication is the retrovirus HIV-1 [24]. A defining characteristic of retroviruses is the integration of a reverse-transcribed double-stranded DNA copy of their genome into the host cell DNA. Since HIV-1 can infect non-dividing cells, it necessitates the import of the viral genome through the intact nuclear envelope [25]. After release into the cytosol of a newly infected target cell, capsids (also called viral cores) are transported towards the nucleus along the microtubule network [26–28]. Upon reaching the nucleus, capsids interact with cytoplasmic components of the nuclear pore complex (NPC) and traverse the NPC channel, with seemingly intact capsids being frequently observed in the nucleoplasm of infected cells [29–32]. Within the nucleus, capsids are eventually directed to nuclear speckles, where they accumulate and undergo genome uncoating, releasing the fully reverse-transcribed viral DNA for subsequent integration into host cell chromatin [30,32–34]. While recent studies shed light on the mechanisms involved in nuclear import of capsids [29,35], the molecular events involved in the targeting of nuclear speckles and in genome uncoating remain poorly understood.

Here, we establish and employ a post-expansion labeling ultrastructure expansion microscopy (U-ExM) protocol to visualize intra-nuclear structures involved in post-entry events during HIV-1 infection, applicable to both cell culture-adapted cell lines and primary immune cells [18,36]. We characterize the robustness of this protocol across different cell types and nuclear compartments, provide references for evaluating sample integrity, and emphasize the advantages of using a post-expansion labeling approach to probe viral structures.

2. Materials and Methods

2.1. Cell Culture and Virus Preparation

Cells were kept in a humidified incubator at 37 °C in a 5% CO_2 atmosphere. HeLa-based TZM-bl cells (NIH AIDS Reagent program #8129-442, also called JC57BL-13) [37,38], and human embryonic kidney cells from the 293T/17 line (here referred to as HEK293T, #CRL-11268TM, purchased from ATCC, Kielpin Lomianki, Poland [39]) were cultured in Glutamax-I Dulbecco's Modified Eagle's Medium (DMEM) (Gibco-Thermo Fischer Scientific, Waltham, MA, USA) supplemented with 10% heat-inactivated fetal bovine serum (FBS, Capricon Scientific, Ebsdorfergrund, Germany), 100 U/mL penicillin, and 100 mg/mL streptomycin (Gibco-Thermo Fischer Scientific, Waltham, MA, USA). U2OS cells endogenously expressing Nup96-SNAP (CLS GmbH, St. Ingbert, Germany, #300444, clone 33 [40]) were cultured in HEK293T growth medium supplemented with 1× MEM NEAA (Gibco-Thermo Fisher Scientific, Waltham, MA, USA). Cell lines were regularly tested for mycoplasma contamination (MycoAlert® mycoplasma detection kit, Lonza Rockland, Basel, Switzerland) and authenticated by STR profiling (Promega PowerPlex 21 Kit; carried out by Eurofins Genomics, Ebersberg, Germany).

Monocyte-derived macrophages (MDMs) were prepared from isolated buffy coats of healthy anonymous human blood donors (purchased from the Heidelberg University Hospital Blood Bank according to the regulations of the local ethics committee, S-023/2022 and S-025/2022, within the SFB 1129 [integrative analysis of the replication and spread of pathogens, funding period 07/2022–06/2026]) by Ficoll density gradient centrifugation using SepMate tubes (StemCell Technologies, Vancouver, BC, Canada). These human peripheral blood mononuclear cells (PBMCs) were cultured in GlutaMax-I Roswell Park Memorial Institute medium 1640 (RPMI 1640, Gibco-ThermoFisher Scientific, Waltham, MA, USA), supplemented with 10% heat-inactivated FBS, 100 U/mL penicillin, 100 mg/mL streptomycin, and 20 mM HEPES. After incubating for 3 h at 37 °C in a 5% CO_2 atmosphere, non-adherent cells were discarded, and the adherent PBMCs were washed with fresh medium. These adherent PBMCs were then differentiated by maintenance in activation medium (GlutaMax-I RPMI 1640 containing 10% heat-inactivated FBS, 100 U/mL penicillin, 100 mg/mL streptomycin, 20 mM HEPES, and 10% human AB serum (Sigma Aldrich, St. Louis, MO, USA)) for 7 days, allowing for their transformation into macrophages. The resulting MDMs were subsequently cultured in the activation medium for up to four weeks, with medium changes occurring every 3 days, as previously described [41].

CD4+ T-cells were isolated from buffy coats obtained from healthy blood donors (DRK-Blutspendedienst, Mannheim, Germany) with approval by the local ethics committee (S-023/2022 and S-025/2022). Negative CD4+ T-cell isolation was performed using the RosetteSep™ Human CD4+ T-Cell Enrichment Cocktail (STEMCELL Technologies Inc., Vancouver, BC, Canada) following the manufacturer's instructions. Briefly, the buffy coats were mixed with the enrichment cocktail and incubated before being layered onto Ficoll-Paque™ PLUS (GE Healthcare Bio-Sciences, Uppsala, Sweden) and centrifuged. After isolation, the cells were washed, treated with ammonium–chloride–potassium lysis buffer to remove red blood cells, and resuspended in complete T-cell medium consisting of GlutaMax-I RPMI 1640 supplemented with 10% fetal bovine serum (Capricon Scientific, Ebsdorfergrund, Germany), 100 U/mL penicillin/100 mg/mL streptomycin (Gibco-ThermoFisher Scientific, Waltham, MA, USA), and 10 ng/mL IL-2 (Biomol GmbH, Hamburg, Germany). The cells were left in a resting state or activated using Dynabeads® Human T-Activator CD3/CD28 (Gibco-ThermoFisher Scientific, Waltham, MA, USA) at a bead-to-cell ratio of 1:5 and cultured in 24-well plates at a cell density between 1 and 2×10^6 cells/mL for 72 h in a humidified incubator at 37 °C with 5% CO_2.

For production of non-infectious, RT-competent HIV-1 (NNHIV) particles, HEK293T cells were seeded into 10 cm dishes at a density of 5×10^6 cells in 6 mL per dish, one day prior to transfection. The following day, the cells were transfected using a standard calcium phosphate procedure with a total of 10 μg of DNA per 10 cm dish. The transfection mixture included the non-infectious proviral plasmid NNHIV env(stop) ANCH [29,30], a Vpr.IN.eGFP expression plasmid containing the corresponding mutations in the IN gene on the NNHIV plasmid [42], and a pCMV-VSV-G expression plasmid (Addgene plasmid #8454, a gift from Bob Weinberg [43]) in the ratio of 7.7:1.3:1.0 μg. After 4–6 h, medium was replaced, and the cells were incubated at 37 °C for an additional two days prior to harvesting the supernatant. The supernatant was then centrifuged at 300× *g* for 5 min to remove cell debris and subsequently filtered through 0.45 μm mixed cellulose ester (MCE) filters. The filtered supernatant was layered onto a 20% (*w*/*v*) sucrose cushion and centrifuged at 107,000× *g* for 1.5 h at 4 °C. The resulting viral particles were resuspended in 30 μL of ice-cooled phosphate-buffered saline (PBS) containing 10% FBS and 10 mM HEPES (pH 7.5), pooled, rapidly frozen in liquid nitrogen, and stored at −80 °C. Viral stocks were quantified using the SYBR Green-based Product Enhanced Reverse Transcription assay (SG-PERT) as described previously [44].

2.2. Cell Seeding and Fixation

MDM and TZM-bl cells were directly seeded onto 6–18 mm glass coverslips, which were placed in multi-well plates, at least 24 h before fixation. Cell density was adjusted to reach 70–90% confluency on the day of fixation. CD4+ T-cells were seeded onto polyethylenimine (PEI)-coated 12 mm coverslips at a density of 300,000 cells per coverslip and incubated for 40 min at 37 °C in a 5% FBS/PBS solution. The PEI coating was applied for 1 h at 37 °C using 100 μL of a 0.5 mg/mL PEI solution. Prior to transferring the cells, the coated coverslips were washed three times with ddH_2O. For fixation, the medium was removed, and the coverslips were washed once with PBS. Fixation was performed for 10 min at room temperature using 4% paraformaldehyde (PFA, Electron Microscopy Sciences, Hatfield, PA, USA, #15719)/0.0075% glutaraldehyde (GA, Sigma Aldrich, Taufkirchen, Germany, G5882-10X1ML) in PBS. Subsequently, the coverslips were washed three times with PBS before either proceeding to U-ExM or being stored in PBS at 4 °C. For imaging during the U-ExM process, nuclei were briefly stained with 1 mg/mL Hoechst 33342 (here referred to only as Hoechst, Thermo Fisher Scientific, Waltham, MA, USA) for 5 min at room temperature in the dark. Afterwards, the coverslips were washed twice with PBS before imaging and subsequent expansion.

For determination of optimal fixation conditions for maintaining cytoplasmic integrity, cells were washed and fixed using either a pH 6.8 PEM buffer (composed of 80 mM PIPES, 5 mM EGTA, and 2 mM $MgCl_2$ in ddH_2O) as described by Leyton–Puig [45] or a pH 7.4 PHEM buffer (containing 60 mM PIPES, 25 mM HEPES, 10 mM EGTA, and 2 mM $MgCl_2$ in ddH_2O) as outlined by Sobue [46] instead of PBS. For fixation with a pH 6.1 cytoskeleton buffer (10 mM MES, 150 mM NaCl, 5 mM EGTA, 5 mM glucose, and 5 mM $MgCl_2$ in ddH_2O) following the protocol of Small [47], cells were rinsed once with pre-warmed cytoskeleton buffer. Cells were then detergent extracted with 0.25% Triton X-100/0.3% GA in cytoskeleton buffer for 90 sec at room temperature, followed by fixation with 2% GA in cytoskeleton buffer for 15 min at room temperature [45,48,49]. After fixation, cells were rinsed once with cytoskeleton buffer and then quenched with 50 mM NH_4Cl in PBS for 10 min at room temperature. Finally, cells were washed twice with PBS before progression to U-ExM as described below.

For cryofixation, samples were treated according to the protocol provided by Laporte et al. [50]. Briefly, cells seeded on 6 mm coverslips were incubated briefly with either HEK293T growth medium or HEPES-buffered Hanks Balanced Salt Solution (HHBSS composed of 20 mM HEPES, 137 mM NaCl, 0.49 mM $MgCl_2$, 1.26 mM $CaCl_2$, 5.4 mM KCl, 0.41 $MgSO_4$, 0.64 mM KH_2PO_4, 3 mM $NaHCO_3$, and 5.5 mM glucose) containing 10% glycerol at pH 7.4. The samples were then plunge-frozen in liquid ethane that had been precooled at liquid nitrogen temperature using a Leica EM GP2 plunge freezer (Leica Microsystems, Wetzlar, Germany). The plunge freezer was operated as indicated by the manufacturer in a sample chamber environment of 25 °C and 80% humidity, a liquid ethane temperature of −183 °C, and using one-sided blotting on Whatman grade 1 filter paper (Sigma Aldrich, St. Louis, MO, USA; #1001-055) with a blotting time of 5 s. Subsequently, the samples were incubated in dry ice-chilled acetone (Sigma Aldrich, Taufkirchen, Germany, #270725-1L, HPLC, ≥99.9%), progressively transitioned to 0 °C before rehydration in a series of ethanol (Fischer Scientific-Thermo Fischer Scientific, Waltham, MA, USA, #E/0650DF/15 1L, ≥99.8%)/water baths with increasing water percentages [50]. Once the samples reached PBS incubation, samples were expanded as described below.

2.3. Ultrastructure Expansion Microscopy (U-ExM)

The protocol established by Gambarotto et al. [36] was implemented with minor modifications to optimize gel handling and reagent usage. A comprehensive protocol, including detailed information on all reagents, can be found in File S1. As previously described, sodium acrylate quality is critical for expansion. Here, we used sodium acrylate powder from AK Scientific (Union City, CA, USA; R624-5g), which was stored at −20 °C and consistently yielded good results in our hands.

The U-ExM protocol involves several steps (Table 1), each with extended incubation times. The approximate durations for each step are as follows: anchoring/crosslinking prevention (3.5 h), gelation (1.5 h), denaturation (2 h), first expansion (either 3× 30 min, or 2× 30 min and once overnight), gel shrinking (30 min), blocking (30 min), primary antibody incubation (2.5 h to overnight), washing (50 min), secondary antibody incubation (2.5 h), washing (50 min), and final expansion and staining with synthetic fluorescent sphingosine BODIPY-Ceramide dyes (BODIPY, Thermo Fisher Scientific, Waltham, MA, USA) staining (1× 30 min, overnight, and 2× 30 min). In some cases, we included an additional staining step using a fluorescently conjugated NHS-ester (ATTO-TEC, Siegen, Germany) after the second antibody incubation and washing, which extended the protocol for an additional 2 h to account for incubation and washing. To mitigate contamination risks during the long incubation times, particularly those involving bovine serum albumin (BSA, fatty acid free from Carl Roth®, Karlsruhe, Germany), we supplemented all solutions with 0.01% NaN_3 (*w*/*v*) for any incubation lasting 30 min or longer.

Table 1. Different steps of the U-ExM protocol and their approximate duration.

	Step	Time
Day 1	Anchoring	3.5 h
	Gelation	1.5 h
	Denaturation	2 h
	First expansion	1.5 h
	PBS shrinking	0.5 h
	Blocking	0.5 h
	Primary antibody incubation	overnight
Day 2	Washing	1 h
	Secondary antibody incubation	2.5 h
	Washing	1 h
	[NHS-ester staining]	[1.5 h]
	[Washing]	[0.5 h]
	Second expansion and BODIPY staining	overnight
Day 3	Final expansion steps	1 h

Although the U-ExM protocol can be completed over three days, we opted for a two-day schedule for the expansion process, with imaging typically conducted on the third and fourth days.

2.3.1. Sample Mounting

Gel pieces were imaged using either Poly-D-Lysine-coated ibidi 8-well glass bottom µ-slides (ibidi GmbH, Gräfeling, Germany) or uncoated 6-well plates (Corning GmbH, Kaiserslautern, Germany). To prepare the ibidi-coated µ-slides, 200–500 µL of a 0.1 mg/mL. Poly-D-Lysine solution (Gibco-Thermo Fisher Scientific, Waltham, MA, USA) was added to each well and incubated for 1 h at 37 °C. Afterwards, the wells were washed three times with ddH_2O. The orientation of cells within the gels was determined using a Nikon eclipse Ts2 widefield inverted benchtop microscope (Nikon Instruments Inc., Tokyo, Japan) by examining either Hoechst or BODIPY staining through a 10×/0.25 air objective. For gel mounting, a piece of gel that fit into a well of an ibidi 8-well µ-slide was carefully excised and gently pressed, cell-side down, into the coated well to ensure leveling and minimize air bubbles. To prevent the gels from drying during imaging, 2–3 drops of ddH_2O were added to the well.

Unmounted, expanded, and stained gel pieces were kept in ddH_2O supplemented with 0.01% NaN_3 for up to one week, allowing for later mounting and imaging.

2.3.2. Expansion Factor Measurement

Since only sections of the original gel were used for imaging, it was not possible to determine the final expansion factor by measuring the final gel diameter, as has been carried out previously [36]. Instead, we measured the diameter of the gel after shrinkage and just prior to excising the gel pieces to determine an initial expansion factor, EF_{PBS}. Rectangular gel pieces were then excised and measured before proceeding with staining and expansion. After the final expansion and prior to mounting, we measured the fully expanded gel piece again to determine a second expansion factor, EF_{H2O}. Consequently, the final expansion factor EF_{Final} was calculated as follows:

$$EF_{Final} = EF_{PBS} \times EF_{H2O} \quad (1)$$

2.4. Fluorescence Microscopy

Imaging of samples during the U-ExM process was conducted using a Nikon eclipse Ts2 widefield inverted benchtop microscope (Nikon, Tokyo, Japan) equipped with a 10×/0.25 Ph1 ADL air objective (working distance (WD) 6.2 mm) and a 385/470/560 filter set. Images were acquired with a CMOS DMK 33UX174 camera (theimagingsource, Bremen, Germany) using the NIS-Elements D software (version 5.02, Nikon, Tokyo, Japan). Exposure time, gain, and gamma values were optimized to enhance contrast and visibility.

For large overviews of gel pieces as well as GelMap coverslips (Utrecht University, Utrecht, Netherlands) [51], we employed the Zeiss CellDiscoverer 7 automated microscope (Zeiss, Oberkochen, Germany). Gels and coverslips were imaged in either 6-well plates or ibidi 8-well glass bottom µ-slides. Images were acquired using either a Plan-Apochromat 5×/0.35/WD 5.1 mm or a Plan-Apochromat 20×/0.7/WD 2.2 mm autocorr air objective with 1×–2× tube lens magnification and a Zeiss Axiocam 712 sCMOS camera for detection. Multichannel images were acquired using light-emitting diodes at 385 nm, 470 nm, 567 nm, and 625 nm for excitation, paired with a quad-bandpass filter (QBP 425/30 + 514/30 + 592/25 + 709/100 nm) for emission. Image tiling was executed with a 10% overlap using the *Definite Focus* module for focus stabilization.

Point laser scanning confocal microscopy was performed on a Zeiss LSM900 microscope (Zeiss, Oberkochen, Germany) equipped with the Airyscan 2 detector. Imaging was performed using either a Plan-Apochromat 20×/0.8/WD 0.55 mm air or a Plan-Apochromat 63×/1.4/WD 0.193 mm oil immersion objective. Similar to previous imaging, tiling was performed with 10% overlap and the *Definite Focus* module for focus stabilization. Multichannel images were acquired sequentially in either frame or stack scanning mode, employing diode lasers at 405 nm, 488 nm, 561 nm, and 640 nm for imaging of fluorophores in the blue, green, orange, and red spectra of the visible light, respectively. Emission detection was configured using variable dichroic mirrors, and detection was optimized for emission spectra of different fluorophores. GaAsP PMT or Airyscan detectors were ultilized, with gain typically set between 700 and 900 V, while the offset remained at 0%. Sampling was optimized for Nyquist criteria (ca. 50 nm laterally and 250 nm axially for the 63×/1.4 oil immersion objective), conducted bidirectionally at maximum scanning speed with 2× line averaging. Post-acquisition processing of 2D or 3D Airyscan images was performed using ZEN Blue (version 3.1, Zeiss, Oberkochen, Germany), applying the default Airyscan Filtering (AF) strength.

STED microscopy was performed on an Expert Line STED system (Abberior Instruments GmbH, Göttingen, Germany) equipped with an SLM-based *easy3D* module and an Olympus IX83 microscope body with a 100× oil immersion objective (NA 1.4; WD 0.13 mm; Olympus UPlanSApo). STED images were acquired using a 640 nm excitation laser and spectral detectors, with detection ranges optimized for fluorophore emission accordingly. A 775 nm STED laser (operating at 10–20% of the maximal power of 3 mW) was used for depletion, with a pixel dwell time of 10 µs and 15 nm xy sampling.

2.5. Image Selection, Processing, and Analysis

For general image visualization and analysis, the FIJI distribution of the open source software ImageJ (version 1.54f) was used [52]. Images were contrast adjusted to enhance visibility and filtered with a mean filter with a radius of 1–2 pixel. Data graphs were generated and statistical analysis performed using GraphPad Prism version 5.00 for Windows (GraphPad Software, San Diego, CA, USA). Images presented were selected as representative examples of 2–3 biological replicates, with at least 3 images (single planes or z-stacks) being acquired per replicate after visual examination of samples. For samples infected with NNHIV, infected cells were identified by the presence of CA- and high CPSF6-containing structures in the nucleus before acquisition of 5–7 images over 2 biological replicates.

2.5.1. Stitching

For tiled images, stitching was performed using Zen Blue version 3.9 without shading correction, using the nuclear channel as a reference.

2.5.2. Bleedthrough Correction

Due to the usage of antibodies and dyes with spectral overlap as well as strong fluorescence intensity in some channels, some bleedthrough was observed in multi-channel imaging during point laser scanning confocal microscopy. To determine if fluorophore bleedthrough was present, images were examined for the presence of duplicate structures in multiple channels. If bleedthrough was present in a channel, a subtraction-based correction approach was used to locally remove bleedthrough in the affected image [53]. Subtraction-based bleedthrough correction can be used in cases where spectral crossover is only present in one direction, i.e., if there is a reference channel without any bleedthrough available, as was the case here.

First, the contribution of the fluorophore that causes bleedthrough was determined in the channel that was affected by bleedthrough. To do so, in the image that was affected by bleedthrough, the mean signal intensity was calculated in an area that only showed bleedthrough. Then, the mean signal intensity was calculated in the same area of the channel where bleedthrough originated. Subsequently, a bleedthrough correction factor was determined by dividing the mean signal intensity of the channel where bleedthrough originated by the mean signal intensity of the channel that was affected by bleedthrough. To create a correction image that represented the bleedthrough, the original channel where the bleedthrough originated was then divided by the calculated correction factor. The resulting lower intensity correction image was then subtracted from the channel that was affected by bleedthrough via the *Image Calculator* module in ImageJ (Figure S1).

2.5.3. Nucleus Segmentation and Nuclear Diameter Measurements

To determine the dimensions of the nucleus before and after expansion, the same set of nuclei in both conditions were segmented. Prior to segmentation, images of nuclei acquired before expansion were processed using a median filter with a radius of 1 pixel. Subsequently, the nuclear channels of both images were segmented using the pre-trained cell segmentation algorithm CP within the cellpose 2.2.1 framework (platform Win32, python version 3.8.16, and torch version 1.12.0) [54]. Label images were exported as PNG files and further processed in ImageJ. To separate nuclei before analysis, an erosion in xy of 1 pixel was carried out using the ImageJ *3D Suite Close Labels* function [55]. The images were then thresholded, and the Feret diameter was measured using the *Particle Analyzer* function in ImageJ with no restrictions for circularity or size.

2.5.4. NPC Dimension Measurements

For evaluating the properties of the Nup153 foci as a proxy for the dimensions of the NPC, a maximum intensity projection of 21 z-slices, each spaced 140 nm apart, starting from the bottom of a U2OS nucleus was created. The image was thresholded (600–65,535) to isolate Nup153 signal foci, and the resulting binary image was analyzed using the *Analyze Particles* function in ImageJ, excluding all objects at the borders of the image and only including objects with a size of 20 pixel^2 and larger. Subsequently, the maximum and minimum Feret diameters were divided by the expansion factor EF_{Final} to account for expansion.

2.5.5. Nup153 and Nup96-SNAP Signal Intensity Measurement

To measure the signal intensity of the Nup153- and Nup96-SNAP channels, a line was drawn through an area of the nuclear envelope containing several distinct Nup153 foci. Signal intensities along the line were measured in both channels using the *plot profile* function of ImageJ. The intensity values were exported and normalized to the minimum and maximal values for each channel. During plotting, the graphs were aligned to the point of highest intensity to compensate for any potential shifts that may have occurred during imaging.

3. Results

Expansion microscopy has gained significant traction within the broader research community; however, its application in virology remains limited, with particularly few studies exploring HIV-1-related phenomena [56,57]. Other studies have investigated general aspects of the broader virology field, such as the spatial distribution of virions and viral proteins in infected cells and tissues, as well as host cell alterations following infection [58–61]. Among the available expansion protocols, the U-ExM method, which offers a 4–5-fold expansion alongside a post-expansion labeling approach, has found broader application in fields such as parasitology, centriole biology, and neurobiology, where it has significantly improved the visualization of target structures [62–66]. Here, we aim to assess the effectiveness of the U-ExM protocol to probe processes during virus infection, with a particular focus on nuclear HIV-1–cell interactions at high resolution.

3.1. The Ultrastructure Expansion Microscopy Protocol Allows Robust Expansion with Good Preservation of Cellular Morphology

We first benchmarked the U-ExM protocol for application in TZM-bl cells, a HeLa-derived cell line frequently used in HIV-1 research [38]. To establish fundamental expansion parameters, we adhered closely to the protocol outlined by Gambarotto in 2021 [36], with minor modifications to optimize sample handling and reagent usage (refer to Materials and Methods as well as File S1 for a detailed protocol). To monitor the expansion process and the cell state during the U-ExM protocol, we seeded TZM-bl cells onto glass coverslips and subjected them to the U-ExM process. We employed nuclear staining using Hoechst at low concentrations directly after fixation to visualize the position and morphology of nuclei. Imaging was conducted following fixation, after incubation with the anchoring solution, immediately after gelation within intact gels sandwiched between the sample coverslip and a rectangular glass slide, as well as in detached gels post-denaturation (Figure 1A,B). Cells were clearly visible at all time points prior to denaturation, exhibiting normal cellular and nuclear morphology and size (Figure 1B). Although the Hoechst signal was lost after denaturation, most likely due to the harsh conditions necessary for sample homogenization, cells remained discernible in transmitted light (Figure 1B). Consistent with the observed size increase of the gel during denaturation, longitudinal cell size increased from ca. 30 μm to up to 100 μm in the gels post-denaturation (Figure 1B). Following the U-ExM protocol, gels were fully expanded for the first time via incubation in deionized water overnight and subsequently shrunk again by incubation in PBS to facilitate sample handling. To minimize reagent usage for staining, we excised 1 cm × 0.5 cm pieces from the center of the gel for blocking and staining. For initial characterization of cellular and nuclear morphology, chromatin was stained using Hoechst, and samples were incubated with BODIPY during the final expansion to visualize lipid membranes, as previously established [67]. Measurement of gel and coverslip sizes before and after expansion indicated an expansion factor of ca. 4.3, in line with previous studies in mammalian cells and other eukaryotes [18,63,67]. After the final expansion, cells showed a normal overall morphology within the gel, with longitudinal cell sizes often exceeding 100 μm, consistent with the calculated expansion factor (Figure 1B). It is noteworthy that samples were more prone to degradation and contamination after expansion; hence, we consistently employed 0.01% sodium azide for any incubation or storage period exceeding one hour.

Next, we assessed whether samples experienced distortion during the expansion process by directly comparing cells before and after expansion. In traditional pre-expansion labeling protocols, structures can be imaged both prior to and following expansion, allowing direct comparisons. However, in the case of U-ExM, direct comparisons are more challenging, as staining typically occurs only after expansion. To identify the same set of cells before and after expansion as well as to visualize any sample deformation, we employed the GelMap system developed by the Kapitein lab [51]. The GelMap consists of a coverslip printed with a defined fluorescently labeled protein pattern comprising a 20 × 20 µm grid along with numerical and letter markers (Figure S2A). For reference images, we acquired wide-field images of fixed TZM-bl cells seeded on an 18 mm GelMap coverslip patterned with a custom R2-myc-his nanobody conjugated to ATTO 647N (Figures 1C and S2B,D). Subsequently, the whole GelMap coverslip was expanded, and a 1 cm × 0.5 cm gel piece was excised from the center of the gel before blocking and staining. Although the ATTO 647N fluorescence was lost during the U-ExM procedure, the GelMap grid could be visualized via α-R2-myc immunofluorescence staining in wide-field microscopy after expansion, allowing for identification of positions that had been imaged before expansion (Figures 1D and S2C,E).

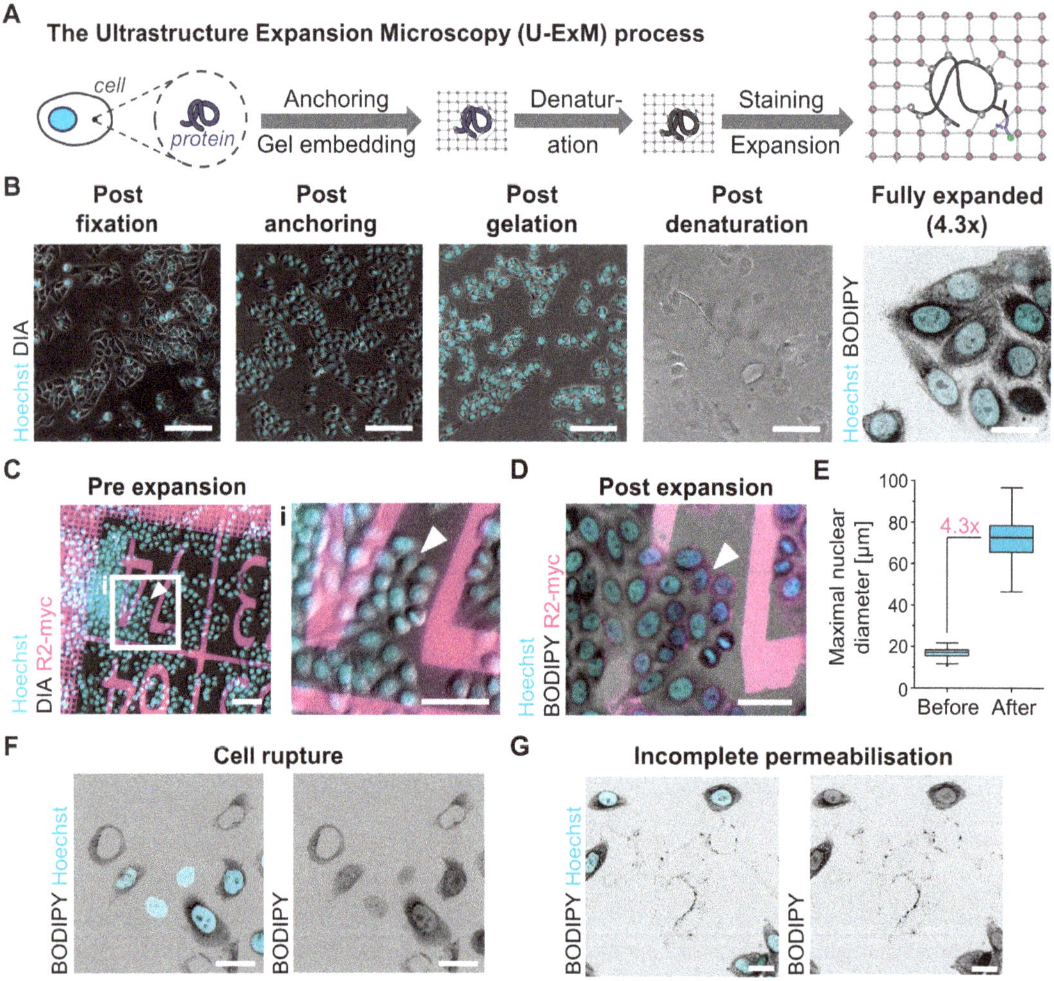

Figure 1. U-ExM enables robust 4.3-fold expansion with minor macroscopic distortions. (**A**) A schematic overview illustrating the different steps involved in the U-ExM procedure. (**B**) TZM-bl cells, fixed with PFA/GA, stained for chromatin (Hoechst, cyan) and membranes (BODIPY, right panel, gray) and imaged at different steps of the U-ExM process. Scale bar: 100 µm. (**C**) TZM-bl cells seeded onto an 18 mm GelMap slide patterned with R2-myc-his-ATTO 647N (magenta), fixed with

PFA/GA, and stained for chromatin (Hoechst, cyan). Scale bar overview, 100 µm; scale bar enlargement, 50 µm. Arrowhead marks the position of the same cluster of cells before and after expansion in (**C**,**D**). (**D**) TZM-bl cells seeded onto an 18 mm GelMap slide, fixed with PFA/GA, expanded by U-ExM, and stained for chromatin (Hoechst, cyan), membranes (BODIPY, gray), and R2-myc (magenta). Scale bar, 200 µm. (**E**) Measurement of the Feret diameter of the nuclei of the same set of TZM-bl cells seeded onto an 18 mm GelMap slide, fixed with PFA/GA and stained for chromatin, before and after expansion; $n = 168$; $\text{median}_{\text{before}} = 17.03$ µm, $\text{median}_{\text{after}} = 72.63$ µm; data plotted according to Tukey box plot. (**F**,**G**) TZM-bl cells stained for chromatin (Hoechst, cyan) and membranes (BODIPY, gray) exhibiting expansion artifacts. Scale bar, 100 µm. All scale bars in expanded samples reflect post-expansion sizes.

To compare the appearance of cells before and after expansion, we segmented the same set of nuclei using the cellpose 2 CP algorithm on images taken before and after expansion, measuring the maximum diameter of these nuclei (Figure S3A,B). Comparison revealed a maximum diameter of ca. 17 µm before and 73 µm after expansion, indicating a nuclear expansion of about 4.3-fold. This finding aligns with our measurements of the expansion factor based on gel size, suggesting that the observed increase in gel size is directly translatable to microscopic structures (Figure 1E). Although the overall distribution of nuclei was retained and cells maintained their relative position, the masks of the segmented nuclei before and after expansion could not be overlaid using rigid transformation or scaling, indicating local distortion (Figures 1C,D and S3C). This was further confirmed by inspection of the protein grid, which displayed small local deformations in the grid lattice and large disruptions in certain areas (Figure S2B–E). Nevertheless, the overall shape and morphology of individual cells appeared to be well conserved (Figures 1C,D, S2B–E and S3A,B).

While most cells showed no apparent abnormalities, expansion artifacts were visible in 5–20% of the cells (Figure 1F,G). These artifacts could be categorized into two groups: (i) cells that appeared to not have been permeabilized during the expansion process (Figure 1G), and (ii) cells exhibiting cytoplasmic rupture (Figure 1F). In severe instances, cytoplasmic rupture resulted in the displacement of visually intact nuclei from empty and occasionally collapsed cytoplasmic shells (Figure 1F). Non-permeabilized cells appeared as opaque, cell-shaped structures lacking staining and could be identified already after denaturation (Figure 1B,G). Although both phenotypes were observed across different samples, they never occurred simultaneously. The abnormal cells were visually distinct from properly expanded cells, allowing for their immediate exclusion from further analysis.

3.2. Optimization of Fixation and Imaging Conditions

Although nuclei showed no visible abnormalities even in severe cases of cell rupture, inspection of the cytoplasmic compartment through Airyscan microscopy occasionally revealed gaps in the BODIPY membrane staining. This finding indicates compromised cytoplasmic integrity (Figure S4A). To optimize the preservation of the cytoplasmic compartment during expansion, we employed various fixation procedures, expanded TZM-bl cells, and performed staining using an α-tubulin antibody alongside a membrane marker (BODIPY) and Hoechst. It is noteworthy that microtubules are prone to fragmentation under inadequate fixation conditions [50,68], while membrane staining was utilized to assess overall compartment integrity. We compared several fixation conditions, including cryofixation [50] in different solutions and chemical fixation with 4% PFA/0.0075% GA using different buffer conditions optimized for cytoskeletal conservation (Table 2, Figures 2A and S4B,C).

Regular chemical fixation using a 4% PFA/0.0075% GA in PBS solution resulted in fragmented microtubules and diffuse membrane signals when observed with Airyscan microscopy. In contrast, alternative fixation conditions, including chemical fixation in PHEM buffer and cryofixation in HHBSS/10% glycerol, showed uninterrupted microtubules and distinct membrane networks (Figures 2A and S4B,C). Notably, these different fixation conditions did not alter the appearance of the nuclear compartment (Figures 2A and S4B,C).

Table 2. Different fixation conditions for optimization of sample preservation.

Fixation Condition	Buffer Solution
Cryofixation	Medium, pH 7.4 Hepes-buffered Hanks' balanced salt solution (HHBSS)/10% glycerol, pH 7.4
Chemical fixation (4% PFA/0.0075% GA)	PBS buffer, pH 7.4 PHEM buffer, pH 7.4 PEM buffer, pH 6.8
(0.25% Triton X-100/ 0.3% GA; 2% GA)	Cytoskeleton buffer, pH 6.1

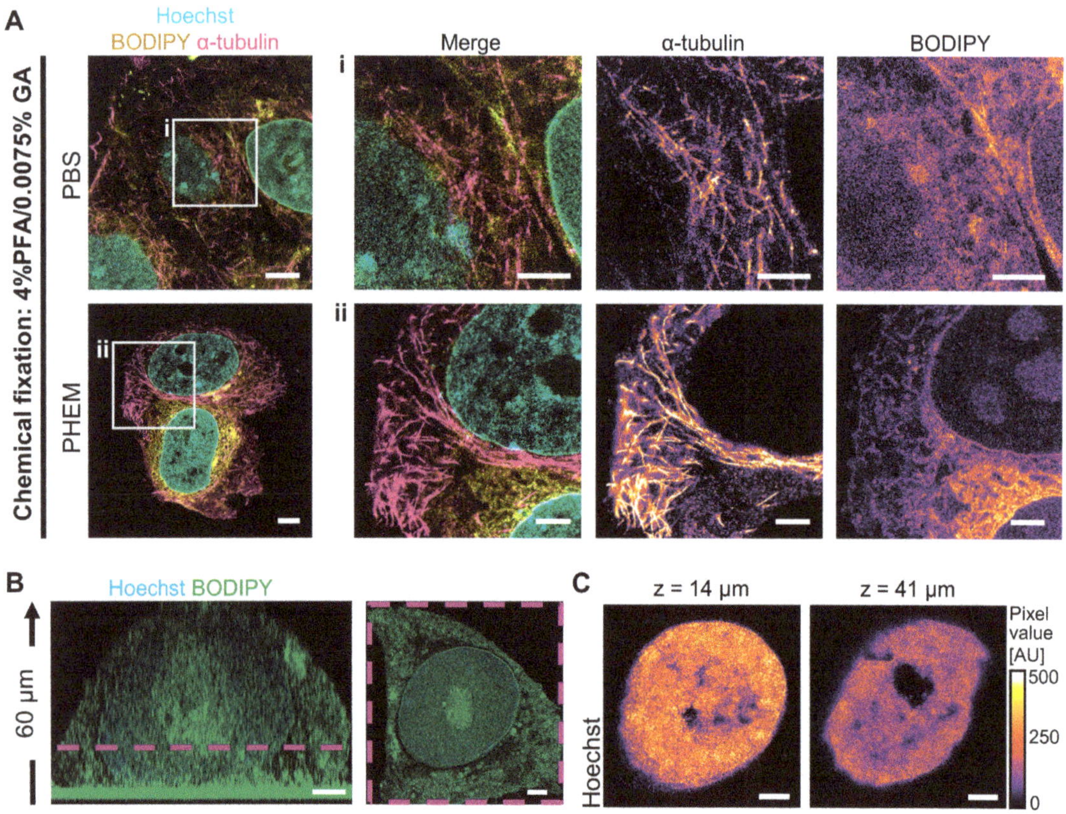

Figure 2. Sample preparation and imaging optimization are critical for expanded samples. (**A**) TZM-bl cells fixed with PFA/GA in either PBS or PHEM buffer, expanded and stained for chromatin (Hoechst, cyan), membranes (BODIPY, yellow), and α-tubulin (magenta). Scale bar overview, 20 μm; scale bar enlargements, 10 μm. (**B**) 3D structure of an expanded TZM-bl cell, stained for chromatin (Hoechst, cyan) and membranes (BODIPY, green). First panel, zy-slice; second panel, xy-slice. The magenta dashed line denotes the position of the xy slice in z. Scale bar, 10 μm. (**C**) Single slices of the same expanded TZM-bl cell as in (**B**), stained for chromatin (Hoechst). Scale bar, 10 μm. All scale bars in expanded samples reflect post-expansion sizes.

Next, we evaluated the 3D sample structure by employing confocal microscopy and optical sectioning. We acquired 3D stacks of whole TZM-bl cells after expansion using Airyscan microscopy, using BODIPY and Hoechst staining to visualize the cellular and nuclear compartments (Figure 2B, Video S1). Reconstruction of cells showed robust 3D expansion without apparent distortion of the cell body or the nucleus (Figure 2B, Video S1). As expected for a 4.3-fold expansion, cells regularly reached heights of over 50 µm (Figure 2B), which resulted in extended imaging times at maximal resolution. In addition, we observed a decrease in fluorescence intensity when acquiring 3D volumes and imaging deeper into the sample (Figure 2B,C, Video S1). This reduction in signal intensity was likely due to a combination of increased scattering and bleaching, which were particularly pronounced with fluorophores within the blue light spectrum, including Alex 405 Plus, Hoechst, and BODIPY (Figure 2B,C, Video S1). Nevertheless, the signal degradation in general stains like Hoechst and BODIPY was largely negligible for the identification of relevant structures (Figure 2C). To prevent signal degradation during imaging of immunostained targets, we also employed fluorophores developed for STED microscopy, including long-stokes shift dyes, which exhibit greater resistance against bleaching [69].

3.3. *U-ExM Achieves Robust Expansion of Different Cell Types*

We subsequently investigated the applicability of the expansion procedure established for the TZM-bl model cell line across different cell types. Here, we focused on cell types relevant for HIV-1 research, particularly primary cells recognized as significant HIV-1 reservoirs [70]. To this end, we fixed and expanded both resting and activated primary CD4+ T-cells as well as monocyte-derived macrophages (MDMs). To evaluate the overall morphology of the expanded cells, we employed a combination of Hoechst and BODIPY staining. Additionally, we used immunofluorescence staining for the inner nuclear pore complex (NPC) protein Nup153 to investigate the NPC distribution and morphology. All cell types were robustly expanded, achieving a ca. 4.3-fold size increase while maintaining intact cell bodies (Figure 3A).

Airyscan microscopy of expanded cell nuclei revealed distinct chromatin domains, characterized by varying degrees of Hoechst staining, likely representing euchromatin (low intensity) and heterochromatin states (high intensity) (Figure 3A). Notably, strong Hoechst staining was prominent at the nuclear periphery and around nucleoli, which are known to serve as heterochromatin hubs [71]. Interestingly, while resting CD4+ T-cells exhibited a high density of condensed chromatin, activated T-cells showed a larger proportion of euchromatin (Figure 3A). This difference likely reflects the distinct chromatin architecture resulting from the varying transcriptional programming between the activated and resting states [72].

Expansion resulted in substantial improvement in rendering the cytoplasmic T-cell compartment, which is typically challenging to visualize by fluorescence microscopy (Figure 3A). All cells displayed a complex membranous network throughout the cytoplasm, with the nucleus containing markedly less BODIPY staining than the cytoplasm (Figure 3A), consistent with the lower lipid and membranous structure contents in the nucleus. In contrast, nucleoli showed a relatively high BODIPY intensity (Figure 3A).

Staining for Nup153 revealed numerous distinct spots decorating the nuclear envelope. The Nup153 foci had a diameter of around 400 nm (corresponding to ca. 100 nm in the pre-expansion state, as expected for nuclear pores [73]) and were primarily localized at the periphery of the chromatin staining, indicating that these foci represent individual NPCs (Figure 3A).

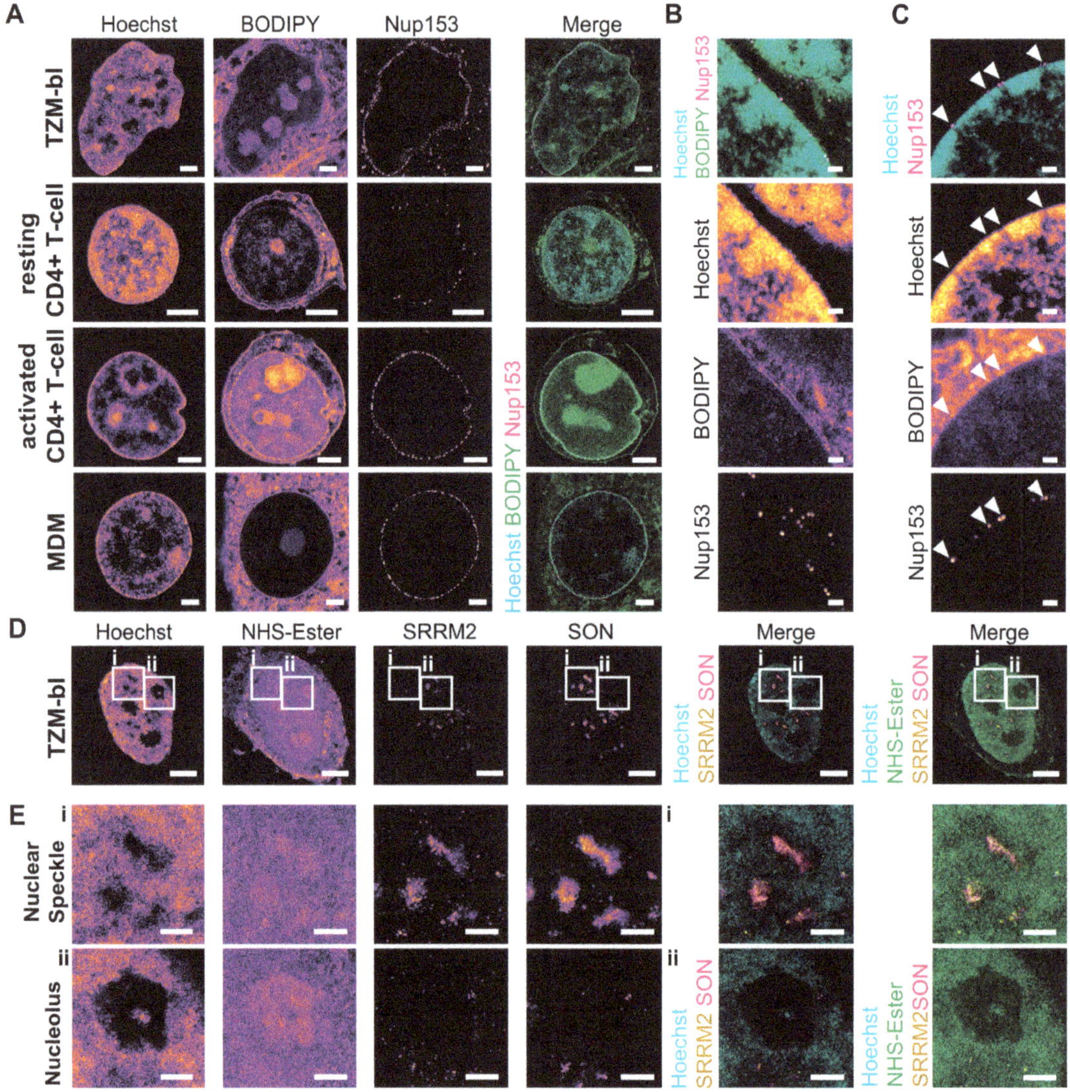

Figure 3. U-ExM allows expansion of various cell types and cellular structures. (**A**) TZM-bl cells, resting and activated CD4+ T-cells, and MDM fixed with PFA/GA, expanded and stained for chromatin (Hoechst, cyan), membranes (BODIPY, green), and Nup153 (magenta). Scale bar, 10 µm. (**B**) Cropped region of the nucleus of an MDM fixed with PFA/GA, expanded and stained for chromatin (Hoechst, cyan), membranes (BODIPY, green), and Nup153 (magenta) showing double membrane layers of the nuclear envelope in membrane staining. Scale bar, 2 µm. (**C**) Cropped region of the nucleus of an MDM fixed with PFA/GA, expanded and stained for chromatin (Hoechst, cyan), membranes (BODIPY, not shown in merge), and Nup153 (magenta), showing heterochromatin-free channels beneath the NPCs in the nucleus. Scale bar, 2 µm; arrowheads mark the position of NPC. (**D**) A TZM-bl cell fixed with PFA/GA, expanded and stained for chromatin (Hoechst, cyan), protein density (NHS-ester, green), SRRM2 (yellow), and SON (magenta), showcasing different intranuclear compartments. Scale bar, 10 µm. (**E**) Enlargements from (**D**) highlighting nuclear speckles and nucleolus. Scale bar, 5 µm. All scale bars in expanded samples reflect post-expansion sizes.

3.4. Different Cellular Structures Are Well Conserved after Expansion

We next focused on the preservation of the nuclear compartment by visualizing different, well-characterized nuclear structures. In non-diving cells, the nucleus acts as a closed compartment, presenting a potential barrier to invading viruses [22,23]. These viruses must hijack the cellular nuclear import machinery, or the NPC, to gain entry into the nucleus [22]. To assess whether U-ExM can be used to investigate virus–host interactions at the nuclear pore, we first probed the ultrastructure of the NPC and nuclear periphery in their unperturbed state. A structurally conserved and well-defined feature of the nuclear periphery is the presence of chromatin-free regions below the NPC, which has already been described in early studies on nuclear architecture using electron microscopy [71]. We fixed and expanded primary MDMs and visualized the different structures of the nuclear periphery by using BODPY to stain the nuclear envelope membrane, Hoechst to visualize the chromatin structure, and Nup153 immunostaining to mark the position of the NPC. Using Airyscan microscopy, we could clearly observe the individual layers of the nuclear envelope in some regions where it was oriented perpendicular to the imaging plane due to the enhanced resolution provided (Figure 3B). Additionally, even in regions where such distinct visualization of the nuclear envelope layers was not achievable, we consistently observed heterochromatin-free regions beneath the NPC, as indicated by Nup153 immunostaining (Figure 3C). These heterochromatin-free regions appeared as narrow, elongated, channel-like structures that eventually connected and merged with larger heterochromatin-free areas further from the nuclear envelope (Figure 3C). These observed structural features align with the widely accepted model of the chromatin landscape at the nuclear periphery, where the interchromatin space is characterized by chromatin-free channels below the NPCs, connecting to a larger, continuous network throughout the nucleus [71].

The interchromatin space of the nucleus also harbors several specialized compartments that may be exploited by viruses for their own purposes [23]. Among these, nuclear speckles are the most abundant. These chromatin-free, membraneless organelles function as storage sites for splicing components and play a crucial role in RNA processing [74]. Another prominent nuclear compartment is the nucleolus, which is primarily involved in ribosome biogenesis [74]. Both nuclear speckles and nucleoli are characterized by low chromatin abundance, high protein density, and formation through phase condensation [74]. They can be visually distinguished by their shape and size: nucleoli are larger and spherical, whereas nuclear speckles are more numerous, smaller, and less defined in structure. Additionally, both compartments can be distinguished by the localization of the nuclear speckle structural proteins SRRM2 and SON, which form the main scaffold of nuclear speckles but are absent in nucleoli [74]. To investigate whether the structural integrity of these compartments is preserved after U-ExM, we fixed and expanded TZM-bl cells, staining them for chromatin using Hoechst and for the two nuclear speckle proteins SRRM2 and SON. We also employed fluorescently conjugated *N*-Hydroxysuccinimide (NHS)-ester to visualize protein distribution within the nuclear compartment, as this reagent reacts with primary amines in proteins, providing a general staining for protein density [10,36]. Airyscan microscopy revealed the presence of multiple chromatin-free regions of different sizes within the nucleus (Figure 3D). As expected, these chromatin-free regions showed higher protein density; however, only the smaller regions colocalized with SON and SRRM2, confirming their identity as nuclear speckles (Figure 3D,E). This finding indicates that the structure and composition of phase-separated nuclear bodies, such as nuclear speckles and nucleoli, are well preserved in U-ExM.

3.5. Nanoscopic Targets Are Well Preserved during Expansion

Since virus–host interactions predominantly occur at a nanoscale, we aimed to investigate whether U-ExM alters the structure of well-defined large macromolecular complexes. The NPC is among the best characterized and structurally defined assemblies within eukaryotic cells. Containing more than 500 individual proteins, the NPC forms a channel with an eight-fold symmetry, consisting of three stacked rings. The diameters of the circular

NPC structures range from 40 to 60 nm in the inner ring's channel to 150 nm in the distal components of the nuclear ring [29,73]. The nuclear basket protein Nup153 is located on the nuclear side of the NPC. Although the structure of the nuclear basket remains poorly defined, single-molecule localization data indicate that it occupies an area with a diameter of ca. 100 nm below the main body of the NPC [73]. When we probed MDM with immunostaining using a primary and secondary antibody cascade against Nup153, single NPC could not be resolved in Airyscan microscopy but could be distinguished using STED microscopy (Figure 4A). In contrast, when applying U-ExM to MDM followed by immunostaining and Airyscan microscopy, we could clearly distinguish individual, circular Nup153 signals, representing NPCs (Figure 4A). The combination of U-ExM and Airyscan microscopy reaches similar resolutions as STED microscopy while also reducing linkage errors during immunostaining, thereby enhancing the separation of individual pore complexes (Figure 4A).

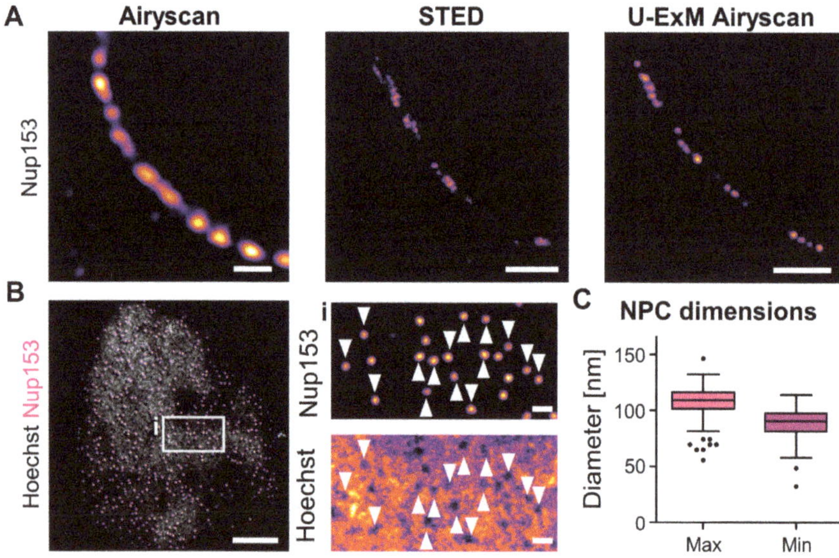

Figure 4. NPC diameter is well conserved after U-ExM. (**A**) Different MDM fixed with PFA/GA and either stained for Nup153 and imaged with Airyscan microscopy or STED microscopy or expanded, stained for Nup153 and imaged with Airyscan microscopy. Scale bar of Airyscan and STED panel, 1 µm; scale bar U-ExM panel, 4 µm. (**B**) Maximum intensity projection (MIP) of the bottom of an expanded U2OS cell nucleus, stained for chromatin (Hoechst, gray) and Nup153 (magenta). Scale bar, 10 µm; scale bar enlargement, 1 µm. Arrowheads mark the position of select NPC. (**C**) Analysis of NUP153 signal to determine NPC dimension, measured as maximum and minimum Feret diameters. Sample size (n = 222); $\text{median}_{\text{max diameter}}$ = 109 nm, $\text{median}_{\text{min diameter}}$ = 91 nm; data blotted according to Tukey box plot with outliers shown as single data points. All scale bars in expanded samples reflect post-expansion sizes.

To further probe the morphology of the NPC, we subjected U2OS cells to U-ExM and performed Nup153 immunostaining alongside Hoechst staining to characterize the nanoscopic environment of the pore. We then acquired Nyquist-optimized 3D volumes of the nuclei close to the sample carrier using Airyscan microscopy (Figure 4B, Video S2). In maximum intensity projections of 3D volumes, NPC appeared as distinct, round structures, again associating with heterochromatin-free regions (Figure 4B). To further explore the structure of the NPC after expansion, we measured the minimum and maximum diameter of the Nup153 signals in the expanded sample (Figures 4C and S5A). Considering the expansion factor of 4.3, we determined that the minimum and maximum diameters of Nup153 ranged from ca. 90–110 nm across a wide range of foci, with an aspect ratio of 1.2 between minimum and maximum diameter. These data indicate that nuclear pores were well conserved at the nanoscale (Figures 4C and S5B).

To push the limit of resolution that can be obtained with widely available imaging methods such as Airyscan microscopy, we stained expanded TZM-bl cells for the distal inner NPC component ELYS. This protein forms a ring structure composed of eight subunits, previously measured to have a diameter of ca. 150 nm using SMLM, making it one of the largest diameters of the NPC [73]. In maximum intensity projections of Nyquist-optimized 3D volumes at the bottom of nuclei stained using a primary and secondary antibody cascade, we successfully visualized some ring-like structures, although most ELYS signals appeared as punctuate structures (Figure 5A). This indicates that U-ExM, in combination with Airyscan microscopy, can resolve the subunit arrangement of structures with a total diameter of as small as 150 nm, even when employing common immunofluorescence strategies. We also probed the spatial separation of the distal cytoplasmic and nuclear components of the NPC after expansion, which are only separated by the length of the NPC itself. Using immunostaining against the peripheral cytoplasmic NPC proteins Nup88 and ELYS, we found a clear separation of both targets across the nuclear envelope in imaging planes perpendicular to the nuclear envelope when using Airyscan microscopy (Figure 5B).

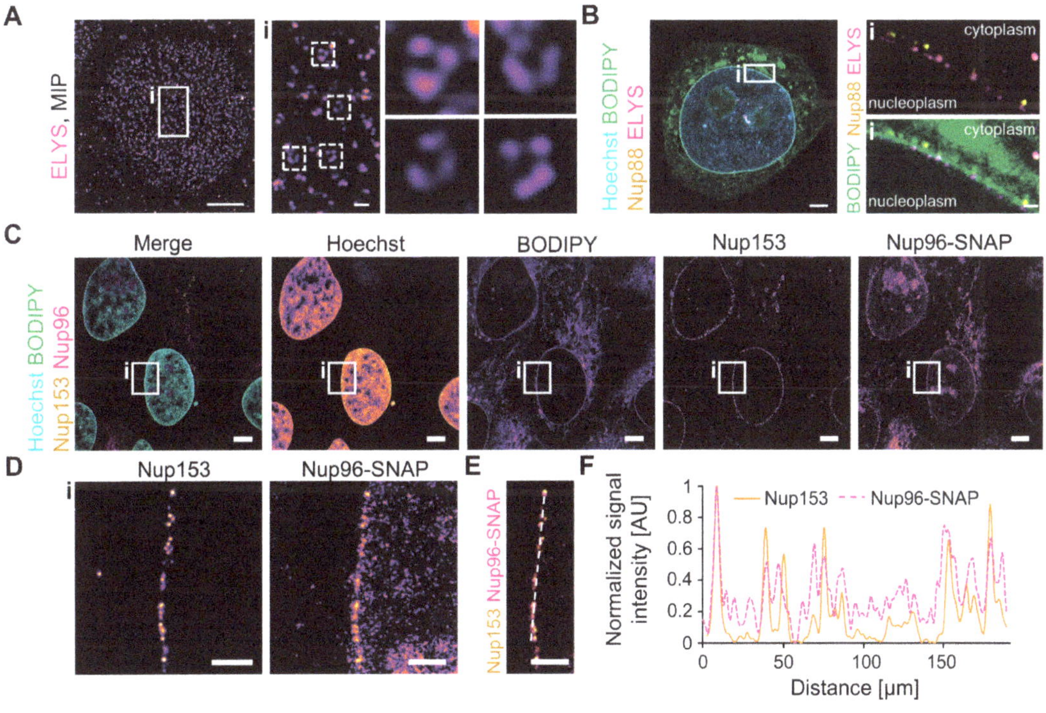

Figure 5. U-ExM allows visualization of nanoscopic structures. (**A**) MIP of a TZM-bl cell fixed with PFA/GA, expanded and stained for ELYS (magenta), showing ring-like structures (enlargements). Scale bar overview, 10 μm; scale bar enlargement, 1 μm. (**B**) Single z-slice of a TZM-bl cell fixed with PFA/GA, expanded and stained for chromatin (Hoechst, cyan), membranes (BODIPY, green), Nup88 (yellow), and ELYS (magenta). Scale bar overview, 10 μm; scale bar enlargement, 1 μm. (**C**) U2OS cells stably expressing Nup96-SNAP fixed with PFA/GA, permeabilized with saponine, and stained with BG-Biotin before expansion and staining for chromatin (Hoechst, cyan), membranes (BODIPY,

green), Nup153 (yellow), and biotin-labeled Nup96 via fluorescently conjugated Streptavidin (magenta). Scale bar, 20 µm. (**D**) Enlargement from (**C**) showing a portion of the nuclear envelope. Scale bar, 5 µm. (**E**) Composite image of the region shown in (**D**). Scale bar: 5 µm; line denotes position of line profile for (**F**) normalized signal intensity of the Nup153 (yellow) and Nup96-SNAP (magenta, dotted line) signal over the line shown in (**E**). Line profiles were aligned to the maximum peak. All scale bars in expanded samples reflect post-expansion sizes.

Although the relative linkage error in U-ExM is reduced compared to both pre-expansion staining expansion workflows and conventional immunofluorescence analysis (ca. 7 nm compared to 30 nm), some uncertainty remains when probing nanoscopic structures. To further minimize linkage errors, several approaches have been developed, including the self-labeling protein tags such as SNAP-, CLIP-, or HaloTag. These tags bind directly and irreversibly to their substrates, resulting in a linkage error equivalent to the size of the tag itself (i.e., ca. 5 nm) [75]. Therefore, we aimed to evaluate the application of such tags in U-ExM using U2OS cells stably expressing the NPC protein Nup96 fused to the SNAP-tag [40]. While the fluorescence of SNAP substrates was lost when ligands were applied prior to expansion, we successfully visualized SNAP-tagged Nup96 by applying a biotin-coupled SNAP substrate to permeabilized cells before U-ExM, followed by detection using fluorescently labeled streptavidin after expansion (Figures 5C and S6). Although streptavidin adds an additional 5 nm linkage error to the detection assembly, it still offers improved localization accuracy compared to antibodies [76]. The combination of biotin-coupled SNAP substrates and fluorescent streptavidin showed more unspecific fluorescence compared to indirect immunostaining against Nup153. However, nuclear envelope-associated Nup96-SNAP signals colocalized well with the Nup153 signal, indicating that this strategy allows the detection of any target that can be linked to biotin with low linkage error (Figures 5C,D and S6).

3.6. U-ExM Can Be Used to Probe Different Steps of the HIV-1 Replication Cycle

Having explored sample preservation during U-ExM with a focus on HIV-1-related structures, we finally aimed to probe different steps in the post-entry phase of the HIV-1 replication cycle. To this end, we first investigated if we could detect HIV-1 capsids or capsid protein (CA)-containing structures in infected cells following expansion. We infected TZM-bl cells with the non-integration competent HIV-1 NL4-3 derivative, NNHIV, which carries both a deletion in Tat and several point mutations in the integrase [29,30]. At 16 h post-infection, the cells were fixed and subjected to U-ExM and indirect immunostaining using two different antisera against CA [41,77], along with immunostaining against Nup153 and BODIPY staining (Figure 6A,B). To avoid bleedthrough, we used secondary antibodies coupled to spectrally well-separated fluorophores (Alexa 405 Plus and Alexa-647) to visualize the signals of both CA antisera. The robust colocalization of signals of both CA antisera confirmed their specificity and functionality in U-ExM (Figure 6A,B). By focusing on individual CA foci, we identified structures likely representing different stages of the HIV-1 replication cycle. Specifically, on the cytoplasmic side of the nuclear envelope, we detected small, circular CA-containing foci of high intensity in close proximity to Nup153 signals, presumably representing capsids interacting with nuclear pores before transit through the NPC (Figure 6A). The diameter of nuclear envelope-associated CA signals was similar to that of Nup153 foci, indicating they may represent individual viral capsids ca. 100 nm in size. Within the nucleus, we found larger, more diffuse areas of CA signals with a lower degree of colocalization between the two CA antisera (Figure 6B,C). These structures presumably represent accumulations of (possibly disintegrating) viral cores, which have previously been observed within nuclear speckles by electron tomography [30].

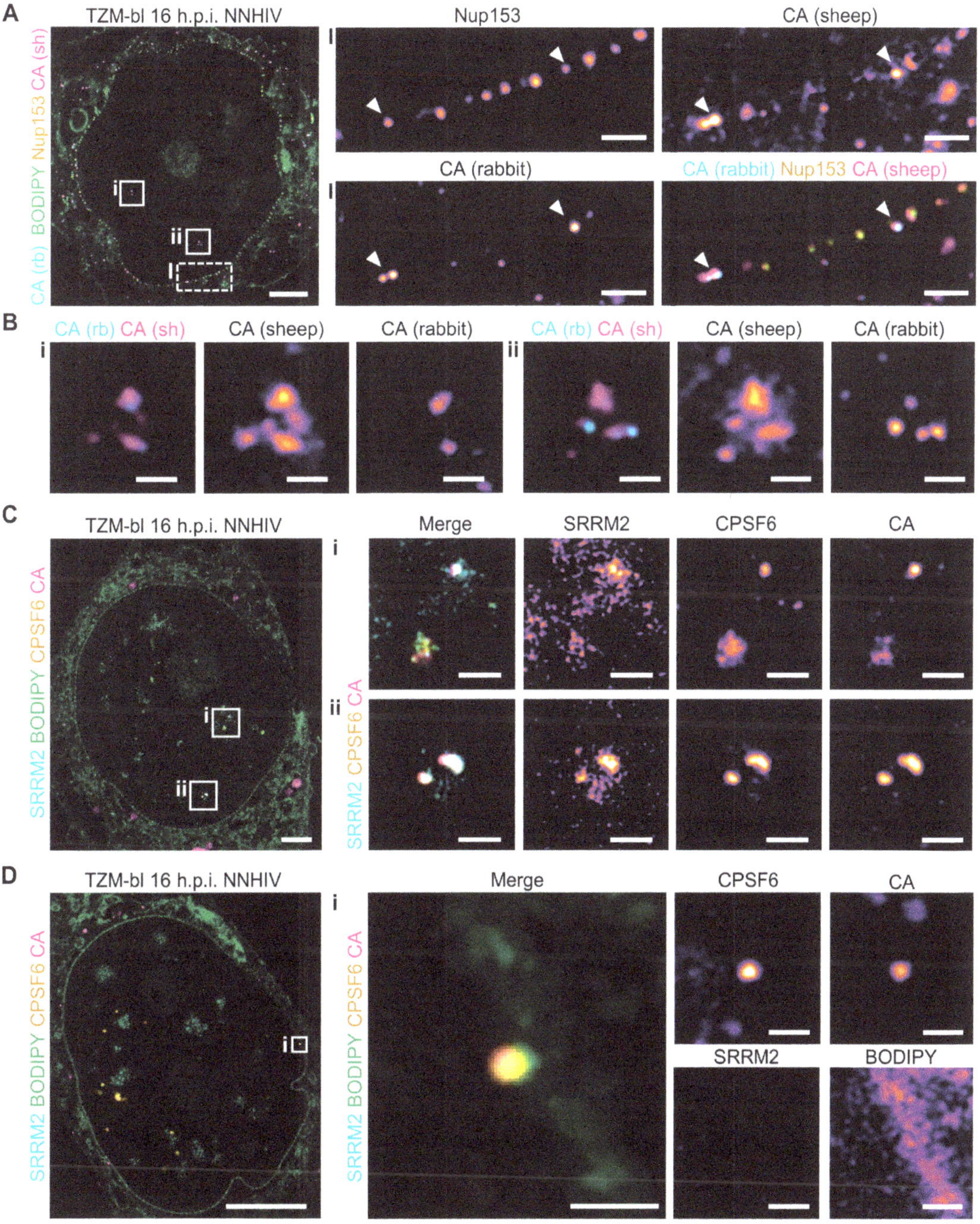

Figure 6. U-ExM can visualize HIV-1 CA-containing structures upon nuclear entry and within the nucleus. (**A**) A TZM-bl cell infected with NNHIV, fixed with PFA/GA at 16 h.p.i., expanded and stained for CA with two different antisera (rabbit α-CA, cyan; sheep α-CA, magenta) as well as Nup153 (yellow) and membranes (BODIPY, green). The sheep α-CA channel was corrected for bleedthrough

from the Nup153 channel. Scale bar overview, 20 μm. Enlargements show HIV-1 CAs (cyan/magenta) interacting with NPC complexes (yellow), marked by arrowheads. Scale bar enlargements, 2 μm. (**B**) Enlargements from (**A**) showing intranuclear CA signals (rabbit α-CA, cyan; sheep α-CA, magenta). Scale bar, 1 μm. (**C**,**D**) TZM-bl cells, infected with NNHIV, fixed with PFA/GA at 16 h.p.i., expanded and stained for the nuclear speckle marker SRRM2 (cyan), membranes (BODIPY, green), CPSF6 (yellow), and CA (sheep α-CA, magenta). Scale bar overview, 20 μm. (**C**) Enlargements show intranuclear CA signals (magenta) showing colocalization with SRRM2 (cyan) and CPSF6 (yellow). The CA channel and the SRRM2 channel were corrected for bleedthrough from the BODIPY channel. Scale bar enlargements, 2 μm. (**D**) Enlargement shows intranuclear CA signals (magenta) with CPSF6 (yellow) close to nuclear envelope marked with BODIPY (green). Scale bar enlargements, 1 μm. All scale bars in expanded samples reflect post-expansion sizes.

To further characterize the nuclear CA-containing structures, we stained samples for membranes using BODIPY, as well as for the nuclear speckle marker SRRM2, the nuclear protein cleavage and polyadenylation-specific factor 6 (CPSF6), and HIV-1 CA. Upon entry into the nucleoplasm, HIV-1 capsids are tightly decorated by CPSF6, which masks CA epitopes in conventional immunostaining [30]. Through indirect immunostaining after U-ExM, we successfully detected both CPSF6 and CA with a high degree of colocalization, appearing as punctate structures presumably representing single viral cores, as well as larger diffuse signals indicative of clusters of (disassembling) capsids (Figure 6B,C). Both structures colocalized with the nuclear speckle marker SRRM2, confirming their identity as nuclear HIV-1 CA-containing structures (and possibly intact viral cores) that have translocated to nuclear speckles (Figure 6C). In one instance, we observed a punctate CA signal close to the nuclear envelope, marked by BODIPY, which presumably represents an HIV-1 capsid that had just traversed the NPC (Figure 6D). Despite its proximity to the nuclear envelope, this CA signal exhibited strong colocalization with CPSF6, consistent with reports of CPSF6 associating with capsids immediately upon nuclear entry [41]. In contrast to CA foci in the inner nucleus, this particular CA focus did not show colocalization with SRRM2 (Figure 6D). These results demonstrate that U-ExM is a valuable tool for identifying and characterizing different stages in the HIV-1 replication cycle.

4. Discussion

In this study, we evaluated the usage of the U-ExM protocol to investigate host–virus interactions, focusing specifically on the post-entry stages of the HIV-1 replication cycle. U-ExM allows a four-fold sample expansion in combination with post-expansion labeling, enabling lower relative linkage errors and increased labeling efficiency. While other protocols achieve around 10-fold expansion, they typically require proteolytic digestion, making them incompatible with post-expansion labeling [13,78]. Despite these limitations, higher expansion factors have been achieved by adaptation of gel recipes for a ca. 10-fold expansion [11] or using iterative expansion microscopy for a 14–26-fold expansion [10,12]. However, adapting gel recipes usually requires careful optimization depending on the sample type, complicating their usage [9,11]. Additionally, 10-fold expansion processes that employ alternative gel chemistries often necessitate specialized equipment to eliminate oxygen during the gel polymerization process [78]. Notably, very recent results show promising advances for the development of novel one-step gelation strategies that allow robust expansion up to 12-fold, which can be used to probe chromatin architecture, NPC structure, as well as virion morphology in 3D [79]. However, the wider feasibility of these approaches remains to be seen. In iterative expansion microscopy, gels are re-embedded into a secondary gel following an initial expansion step, enabling a subsequent round of expansion [10,12]. While these approaches enable visualization of molecular details at a resolution below 20 nm, they are considerably more time-consuming and complex than one-step expansion protocols. In addition, a second expansion may compromise the robustness of the protocol, leading to greater variability in final expansion compared to one-step expansion [12].

By integrating U-ExM with Airyscan microscopy, which delivers improved resolution while being more user-friendly than STED or SMLM [20], we successfully resolved the subunit arrangement of structures with diameters under 150 nm, despite the usage of indirect immunofluorescence staining (Figure 5A,B). Although U-ExM inherently results in low relative linkage errors when using primary and secondary antibodies, localization precision can be further improved by the use of self-labeling tags, such as the SNAP-tag [75] (Figures 5C–F and S6). Our findings demonstrate that by employing biotin-coupled reporters, we can effectively detect self-labeling protein tags after expansion. This strategy has been successfully applied to various targets [80,81] and can be adapted for the detection of any biomolecule tagged with biotin.

Although biotin labeling represents an approach that minimizes linkage errors, it may require cell permeabilization to introduce biotin-coupled moieties and appears to be more prone to background signal compared to traditional immunostaining (Figures 5C–F and S6). This increased background is likely due to the presence of endogenous biotinylated molecules. Consequently, antibody labeling remains the most accessible avenue for target detection in U-ExM. In our study, we evaluated a panel of antibodies against viral and cellular targets for use in U-ExM and found that most were compatible with post-expansion staining, consistent with previous reports [36]. However, we faced challenges in establishing reliable conditions for antibody staining of GFP tags and lamin, despite other studies successfully visualizing GFP after U-ExM [36,82].

We observed that the nuclear compartment was well preserved after U-ExM, while the cytoplasmic compartment needed optimization of fixation for adequate sample preservation, a phenomenon that has been well documented [50]. While cryofixation effectively preserves sample integrity, it requires specialized equipment and prolongs the U-ExM sample preparation process by an additional day [50]. Additionally, previous studies indicate that cryofixation may need further optimization to enhance antibody-mediated detection of NPC structures in expansion microscopy [12]. We found that substituting PBS with different buffer solutions optimized for sample preservation markedly improved post-expansion results, achieving outcomes comparable to those with cryofixation (Figures 2A and S4B,C).

Using U-ExM, we probed the HIV-1 nuclear replication cycle and readily detected HIV-1 viral cores in different stages of the post-entry process (Figure 6). Within the nucleus, individual particles exhibited strong colocalization with the nuclear protein CPSF6, even when positioned adjacent to the nuclear envelope, suggesting they had just entered the nucleus (Figure 6C,D). This finding is consistent with reports showing that capsids are immediately decorated with CPSF6 upon arrival in the nucleoplasm [41,83]. Additionally, we identified more diffuse CA signal areas within the nucleus that colocalized with the nuclear speckle marker SRRM2, presumably representing accumulated and possibly degrading viral cores (Figure 6B,C). These observations support other reports indicating that multiple, apparently morphologically intact capsids gather within nuclear speckles prior to viral genome uncoating and integration into speckle-associated domains [30–32,34].

Of note, we could readily observe both CA and CPSF6 signals in regions that presumably represented intact viral cores (Figure 6C,D). Previously, due to the masking of CA epitopes by the CPSF6 coat, simultaneous detection of CA and CPSF6 was often only possible in classical immunofluorescence after chemical extraction [30]. Here, however, both targets were accessible in parallel with high fidelity, likely due to the exposure of epitopes during expansion. To our knowledge, this is the first clear demonstration of epitope unmasking during U-ExM. This decrowding of dense environments is especially valuable, as it not only enables investigation of viral structures at higher resolution but also unlocks new possibilities for the detection of features that have previously been masked.

Due to their size, singular capsids cannot be faithfully resolved in conventional fluorescence microscopy. At the same time, methods that allow visualization of capsid with appropriate resolution, such as STED and SMLM, are technically challenging and time-consuming—especially when scanning larger volumes—which particularly complicates detection of rare events. In this context, we demonstrate that using U-ExM in combination with Airyscan microscopy allows for clear differentiation of individual capsids as punctate CA signals (Figure 6). This combination facilitates targeted investigations of virus–host interactions at the nanoscale. At the same time, Airyscan microscopy of expanded samples allows fast image acquisition and scanning of large 3D volumes. By combining U-ExM with Airyscan microscopy or other microscopy techniques that achieve similar resolution [84,85], we can identify rare events with high resolution, opening up new avenues for detection of specific host–virus interactions at super-resolution.

5. Conclusions

Here, using the visualization of HIV-1 post-entry events as an example, we illustrate that expansion microscopy is a promising method for the investigation of virus-cell interactions, especially within the nuclear compartment. U-ExM is a robust and versatile protocol applicable to various cell types and structures, effectively bridging the gap between fluorescence and electron microscopy while enhancing the resolution without the need for specialized equipment. As with any super-resolution technique, labeling becomes critical in expansion microscopy due to the linkage errors conferred by different molecular visualization tools. Although post-expansion labeling protocols like U-ExM improve localization accuracy, the impact of antibody size on linkage error remains significant. Thus, we recommend adopting more direct labeling methods, such as self-labeling tags, moving forward [75]. Additionally, alternative labeling strategies that minimize linkage errors, such as directly labeled camelid antibodies [86] or genetic code expansion with click labeling [87], could enhance the localization accuracy in expansion microscopy. By further increasing the expansion factor and integrating super-resolution imaging methods such as Airyscan microscopy, structured illumination microscopy (SIM), and STED, we can explore virus–host interactions at unprecedented scales.

Supplementary Materials: The following supporting information can be downloaded at: https://www.mdpi.com/article/10.3390/v16101610/s1. File S1: U-ExM protocol. Figure S1: Overview of bleedthrough correction process. Figure S2: Distortion during U-ExM. Figure S3: Pre- and post-expansion measurements of the same set of nuclei. Figure S4: Different fixation conditions affect the integrity of the cytoplasm during U-ExM. Figure S5: Measurement of NPC diameter. Figure S6: SNAP-tagged targets can be visualized in cells permeabilized with Triton-X100. Video S1: 3D structure of an expanded TZM-bl cell as seen in Figure 2B. Video S2: 3D volume of Nup153 signals and Hoechst staining as seen in Figure 4B. Reference [88] are cited in the supplementary materials.

Author Contributions: Conceptualization, G.M.H., B.M., H.-G.K. and S.K.; methodology, A.P., G.M.H. and S.K.; software, G.M.H. and S.K.; validation, A.P., L.L.R., M.H. and G.M.H.; formal analysis, S.K.; investigation, A.P., G.M.H., L.L.R., M.H. and S.K.; resources, M.A.-Ö., V.S.-B., A.-M.H., V.L., B.M. and H.-G.K.; data curation, A.P., G.M.H. and S.K.; writing—original draft preparation, G.M.H., B.M., H.-G.K. and S.K.; writing—review and editing, G.M.H., A.P., B.M., V.L., H.-G.K. and S.K.; visualization, G.M.H. and S.K.; supervision, A.P., V.L., B.M., H.-G.K. and S.K.; project administration, S.K.; funding acquisition, V.L., B.M. and H.-G.K. All authors have read and agreed to the published version of the manuscript.

Funding: This work was funded by the Deutsche Forschungsgemeinschaft (DFG, German Research Foundation)—Projektnummer 240245660—SFB 1129 (project 5 for H.G.K. and project 6 for B.M.) and grant number KR 906/7-1 (H-G.K.) by Deutsches Zentrum fuer Infektionsforschung (DZIF), grant number TTU 04.710 (V.L. and H-G.K.).

Data Availability Statement: The data presented in this study are available upon request from the corresponding author.

Acknowledgments: We are grateful to the Fackler group (University Hospital Heidelberg, Germany) and the Lusic group (University Hospital Heidelberg, Germany) for providing resting and activated CD4+ T-cells for expansion. We would also like to thank Annika Binder (Frischknecht group, University Hospital Heidelberg, Germany), Caroline Simon and Felix Mikus (Dey group, EMBL Heidelberg, Germany), as well as Yannik Voss (Frischknecht lab, University Hospital Heidelberg,

Germany) and Julien Guizetti (Guizetti lab, University Hospital Heidelberg, Germany) for their help and the many fruitful discussions about expansion microscopy. We also thank De-en Sun (Wombacher group and Johnsson group, Max Planck Institute for Medical Research, Germany) for his helpful suggestions regarding visualization of SNAP-tags in expansion microscopy. Finally, we would like to acknowledge the microscopy support from the Infectious Diseases Imaging Platform (IDIP) at the Center for Integrative Infectious Disease Research, Heidelberg, Germany.

Conflicts of Interest: The authors declare no conflicts of interest. The funders had no role in the design of this study, in the collection, analysis, or interpretation of data, in the writing of the manuscript, or in the decision to publish the results.

References

1. Briggs, J.A.G.; Wilk, T.; Welker, R.; Kräusslich, H.G.; Fuller, S.D. Structural organization of authentic, mature HIV-1 virions and cores. *EMBO J.* **2003**, *22*, 1707. [CrossRef] [PubMed]
2. Prakash, K.; Diederich, B.; Heintzmann, R.; Schermelleh, L. Super-resolution microscopy: A brief history and new avenues. *Philos. Trans. R. Soc. A* **2022**, *380*, 20210110. [CrossRef] [PubMed]
3. Robb, N.C. Virus morphology: Insights from super-resolution fluorescence microscopy. *Biochim. Biophys. Acta (BBA)—Mol. Basis Dis.* **2022**, *1868*, 166347. [CrossRef] [PubMed]
4. Müller, B.; Heilemann, M. Shedding new light on viruses: Super-resolution microscopy for studying human immunodeficiency virus. *Trends Microbiol.* **2013**, *21*, 522–533. [CrossRef]
5. Laketa, V. Microscopy in Infectious Disease Research—Imaging Across Scales. *J. Mol. Biol.* **2018**, *430*, 2612–2625. [CrossRef]
6. Chen, H.; Yan, G.; Wen, M.H.; Brooks, K.N.; Zhang, Y.; Huang, P.S.; Chen, T.Y. Advancements and Practical Considerations for Biophysical Research: Navigating the Challenges and Future of Super-resolution Microscopy. *Chem. Biomed. Imaging* **2024**, *2*, 331–344. [CrossRef]
7. Chen, F.; Tillberg, P.W.; Boyden, E.S. Expansion Microscopy. *Science* **2015**, *347*, 543. [CrossRef]
8. Hümpfer, N.; Thielhorn, R.; Ewers, H. Expanding boundaries—A cell biologist's guide to expansion microscopy. *J. Cell Sci.* **2024**, *137*, jcs260765. [CrossRef]
9. Wen, G.; Leen, V.; Rohand, T.; Sauer, M.; Hofkens, J. Current Progress in Expansion Microscopy: Chemical Strategies and Applications. *Chem. Rev.* **2023**, *123*, 3299–3323. [CrossRef]
10. M'Saad, O.; Bewersdorf, J. Light microscopy of proteins in their ultrastructural context. *Nat. Commun.* **2020**, *11*, 3850. [CrossRef]
11. Klimas, A.; Gallagher, B.R.; Wijesekara, P.; Fekir, S.; DiBernardo, E.F.; Cheng, Z.; Stolz, D.B.; Cambi, F.; Watkins, S.C.; Brody, S.L.; et al. Magnify is a universal molecular anchoring strategy for expansion microscopy. *Nat. Biotechnol.* **2023**, *41*, 858–869. [CrossRef]
12. Louvel, V.; Haase, R.; Mercey, O.; Laporte, M.H.; Eloy, T.; Baudrier, É.; Fortun, D.; Soldati-Favre, D.; Hamel, V.; Guichard, P. iU-ExM: Nanoscopy of organelles and tissues with iterative ultrastructure expansion microscopy. *Nat. Commun.* **2023**, *14*, 7893. [CrossRef] [PubMed]
13. Damstra, H.G.; Mohar, B.; Eddison, M.; Akhmanova, A.; Kapitein, L.C.; Tillberg, P.W. Visualizing cellular and tissue ultrastructure using Ten-fold Robust Expansion Microscopy (TREx). *eLife* **2022**, *11*, e73775. [CrossRef] [PubMed]
14. Chozinski, T.J.; Halpern, A.R.; Okawa, H.; Kim, H.J.; Tremel, G.J.; Wong, R.O.; Vaughan, J.C. Expansion microscopy with conventional antibodies and fluorescent proteins. *Nat. Methods* **2016**, *13*, 485–488. [CrossRef] [PubMed]
15. Tillberg, P.W.; Chen, F.; Piatkevich, K.D.; Zhao, Y.; Yu, C.C.; English, B.P.; Gao, L.; Martorell, A.; Suk, H.J.; Yoshida, F.; et al. Protein-retention expansion microscopy of cells and tissues labeled using standard fluorescent proteins and antibodies. *Nat. Biotechnol.* **2016**, *34*, 987–992. [CrossRef]
16. Ku, T.; Swaney, J.; Park, J.Y.; Albanese, A.; Murray, E.; Cho, J.H.; Park, Y.G.; Mangena, V.; Chen, J.; Chung, K. Multiplexed and scalable super-resolution imaging of three-dimensional protein localization in size-adjustable tissues. *Nat. Biotechnol.* **2016**, *34*, 973–981. [CrossRef]
17. Zwettler, F.U.; Reinhard, S.; Gambarotto, D.; Bell, T.D.; Hamel, V.; Guichard, P.; Sauer, M. Molecular resolution imaging by post-labeling expansion single-molecule localization microscopy (Ex-SMLM). *Nat. Commun.* **2020**, *11*, 3388. [CrossRef]
18. Gambarotto, D.; Zwettler, F.U.; Le Guennec, M.; Schmidt-Cernohorska, M.; Fortun, D.; Borgers, S.; Heine, J.; Schloetel, J.G.; Reuss, M.; Unser, M.; et al. Imaging cellular ultrastructures using expansion microscopy (U-ExM). *Nat. Methods* **2019**, *16*, 71–74. [CrossRef]
19. Hamel, V.; Guichard, P. Improving the resolution of fluorescence nanoscopy using post-expansion labeling microscopy. *Methods Cell Biol.* **2021**, *161*, 297–315. [CrossRef]
20. Huff, J. The Airyscan detector from ZEISS: Confocal imaging with improved signal-to-noise ratio and super-resolution. *Nat. Methods* **2015**, *12*, i–ii. [CrossRef]
21. Hell, S.W.; Wichmann, J. Breaking the diffraction resolution limit by stimulated emission: Stimulated-emission-depletion fluorescence microscopy. *Opt. Lett.* **1994**, *19*, 780. [CrossRef] [PubMed]
22. Lucic, B.; de Castro, I.J.; Lusic, M. Viruses in the Nucleus. *Cold Spring Harb. Perspect. Biol.* **2021**, 13. [CrossRef]
23. Charman, M.; Weitzman, M.D. Replication Compartments of DNA Viruses in the Nucleus: Location, Location, Location. *Viruses* **2020**, *12*, 151. [CrossRef] [PubMed]

24. Muller, T.G.; Zila, V.; Muller, B.; Krausslich, H.G. Nuclear Capsid Uncoating and Reverse Transcription of HIV-1. *Annu. Rev. Virol.* **2022**, *9*, 261–284. [CrossRef] [PubMed]
25. Yamashita, M.; Emerman, M. Retroviral infection of non-dividing cells: Old and new perspectives. *Virology* **2006**, *344*, 88–93. [CrossRef]
26. McDonald, D.; Vodicka, M.A.; Lucero, G.; Svitkina, T.M.; Borisy, G.G.; Emerman, M.; Hope, T.J. Visualization of the intracellular behavior of HIV in living cells. *J. Cell Biol.* **2002**, *159*, 441–452. [CrossRef] [PubMed]
27. Badieyan, S.; Lichon, D.; Andreas, M.P.; Gillies, J.P.; Peng, W.; Shi, J.; DeSantis, M.E.; Aiken, C.R.; Böcking, T.; Giessen, T.W.; et al. HIV-1 binds dynein directly to hijack microtubule transport machinery. *bioRxiv* **2023**. [CrossRef]
28. Naghavi, M.H. HIV-1 capsid exploitation of the host microtubule cytoskeleton during early infection. *Retrovirology* **2021**, *18*, 19. [CrossRef]
29. Zila, V.; Margiotta, E.; Turoňová, B.; Müller, T.G.; Zimmerli, C.E.; Mattei, S.; Allegretti, M.; Börner, K.; Rada, J.; Müller, B.; et al. Cone-shaped HIV-1 capsids are transported through intact nuclear pores. *Cell* **2021**, *184*, 1032–1046.e18. [CrossRef]
30. Müller, T.G.; Zila, V.; Peters, K.; Schifferdecker, S.; Stanic, M.; Lucic, B.; Laketa, V.; Lusic, M.; Müller, B.; Kräusslich, H.G. Hiv-1 uncoating by release of viral cdna from capsid-like structures in the nucleus of infected cells. *eLife* **2021**, *10*, e64776. [CrossRef]
31. Burdick, R.C.; Li, C.; Munshi, M.; Rawson, J.M.O.; Nagashima, K.; Hu, W.S.; Pathak, V.K. HIV-1 uncoats in the nucleus near sites of integration. *Proc. Natl. Acad. Sci. USA* **2020**, *117*, 5486–5493. [CrossRef]
32. Schifferdecker, S.; Zila, V.; Müller, T.G.; Sakin, V.; Anders-Össwein, M.; Laketa, V.; Kräusslich, H.G.; Müller, B. Direct Capsid Labeling of Infectious HIV-1 by Genetic Code Expansion Allows Detection of Largely Complete Nuclear Capsids and Suggests Nuclear Entry of HIV-1 Complexes via Common Routes. *mBio* **2022**, *13*, e01959-22. [CrossRef]
33. Francis, A.C.; Marin, M.; Singh, P.K.; Achuthan, V.; Prellberg, M.J.; Palermino-Rowland, K.; Lan, S.; Tedbury, P.R.; Sarafianos, S.G.; Engelman, A.N.; et al. HIV-1 replication complexes accumulate in nuclear speckles and integrate into speckle-associated genomic domains. *Nat. Commun.* **2020**, *11*, 3505. [CrossRef]
34. Rensen, E.; Mueller, F.; Scoca, V.; Parmar, J.J.; Souque, P.; Zimmer, C.; Di Nunzio, F. Clustering and reverse transcription of HIV-1 genomes in nuclear niches of macrophages. *EMBO J.* **2021**, *40*, e105247. [CrossRef]
35. Kreysing, J.P.; Heidari, M.; Zila, V.; Cruz-Leon, S.; Obarska-Kosinska, A.; Laketa, V.; Welsch, S.; Koefinger, J.; Turonova, B.; Hummer, G.; et al. Passage of the HIV capsid cracks the nuclear pore. *bioRxiv* **2024**. [CrossRef]
36. Gambarotto, D.; Hamel, V.; Guichard, P. Ultrastructure expansion microscopy (U-ExM). *Methods Cell Biol.* **2021**, *161*, 57–81. [CrossRef]
37. Platt, E.J.; Wehrly, K.; Kuhmann, S.E.; Chesebro, B.; Kabat, D. Effects of CCR5 and CD4 cell surface concentrations on infections by macrophagetropic isolates of human immunodeficiency virus type 1. *J. Virol.* **1998**, *72*, 2855–2864. [CrossRef]
38. Wei, X.; Decker, J.M.; Liu, H.; Zhang, Z.; Arani, R.B.; Kilby, J.M.; Saag, M.S.; Wu, X.; Shaw, G.M.; Kappes, J.C. Emergence of Resistant Human Immunodeficiency Virus Type 1 in Patients Receiving Fusion Inhibitor (T-20) Monotherapy. *Antimicrob. Agents Chemother.* **2002**, *46*, 1896. [CrossRef]
39. Pear, W.S.; Nolan, G.P.; Scott, M.L.; Baltimore, D. Production of high-titer helper-free retroviruses by transient transfection. *Proc. Natl. Acad. Sci. USA* **1993**, *90*, 8392–8396. [CrossRef]
40. Thevathasan, J.V.; Kahnwald, M.; Cieśliński, K.; Hoess, P.; Peneti, S.K.; Reitberger, M.; Heid, D.; Kasuba, K.C.; Hoerner, S.J.; Li, Y.; et al. Nuclear pores as versatile reference standards for quantitative superresolution microscopy. *Nat. Methods* **2019**, *16*, 1045. [CrossRef]
41. Bejarano, D.A.; Peng, K.; Laketa, V.; Börner, K.; Jost, K.L.; Lucic, B.; Glass, B.; Lusic, M.; Müller, B.; Kräusslich, H.G. HIV-1 nuclear import in macrophages is regulated by CPSF6-capsid interactions at the nuclear pore complex. *eLife* **2019**, *8*, e41800. [CrossRef]
42. Albanese, A.; Arosio, D.; Terreni, M.; Cereseto, A. HIV-1 Pre-Integration Complexes Selectively Target Decondensed Chromatin in the Nuclear Periphery. *PLoS ONE* **2008**, *3*, 2413. [CrossRef]
43. Stewart, S.A.; Dykxhoorn, D.M.; Palliser, D.; Mizuno, H.; Yu, E.Y.; An, D.S.; Sabatini, D.M.; Chen, I.S.; Hahn, W.C.; Sharp, P.A.; et al. Lentivirus-delivered stable gene silencing by RNAi in primary cells. *RNA* **2003**, *9*, 493–501. [CrossRef] [PubMed]
44. Pizzato, M.; Erlwein, O.; Bonsall, D.; Kaye, S.; Muir, D.; McClure, M.O. A one-step SYBR Green I-based product-enhanced reverse transcriptase assay for the quantitation of retroviruses in cell culture supernatants. *J. Virol. Methods* **2009**, *156*, 1–7. [CrossRef] [PubMed]
45. Leyton-Puig, D.; Kedziora, K.M.; Isogai, T.; Den Van Broek, B.; Jalink, K.; Innocenti, M. PFA fixation enables artifact-free super-resolution imaging of the actin cytoskeleton and associated proteins. *Biol. Open* **2016**, *5*, 1001. [CrossRef]
46. Sobue, K.; Fujio, Y.; Kanda, K. Tumor promoter induces reorganization of actin filaments and calspectin (fodrin or nonerythroid spectrin) in 3T3 cells. *Proc. Natl. Acad. Sci. USA* **1988**, *85*, 482. [CrossRef]
47. Small, J.V.; Rinnerthaler, G.; Hinssen, H. Organization of actin meshworks in cultured cells: The leading edge. *Cold Spring Harb. Symp. Quant. Biol.* **1982**, *46*, 599–611. [CrossRef]
48. Lickert, S.; Sorrentino, S.; Studt, J.D.; Medalia, O.; Vogel, V.; Schoen, I. Morphometric analysis of spread platelets identifies integrin αIIbβ3-specific contractile phenotype. *Sci. Rep.* **2018**, *8*, 5428. [CrossRef]
49. Xu, K.; Babcock, H.P.; Zhuang, X. Dual-objective STORM reveals three-dimensional filament organization in the actin cytoskeleton. *Nat. Methods* **2012**, *9*, 185. [CrossRef]
50. Laporte, M.H.; Klena, N.; Hamel, V.; Guichard, P. Visualizing the native cellular organization by coupling cryofixation with expansion microscopy (Cryo-ExM). *Nat. Methods* **2022**, *19*, 216–222. [CrossRef]

51. Damstra, H.G.J.; Passmore, J.B.; Serweta, A.K.; Koutlas, I.; Burute, M.; Meye, F.J.; Akhmanova, A.; Kapitein, L.C. GelMap: Intrinsic calibration and deformation mapping for expansion microscopy. *Nat. Methods* **2023**, *20*, 1573. [CrossRef] [PubMed]
52. Schindelin, J.; Arganda-Carreras, I.; Frise, E.; Kaynig, V.; Longair, M.; Pietzsch, T.; Preibisch, S.; Rueden, C.; Saalfeld, S.; Schmid, B.; et al. Fiji: An open-source platform for biological-image analysis. *Nat. Methods* **2012**, *9*, 676–682. [CrossRef]
53. Zimmermann, T. Spectral Imaging and Linear Unmixing in Light Microscopy. In *Microscopy Techniques*; Springer: Berlin/Heidelberg, Germany, 2005. [CrossRef]
54. Pachitariu, M.; Stringer, C. Cellpose 2.0: How to train your own model. *Nat. Methods* **2022**, *19*, 1634–1641. [CrossRef] [PubMed]
55. Ollion, J.; Cochennec, J.; Loll, F.; Escudé, C.; Boudier, T. TANGO: A generic tool for high-throughput 3D image analysis for studying nuclear organization. *Bioinformatics* **2013**, *29*, 1840–1841. [CrossRef] [PubMed]
56. Acke, A.; Van Belle, S.; Louis, B.; Vitale, R.; Rocha, S.; Voet, T.; Debyser, Z.; Hofkens, J. Expansion microscopy allows high resolution single cell analysis of epigenetic readers. *Nucleic Acids Res.* **2022**, *50*, e100. [CrossRef]
57. Amodio, A.; Cassani, M.; Mummolo, L.; Cortez-Jugo, C.; Bhangu, S.K.; Symons, J.; Ahlenstiel, C.L.; Forte, G.; Ricci, F.; Kelleher, A.D.; et al. Nanoscale probing and imaging of HIV-1 RNA in cells with a chimeric LNA–DNA sensor. *Nanoscale* **2022**, *14*, 3049–3061. [CrossRef]
58. Mascheroni, L.; Scherer, K.M.; Manton, J.D.; Ward, E.; Dibben, O.; Kaminski, C.F. Combining sample expansion and light sheet microscopy for the volumetric imaging of virus-infected cells with super-resolution. *Biomed. Opt. Express* **2020**, *11*, 5032–5044. [CrossRef]
59. Nijenhuis, W.; Damstra, H.G.J.; van Grinsven, E.J.; Iwanski, M.K.; Praest, P.; Soltani, Z.E.; van Grinsven, M.M.P.; Brunsveld, J.E.; de Kort, T.; Rodenburg, L.W.; et al. Optical nanoscopy reveals SARS-CoV-2-induced remodeling of human airway cells. *bioRxiv* **2021**. [CrossRef]
60. Rozario, A.M.; Zwettler, F.; Duwé, S.; Hargeaves, R.B.; Brice, A.; Dedecker, P.; Sauer, M.; Moseley, G.W.; Whelan, D.R.; Bell, T.D.M. 'Live and Large': Super-Resolution Optical Fluctuation Imaging (SOFI) and Expansion Microscopy (ExM) of Microtubule Remodelling by Rabies Virus P Protein. *Aust. J. Chem.* **2020**, *73*, 686–692. [CrossRef]
61. Matschke, J.; Hartmann, K.; Pfefferle, S.; Wang, Y.; Valdes, P.A.; Thies, E.; Schweizer, M.; Lütgehetmann, M.; Schmiedel, S.; Bernreuther, C.; et al. Inefficient tissue immune response against MPXV in an immunocompromised mpox patient. *J. Med. Virol.* **2024**, *96*, e29811. [CrossRef]
62. Chan, A.W.; Broncel, M.; Yifrach, E.; Haseley, N.R.; Chakladar, S.; Andree, E.; Herneisen, A.L.; Shortt, E.; Treeck, M.; Lourido, S. Analysis of CDPK1 targets identifies a trafficking adaptor complex that regulates microneme exocytosis in Toxoplasma. *eLife* **2023**, *12*, RP85654. [CrossRef] [PubMed]
63. Bertiaux, E.; Balestra, A.C.; Bournonville, L.; Louvel, V.; Maco, B.; Soldati-Favre, D.; Brochet, M.; Guichard, P.; Hamel, V. Expansion microscopy provides new insights into the cytoskeleton of malaria parasites including the conservation of a conoid. *PLoS Biol.* **2021**, *19*, e3001020. [CrossRef]
64. Schweizer, N.; Haren, L.; Dutto, I.; Viais, R.; Lacasa, C.; Merdes, A.; Lüders, J. Sub-centrosomal mapping identifies augmin-γTuRC as part of a centriole-stabilizing scaffold. *Nat. Commun.* **2021**, *12*, 6042. [CrossRef]
65. Zwettler, F.U.; Spindler, M.C.; Reinhard, S.; Klein, T.; Kurz, A.; Benavente, R.; Sauer, M. Tracking down the molecular architecture of the synaptonemal complex by expansion microscopy. *Nat. Commun.* **2020**, *11*, 3222. [CrossRef]
66. Le Borgne, P.; Greibill, L.; Laporte, M.H.; Lemullois, M.; Bouhouche, K.; Temagoult, M.; Rosnet, O.; Le Guennec, M.; Lignières, L.; Chevreux, G.; et al. The evolutionary conserved proteins CEP90, FOPNL, and OFD1 recruit centriolar distal appendage proteins to initiate their assembly. *PLoS Biol.* **2022**, *20*, e3001782. [CrossRef]
67. Liffner, B.; Absalon, S. Expansion microscopy reveals plasmodium falciparum blood-stage parasites undergo anaphase with a chromatin bridge in the absence of mini-chromosome maintenance complex binding protein. *Microorganisms* **2021**, *9*, 2306. [CrossRef] [PubMed]
68. Leduc, C.; Salles, A.; Shorte, S.L.; Etienne-Manneville, S. Imaging Intermediate Filaments and Microtubules with 2-dimensional Direct Stochastic Optical Reconstruction Microscopy. *J. Vis. Exp.* **2018**, *2018*, 57087. [CrossRef]
69. Wurm, C.A.; Kolmakov, K.; Göttfert, F.; Ta, H.; Bossi, M.; Schill, H.; Berning, S. Novel red fluorophores with superior performance in STED microscopy. *Opt. Nanoscopy* **2012**, *1*, 7. [CrossRef]
70. García, M.; Buzón, M.J.; Benito, J.M.; Rallón, N. Peering into the HIV reservoir. *Rev. Med. Virol.* **2018**, *28*, e1981. [CrossRef]
71. Cremer, T.; Cremer, M.; Hübner, B.; Strickfaden, H.; Smeets, D.; Popken, J.; Sterr, M.; Markaki, Y.; Rippe, K.; Cremer, C. The 4D nucleome: Evidence for a dynamic nuclear landscape based on co-aligned active and inactive nuclear compartments. *FEBS Lett.* **2015**, *589*, 2931–2943. [CrossRef]
72. Bediaga, N.G.; Coughlan, H.D.; Johanson, T.M.; Garnham, A.L.; Naselli, G.; Schröder, J.; Fearnley, L.G.; Bandala-Sanchez, E.; Allan, R.S.; Smyth, G.K.; et al. Multi-level remodelling of chromatin underlying activation of human T cells. *Sci. Rep.* **2021**, *11*, 528. [CrossRef] [PubMed]
73. Sabinina, V.J.; Hossain, M.J.; Hériché, J.K.; Hoess, P.; Nijmeijer, B.; Mosalaganti, S.; Kueblbeck, M.; Callegari, A.; Szymborska, A.; Beck, M.; et al. Three-dimensional superresolution fluorescence microscopy maps the variable molecular architecture of the nuclear pore complex. *Mol. Biol. Cell* **2021**, *32*, 1523–1533. [CrossRef] [PubMed]
74. Hirose, T.; Ninomiya, K.; Nakagawa, S.; Yamazaki, T. A guide to membraneless organelles and their various roles in gene regulation. *Nat. Rev. Mol. Cell Biol.* **2022**, *24*, 288–304. [CrossRef]

75. Erdmann, R.S.; Baguley, S.W.; Richens, J.H.; Wissner, R.F.; Xi, Z.; Allgeyer, E.S.; Zhong, S.; Thompson, A.D.; Lowe, N.; Butler, R.; et al. Labeling Strategies Matter for Super-Resolution Microscopy: A Comparison between HaloTags and SNAP-tags. *Cell Chem. Biol.* **2019**, *26*, 584. [CrossRef]
76. Lukinavičius, G.; Alvelid, J.; Gerasimaitė, R.; Rodilla-Ramirez, C.; Nguyễn, V.T.; Vicidomini, G.; Bottanelli, F.; Han, K.Y.; Testa, I. Stimulated emission depletion microscopy. *Nat. Rev. Methods Primers* **2024**, *4*, 56. [CrossRef]
77. Hanne, J.; Göttfert, F.; Schimer, J.; Anders-Össwein, M.; Konvalinka, J.; Engelhardt, J.; Müller, B.; Hell, S.W.; Kräusslich, H.G. Stimulated Emission Depletion Nanoscopy Reveals Time-Course of Human Immunodeficiency Virus Proteolytic Maturation. *ACS Nano* **2016**, *10*, 8215–8222. [CrossRef]
78. Truckenbrodt, S.; Maidorn, M.; Crzan, D.; Wildhagen, H.; Kabatas, S.; Rizzoli, S.O. ×10 expansion microscopy enables 25-nm resolution on conventional microscopes. *EMBO Rep.* **2018**, *19*, 45836. [CrossRef] [PubMed]
79. Norman, R.X.; Chen, Y.C.; Recchia, E.E.; Loi, J.; Rosemarie, Q.; Lesko, S.L.; Patel, S.; Sherer, N.; Takaku, M.; Burkard, M.E.; et al. One step 4× and 12× 3D-ExM: Robust super-resolution microscopy in 1 cell biology. *bioRxiv* **2024**. [CrossRef]
80. Sun, D.E.; Fan, X.; Shi, Y.; Zhang, H.; Huang, Z.; Cheng, B.; Tang, Q.; Li, W.; Zhu, Y.; Bai, J.; et al. Click-ExM enables expansion microscopy for all biomolecules. *Nat. Methods* **2020**, *18*, 107–113. [CrossRef]
81. Shi, X.; Li, Q.; Dai, Z.; Tran, A.A.; Feng, S.; Ramirez, A.D.; Lin, Z.; Wang, X.; Chow, T.T.; Chen, J.; et al. Label-retention expansion microscopy. *J. Cell Biol.* **2021**, *220*, e202105067. [CrossRef]
82. Liffner, B.; Cepeda Diaz, A.K.; Blauwkamp, J.; Anaguano, D.; Frolich, S.; Muralidharan, V.; Wilson, D.W.; Dvorin, J.D.; Absalon, S. Atlas of Plasmodium falciparum intraerythrocytic development using expansion microscopy. *eLife* **2023**, *12*, RP88088. [CrossRef] [PubMed]
83. Zila, V.; Müller, T.G.; Laketa, V.; Müller, B.; Kräusslich, H.G. Analysis of CA Content and CPSF6 Dependence of Early HIV-1 Replication Complexes in SupT1-R5 Cells. *mBio* **2019**, *10*, e02501-19. [CrossRef] [PubMed]
84. Heintzmann, R.; Huser, T. Super-Resolution Structured Illumination Microscopy. *Chem. Rev.* **2017**, *117*, 13890–13908. [CrossRef] [PubMed]
85. Delattre, S. Igniting New Confocal Imaging Potential—Nikon AX R Series with NSPARC. *Micros. Today* **2023**, *31*, 23–27. [CrossRef]
86. Ries, J.; Kaplan, C.; Platonova, E.; Eghlidi, H.; Ewers, H. A simple, versatile method for GFP-based super-resolution microscopy via nanobodies. *Nat. Methods* **2012**, *9*, 582–584. [CrossRef]
87. Müller, T.G.; Sakin, V.; Müller, B. A Spotlight on Viruses—Application of Click Chemistry to Visualize Virus-Cell Interactions. *Molecules* **2019**, *24*, 481. [CrossRef] [PubMed]
88. Hanne, J.; Zila, V.; Heilemann, M.; Müller, B.; Kräusslich, H.G. Super-resolved insights into human immunodeficiency virus biology. *FEBS Lett.* **2016**, *590*, 1858–1876. [CrossRef] [PubMed]

viruses

MDPI

Article

Multiplex Microscopy Assay for Assessment of Therapeutic and Serum Antibodies against Emerging Pathogens

Nuno Sartingen [1], Vanessa Stürmer [1], Matthias Kaltenböck [1], Thorsten G. Müller [1], Paul Schnitzler [1], Anna Kreshuk [2], Hans-Georg Kräusslich [1,3] Uta Merle [4], Frauke Mücksch [1], Barbara Müller [1], Constantin Pape [5,6] and Vibor Laketa [1,3,*]

1 Department of Infectious Diseases, Virology, Medical Faculty Heidelberg, Heidelberg University, 69120 Heidelberg, Germany; nuno.sartingen@stud.uni-heidelberg.de (N.S.); vanessa.stuermer@embl.de (V.S.); matthiaskaltenboeck@gmail.com (M.K.); muellert@ie-freiburg.mpg.de (T.G.M.); paul.schnitzler@med.uni-heidelberg.de (P.S.); hans-georg.kraeusslich@med.uni-heidelberg.de (H.-G.K.); frauke.muecksch@med.uni-heidelberg.de (F.M.); barbara.mueller@med.uni-heidelberg.de (B.M.)
2 European Molecular Biology Laboratory, 69117 Heidelberg, Germany; anna.kreshuk@embl.de
3 German Center for Infection Research (DZIF), Partner Site Heidelberg, 69120 Heidelberg, Germany
4 Department of Internal Medicine IV, University Hospital Heidelberg, 69120 Heidelberg, Germany; uta.merle@med.uni-heidelberg.de
5 Institute of Computer Science, Göttingen University, 37073 Göttingen, Germany; constantin.pape@informatik.uni-goettingen.de
6 Cluster of Excellence 'Multiscale Bioimaging: from Molecular Machines to Networks of Excitable Cells' (MBExC), Göttingen University, 37073 Göttingen, Germany
* Correspondence: vibor.laketa@med.uni-heidelberg.de

Abstract: The emergence of novel pathogens, exemplified recently by the severe acute respiratory syndrome coronavirus 2 (SARS-CoV-2), highlights the need for rapidly deployable and adaptable diagnostic assays to assess their impact on human health and guide public health responses in future pandemics. In this study, we developed an automated multiplex microscopy assay coupled with machine learning-based analysis for antibody detection. To achieve multiplexing and simultaneous detection of multiple viral antigens, we devised a barcoding strategy utilizing a panel of HeLa-based cell lines. Each cell line expressed a distinct viral antigen, along with a fluorescent protein exhibiting a unique subcellular localization pattern for cell classification. Our robust, cell segmentation and classification algorithm, combined with automated image acquisition, ensured compatibility with a high-throughput approach. As a proof of concept, we successfully applied this approach for quantitation of immunoreactivity against different variants of SARS-CoV-2 spike and nucleocapsid proteins in sera of patients or vaccinees, as well as for the study of selective reactivity of monoclonal antibodies. Importantly, our system can be rapidly adapted to accommodate other SARS-CoV-2 variants as well as any antigen of a newly emerging pathogen, thereby representing an important resource in the context of pandemic preparedness.

Keywords: multiplex microscopy; machine learning; serology; emerging pathogens; monoclonal antibodies; SARS-CoV-2

1. Introduction

The emergence of the novel pathogenic severe acute respiratory syndrome coronavirus 2 (SARS-CoV-2) in 2019 [1,2] and its rapid pandemic spread had profound repercussions on societies worldwide [3–5]. To mitigate the viral transmission and reduce infection rates, governments implemented extensive lockdown measurements and social restrictions. In an effort to restrain the pandemic, the swift development of SARS-CoV-2 vaccines led to a massive vaccination effort with more than 5 billion people vaccinated worldwide as of July 15, 2024 [6–8]. However, the continuous evolution of new viral variants led to immune evasion, affecting viral infectivity and pathogenicity [9], and compromised the effectiveness of the first vaccines based on the wild-type Wuhan-Hu-1 sequence of the viral spike protein [10,11]. The spike protein also had a central role in the development of anti-viral treatments based on therapeutic monoclonal antibodies (mAbs) [12,13]. However, with the emergence of novel variants, the initially developed therapeutic mAbs lost their anti-viral efficacy due to mutations in the spike protein [14–16]. These lessons learned during the COVID-19 pandemic need to be leveraged to prepare our response to any future pandemics characterized by quickly evolving pathogen variants. In the context of pandemic preparedness, it will be essential to rapidly (i) evaluate the status of the immunity of the population and assure the effectiveness of vaccines and (ii) support the development of new mAb-based anti-viral treatments.

To achieve these objectives, it is essential to develop comprehensive serological assays that can assess antibody binding to different viral proteins and are rapidly adaptable to include emerging variants of concern. In the broader context of public health management and infectious disease control, serological assays are an essential complement to PCR testing. Antibody detection and quantification played a fundamental role in decision-making and policy guidance aimed at restraining the spread of SARS-CoV-2 [17,18] and will be equally important in constraining any pandemic in the future. Currently, commercially available enzyme-linked immunosorbent assays or (electro) chemiluminescent immunoassays (ELISA, (E)CLIA) represent the gold standard for antibody detection. However, adapting them rapidly to new viral variants in an evolving pandemic situation presents challenges, primarily due to the need for protein expression, purification and optimization of biochemical properties. This challenge is especially pronounced for highly glycosylated proteins expressed in cellular and viral membranes [19–22]. Moreover, standard ELISA assays cannot assess binding to multiple viral proteins simultaneously, effectively limiting their throughput and information content. Multiplex serology for simultaneous detection of antibodies against different SARS-CoV-2 proteins based on protein microarrays [23] or fluorescent beads [24] has been applied during the COVID-19 pandemic. However, these approaches also rely on the availability of purified proteins. Additionally, the recent SARS-CoV-2 pandemic highlighted the importance of alternative serological assays in situations where commercial ELISA supplies are scarce or technically infeasible.

In response to the outlined challenges, we describe the development of an automated multiplex microscopy and machine learning-based serology assay for simultaneous differential antibody detection, focusing on SARS-CoV-2 as a proof of concept. For this, we build upon our previous work, where we established a serology assay for SARS-CoV-2 antibody detection in human samples [25] and employed it as a complementary assay in an early large-scale seroprevalence study [26]. The multiplexed approach presented here provides differentiated data about antibody reactivity against various viral antigens at once and can be conducted within a biosafety level 1 laboratory. By applying this method to screen a cohort of sera from convalescent individuals and vaccinees, we successfully demonstrated its efficacy in determining immunoreactivity against the SARS-CoV-2 nucleocapsid protein and different variants of SARS-CoV-2 spike proteins. Additionally, it enabled the evaluation of selective reactivity of (therapeutic) monoclonal antibodies (mAbs). Notably, the system described here can be swiftly adapted to assess immunoreactivity to new SARS-CoV-2 variants or other antigens from emerging pathogens, making it a valuable tool for pandemic preparedness. It represents a valuable tool in our portfolio of serological assays with the potential to complement and expand diagnostic procedures as well as facilitate the screening and testing of therapeutic antibodies.

2. Materials and Methods

2.1. Human Material

Negative control serum samples ($n = 233$) were collected for various serological testing in the routine laboratory of the Center of Infectious Diseases, University Hospital Heidelberg between 2015 and 2019, before the start of the SARS-CoV-2 outbreak. Samples used corresponded to pseudonymized material from the Center of Infectious Diseases, Heidelberg. SARS-CoV-2 positive sera were collected from 41 RT-PCR-confirmed symptomatic COVID-19 patients treated at the University Hospital Heidelberg after giving their written informed consent (ethics votum no S-148/2020, University Hospital Heidelberg). Days post-infection were defined based on the anamnesis carried out upon admission. Serum samples were stored at −20 °C until use.

2.2. Plasmids

Plasmids were cloned using standard molecular biology techniques and verified by commercial Sanger sequencing (Eurofins Genomics, Ebersberg, Germany). pWPI lentiviral plasmids [27] served as a vector backbone for the insertion of genes of interest. Genes of interest were integrated into the pWPI sequence using classical cloning and Gibson assembly. The assembly was performed using the NEB HiFi Mastermix (New England Biolabs, Ipswich, MA, USA) with 30 base–pair overlap regions. PCR was performed using Q5 High-Fidelity DNA Polymerase (New England Biolabs), according to the manufacturer's instructions with primers purchased from Eurofins Genomics. E. coli DH5α and Stbl2 [28] (Thermo Fisher Scientific, Waltham, MA, USA) were used for amplification of standard plasmids or LTR-containing plasmids, respectively. The list of plasmids used can be found in Table S1. Schematic representation of the generated plasmids is shown in Figure S1.

2.3. Cell Culture

HeLa cells and Hek293T cells were authenticated using STR profiling (Eurofins Genomics) and monitored for mycosplasma contamination using the MycoAlert mycoplasma detection kit (Lonza, Basel, Switzerland). Cells were cultured at 37 °C and 5% CO_2 in Dulbecco's Modified Eagle's Medium (DMEM; Thermo Fisher Scientific) containing 4.5 g/L D-glucose and L-glutamine supplemented with 10% fetal calf serum (FCS; Sigma-Aldrich, St. Louis, MO, USA), 100 U/mL penicillin and 100 μg/mL streptomycin (PAN Biotech, Aidenbach, Germany).

2.4. Generation of Cell Lines

A total of 4×10^5 Hek293T cells were seeded per well into 6 well plates and incubated overnight. Lentiviral vector particles were produced by co-transfection of packaging plasmid psPAX2 (Addgene #12260), the respective lentiviral transfer vector pWPI, the envelope expression plasmid pCMV-VSV-G (Addgene #8454) and pAdvantage (Promega, Madison, WI, USA) in a ratio of 1.5:1.0:0.5:0.2 μg into HEK293T cells using polyethylenimine (PEI; 3 μL of 1 mg/mL PEI per μg DNA). At 48 h post-transfection, the supernatant was harvested and filtered through 0.45 μm mixed cellulose ester (MCE) filters, and 500 μL of lentiviral vector particle containing supernatant was added to HeLa cells (5×10^4 cells/well seeded the day before in 12 well plates). After 48 h, the transduced cells were selected using 5 mg/mL blasticidin or 1 mg/mL puromycin and subsequently expanded to a 25 mL or 75 mL cell culture flask depending on the number of cells. After three or five days (for blasticidin and puromycin, respectively) under antibiotic selection, cells were transduced with the respective second construct. Cells were first transduced with lentiviral vector particles encoding the subcellular fluorescent localization marker, followed by transduction with corresponding viral proteins. Cells were cultured under ongoing selection pressure with both antibiotics. Cell lines were cultured separately and mixed in equal numbers at the start of the experiment. The full list of cell lines used and created in this work can be found in Table S2.

2.5. Fluorescence-Activated Cell Sorting

After successful double transduction, fluorescence-activated cell sorting (FACS) was employed for additional double sorting of each generated cell line. Cells were suspended in phosphate-buffered saline (PBS). After addition of undiluted accutase (StemPro™ Accutase® Cell Dissociation Reagent, Thermo Fisher Scientific) as per the manufacturer's instructions, cells were incubated for 15 min at 37 °C. Subsequently, cells were resuspended in PBS and harvested by centrifugation (5 min, 1500 rpm, Heraeus Megafuge 40R (Heraeus, Hanau, Germany)). Sorting of barcoded cell lines was performed based on the expression of the mScarlet-marker. For sorting of spike-expressing cells, additional staining targeting surface-expressed spike proteins was performed. Cells were resuspended in 3 ml PBS and 2% fetal calf serum as blocking buffer and incubated for 15 min. After incubation, a total of 4×10^6 cells per sample were resuspended in a volume of 2 ml PBS and primary anti-IgG1-Wuhan-Hu-1-spike antibody (Biozol, Eching, Germany, Cat# GTX632604) was applied in a dilution of 1:1000. Samples were incubated on ice for 30 min in the dark. Samples were washed with PBS and incubated with polyclonal Alexa Fluor 488 secondary antibody in a dilution of 1:2000 on ice for 20 min in the dark. After washing the samples with PBS, double-positive cells for the mScarlet (568 nm) and viral protein signal (488 nm) were sorted. The nucleocapsid-3xNLS-mSc cell line was only sorted for the mScarlet expression as nucleocapsid is exclusively expressed in the cytosol.

2.6. Monoclonal Antibodies

REGN10933 (Casirivimab) and REGN10987 (Imdevimab) were procured in an infusion solution, containing both antibodies at a concentration of 120 mg/mL (Roche, Basel, Switzerland; Swissmedic marketing authorization number 68329). The antibodies were diluted according to the manufacturer's instructions before use in subsequent experimental procedures. Monoclonal anti-RBD antibodies analyzed in paragraph 3.3 were described previously [29]. Briefly, RBD-specific B cells were isolated from study participants who had been vaccinated with three doses of an mRNA vaccine and reported SARS-CoV-2 Omicron BA.1 breakthrough infection (variant deduced based on prevalent variant circulating in New York City at the time of infection). Antibody sequences were amplified from reverse-transcribed cellular RNA, sequenced, cloned and expressed as recombinant human IgG. The full list of mAbs can be found in Table S3.

2.7. Immunofluorescent Staining

Staining was performed in a black-wall glass bottom 96-well plate (Cellvis, Mountain View, CA, USA 96-well glass bottom plate with high-performance #1.5 cover glass). The various HeLa cell lines were seeded at a density of 1.5×10^4 cells per well 16–18 h before staining (5 cell lines = 3×10^3 per cell line/well). After washing with PBS, cells were fixed with 4% paraformaldehyde (PFA) for 20 min followed by 3× washing with PBS. Cells were then permeabilized with 0.5% Triton X100 (Sigma-Aldrich) for 5 min and washed 3 times with PBS containing 0.02% Tween 20 (Sigma-Aldrich) (PBS+). Blocking was carried out with 2% powdered milk (Roth) in PBS+ for 20 min followed by 3 washes with PBS+. Subsequently, cells were incubated with primary antibodies (mouse anti-spike antibody Biozol GTX632604 1:2000 in PBS+, mouse anti-nucleocapsid Sino Biological, Beijing, China, 40143-MM05 1:4000 in PBS+, mAbs with different indicated dilutions) or patient sera (1:200 in PBS+) for 30 min in the dark. Following three washing steps, samples were incubated with corresponding secondary antibodies for 20 min at room temperature in the dark (goat anti-human IgG-AlexaFluor 488 (Invitrogen, Thermo Fisher Scientific) 1:2000, goat anti-mouse IgG 647 1:500). After two washing steps with PBS+, cells were stained with Hoechst 33342 (Thermo Fisher Scientific) at a concentration of 0.2 μg/mL in PBS for 5 min before undergoing three washes with PBS. Samples were stored at 4 °C until imaging. All washing steps in a 96-well format were performed using the HydroFlex microplate washer (Tecan, Männedorf, Switzerland). Living cells were incubated with mAbs (200 ng/mL and 2 ug/mL) for 15 min directly in growth medium at room temperature, followed by two washing steps with PBS and fixation with 4% PFA for 20 min. Further staining was performed as described above.

2.8. Microscopy

Images were acquired on the Zeiss CellDiscoverer 7 automated microscope (Zeiss, Oberkochen, Germany) using a Plan-Apochromat 20×/0.95 Air objective with 1× tube lens magnification and a Zeiss Axiocam 712 sCMOS camera for detection. Multichannel images were acquired using 385 nm (Nucleus channel) and 625 nm (Antigen expression channel) light-emitting diodes for excitation and a quad-bandpass filter QBP 425/30 + 514/30 + 592/25 + 709/100 for emission. "IgG channel" and "Barcode channel" were acquired using a 470 nm LED together with 515/25 bandpass emission filter and 567 nm LED with 595/30 bandpass emission filter, respectively. A total of 9–12 images were acquired in each well, spaced in a regular 3 × 3 or 3 × 4 pattern centered around the middle of the well or randomly distributed. A definite focus system was used for axial plane stabilization during imaging.

2.9. Enzyme-Linked Immuno-Sorbent Assay (ELISA)

ELISA measurements to determine reactivity against the S1 domain of the viral spike protein were conducted using the Euroimmun Anti-SARS-CoV-2-ELISA (IgG) test kits (Euroimmun, Lübeck, Germany; and EI 2606-9601 451 G). The assays were run on a Euroimmun Analyzer I instrument following the manufacturer's instructions. Optical densities measured for the samples were normalized using the value obtained for a calibrator sample provided in the test kit. The interpretation of the semi-quantitative ratiometric values were obtained following the manufacturer's protocol: values < 0.8 RU/mL were classified as negative, 0.8–1.1 RU/mL as borderline, and values of 1.1 RU/mL or higher as positive.

2.10. Image Analysis

Cell segmentation was adapted from Pape et al. [25] and was based on a seeded watershed using segmented nuclei as seeds and boundary predictions as the heightmap. For nucleus segmentation, the pretrained "Fluo" model of StarDist [30] was used. It was qualitatively checked and shows high prediction quality on our data. For boundary prediction, we used a U-Net [31] applied to the IgG channel, starting from the model trained in Pape et al. [25]. However, due to the different appearance of this channel, the model exhibited diminished prediction quality, resulting in imprecise cell boarders after the watershed. We retrained this model on data from multiple plates using the initial cell segmentation results as pseudo-ground-truth. Despite imperfections in these labels, the resulting network showed clearly improved predictions, generating cell segmentation results that were visually deemed of sufficient quality.

Cell classification was performed with a ResNet34 [32] that was applied to the barcode (mScarlet) channel, the Nucleus channel and the binarized cell segmentation mask centered around each cell and resized to the median cell size. The signal outside of the segmentation mask was removed to avoid crosstalk from the signal of other cells overlapping the bounding box of the current cell. This network was initially trained on several wells that contained isolated cell lines so that the expressed barcode was known for each cell without the need for human labeling. However, this network showed some inaccuracies when applied to images containing mixtures of barcodes. Consequently, we manually annotated 22 additional images by correcting the classification predictions of the initial network with MoBIE [33]. The network was further trained, and its accuracy was evaluated on separate splits of these data.

Initially, images were subjected to manual quality control to remove images with acquisition problems, such as out-of-focus images. The camera background was removed and spectral bleed-through correction was applied prior to cell segmentation and classification. Quality control was performed before obtaining the vp-score and Pearson's correlation. The quality control consisted of the following steps:

1. Excluding images with less than 35 or more than 600 cells.
2. Removing 25% of cells of each cell class with the lowest intensity in spike and nucleocapsid expression in the Antigen expression channel.

3. Removing the cells that were classified as mScarlet-Lamin (non-expressing for viral protein) but showed an expression profile in the "Antigen expression" channel.
4. Removing the cells classified as spike-expressing cells (wt, delta, omicron) with a higher Antigen-expression value than 5000.
5. Removing the cells with a very small or large spatial size that likely correspond to segmentation errors.
6. Removing the cells with low expression in the barcode (mScarlet) channel defined by absolute intensity values for each cell line.
7. Removing the cells with a saturated expression level in the H2A Barcode channel.

Antibody binding was quantified based on immunofluorescence signal intensity in the IgG channel within the labeled masks. Signal intensity was measured in each pixel within the labeled mask to calculate the median signal intensity per cell. From this, median fluorescence intensity values across all cells expressing the same viral protein were calculated. The specific (above background) antibody binding (called "vp-score") was calculated as the ratio between median fluorescence intensity in IgG channels of viral protein-expressing cells and cells expressing solely mSc-Lamin.

$$vp - score = \frac{Median\ intensity\ IgG\ (viral\ protein)}{Median\ intensity\ IgG\ (no\ viral\ protein)}$$

The vp-score for spike ≥ 1.3 and the vp-score for nucleocapsid ≥ 1.5, together with Pearson's correlation ≥ 0.6, were empirically determined and considered indicative of specific binding in experiments with human sera. To compare the relative binding to different spike variants, we normalized the vp-scores to expression levels of each variant determined in the "Antigen expression channel". The raw image data and all analysis results (cell segmentation, cell classification, quality control and further measurements) were visualized within the Fiji plugin MoBIE v4.2.3, enabling further visual quality control and an interpretation of the analysis results. The image analysis pipeline is available as open-source software at https://github.com/kreshuklab/covid-if-2 (accessed on 4 August 2024).

2.11. Statistical Analysis

Statistical tests were performed using Prism v5.01 (GraphPad Software Inc., Boston, MA, USA). Data were plotted using Prism v5.01 or the Python statistical data visualization package matplotlib v3.7.0, seaborn v0.12.2 and pandas v1.5.3. Median intensities were consistently used as the basis for all intensity measurements. Specific methodologies, including statistical tests and calculations, are provided in the figure legends.

3. Results

3.1. Setup of the Multiplex Microscopy Serology Assay

A multiplex microscopy immunofluorescence assay was designed to simultaneously measure SARS-CoV-2 antibodies binding to a panel of viral proteins. This was achieved by generating a panel of cell lines, each engineered to express a different viral protein, and combining these cell lines in a single well for multiplex immunofluorescence analysis (Figure 1A). To allow for the identification of individual cell lines in the mixed population, we devised a barcoding strategy using mScarlet fluorescent protein targeted to specific subcellular compartments. Each viral protein variant expressed was paired with a unique barcode identifier (Figure 1A). Since barcodes are defined by localization patterns rather than by color, a single fluorescence channel is sufficient for the classification of multiple cell lines, each expressing a distinct viral protein.

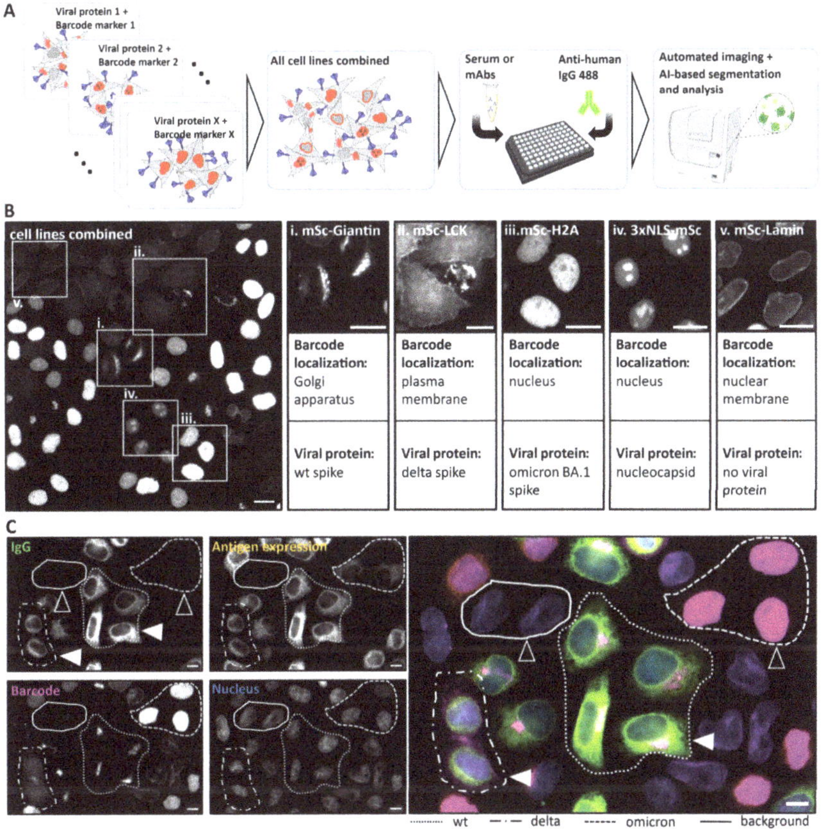

Figure 1. Principle of the multiplex microscopy serology assay. (**A**) Scheme of the multiplex assay workflow. Red structures represent fluorescent barcode proteins at different intracellular localizations, blue arrowhead-shaped structures represent viral proteins. (**B**) Cell lines transduced with different barcoding constructs and corresponding viral proteins were pooled and an overview image in the mScarlet channel was acquired. White boxes labeled with roman numbers mark the magnified areas showing cells with different subcellular localization patterns. Barcode localization and the corresponding viral protein are indicated below each image. (**C**) Images show results obtained using REGEN-COV therapeutic Ab on combined cell lines. Individual channels —"IgG channel" (represents REGEN-COV Ab binding, green), "Barcode channel" (represents viral antigen identity, magenta), "Antigen expression channel" (obtained by pan-specific anti-spike antibody staining and represents the expression levels of different spike variants, orange), "Nucleus channel" (obtained by Hoechst staining, blue) and an overlay image are shown. Cells expressing different variants are marked with different dashed line patterns (shown at the bottom of the image panel). Filled arrowheads indicate cells where REGEN-COV mAb binding can be detected, the empty arrowheads indicate those lacking detectable mAb binding. Scale bar = 10 μm.

To select fluorescent barcode markers that allowed robust discrimination between different cell lines, we screened a panel of mScarlet-tagged peptides and proteins with known subcellular localization [34] (Figure S2A). The criteria for suitability in our multiplex assay were the specificity of characteristic subcellular localization and stability of the distribution pattern upon fixation. We transfected HeLa cells with individually tagged mScarlet expression plasmids and imaged the cells before and after fixation (Figure S2B). Proteins targeting actin (LifeAct-mSc), microtubules (mSc-MapTau), focal adhesions (mSc-

PXN), and peroxisomes (mSc-peroxisome) exhibited diffuse cytosolic localization after fixation and were deemed unsuitable for our assay (Figure S2B). The proteins mSc-Giantin (Golgi), LCK-mSc (plasma membrane), mSc-Lamin (nuclear envelope), 3xNLS-mSc (nucleus), H2A-mSc (nucleus), 4xmts-mSc (mitochondria), ER-mSc (endoplasmic reticulum) and Lamp1-mSc (Lysosomes) displayed characteristic localization patterns well retained upon fixation and were thus considered suitable for machine learning-based automated classification in our multiplex assay (Figure S2B).

Based on these results, we generated a panel of stably transfected HeLa cell lines, each expressing an mScarlet-tagged peptide or protein serving as a distinct subcellular barcode (Figure 1B). These were subsequently paired with the specific SARS-CoV-2 protein or protein variant. Specifically, we chose three variants of the spike (S) protein representative of different phases of the pandemic: Wuhan-Hu-1 (wild type), B.1.617.2 (delta) and BA.1 (omicron)). A Δ18-C-terminal truncated version of the spike protein that has enhanced expression efficiency but preserves the immunogenic potential of the wild type was used [35,36]. Additionally, the conserved nucleocapsid (N) protein, highly immunogenic and commonly used in diagnostic tests to indicate past SARS-CoV-2 infection across circulating variants [37,38], was included in our analysis. As an internal control for nonspecific staining, we employed a cell line expressing a subcellular localization marker (mScarlet-Lamin) without any SARS-CoV-2 antigen (Figure 1B). Cell lines were generated through sequential lentiviral-based transduction. The generated polyclonal cell lines underwent double Fluorescence-Activated Cell Sorting (FACS) to ensure the presence and appropriate expression levels of both proteins in each cell population (Figure S3).

To assess the assay's ability to detect differential antibody binding, we employed a therapeutic mAb cocktail, REGEN-COV. This FDA-approved therapeutic comprises two non-competing human receptor-binding domain (RBD)-specific IgG antibodies—REGN10933 (Casirivimab) and REGN10987 (Imdevimab) [39,40]. Previous studies have demonstrated that these antibodies effectively bind and neutralize wild-type and delta variants of the spike protein, while proving ineffective against the omicron variant [14,41], rendering REGEN-COV an ideal candidate for our proof-of-principle experiments. The appropriate combination of barcoded HeLa cell lines was incubated with the mAb cocktail. Consistent with previous studies, mAb binding was only apparent on cells expressing wild-type and delta spike protein variants, identified by the expression of mSc-Giantin (Golgi) and LCK-mSc (plasma membrane), respectively (Figure 1C). In contrast, no binding was observed on cells expressing the omicron variant (barcoded by mSc-H2A signal in the nucleus) or those devoid of viral protein expression (barcoded by mSc-Lamin signal at the nuclear lamina) (Figure 1C). To ensure that all cells indeed express the viral variants and that the absence of omicron detection was not due to the lack of omicron variant expression, we stained the sample with a pan-specific anti-SARS-CoV-2 spike antibody (clone 1A9, Biozol (Eching, Germany) Cat.: GTX632604), targeting the conserved region (amino acids 1029–1192 of S2 subunit) of the spike protein. This confirmed the expression of all respective viral proteins ("Antigen expression", Figure 1C). These results corroborate prior findings about the REGEN-COV binding across different SARS-CoV-2 variants [14,42,43], validating the assay's capacity to detect differential antibody binding.

3.2. Machine Learning-Based Image Analysis Pipeline

To quantify results and facilitate high-throughput processing, we developed a fully automated analysis pipeline incorporating machine learning techniques. The analysis employed images recorded in four spectral channels captured using automated widefield microscopy—the "Nucleus channel" (for labeling nuclei, essential for quality control and cell segmentation), the "IgG channel" (for assessing specific antibody binding to various viral proteins), the "Barcode (mScarlet) channel" (for cell identification), and the "Antigen expression channel" (to control for viral protein expression levels using pan-specific spike antibody and/or nucleocapsid antibody) (Figure 2A).

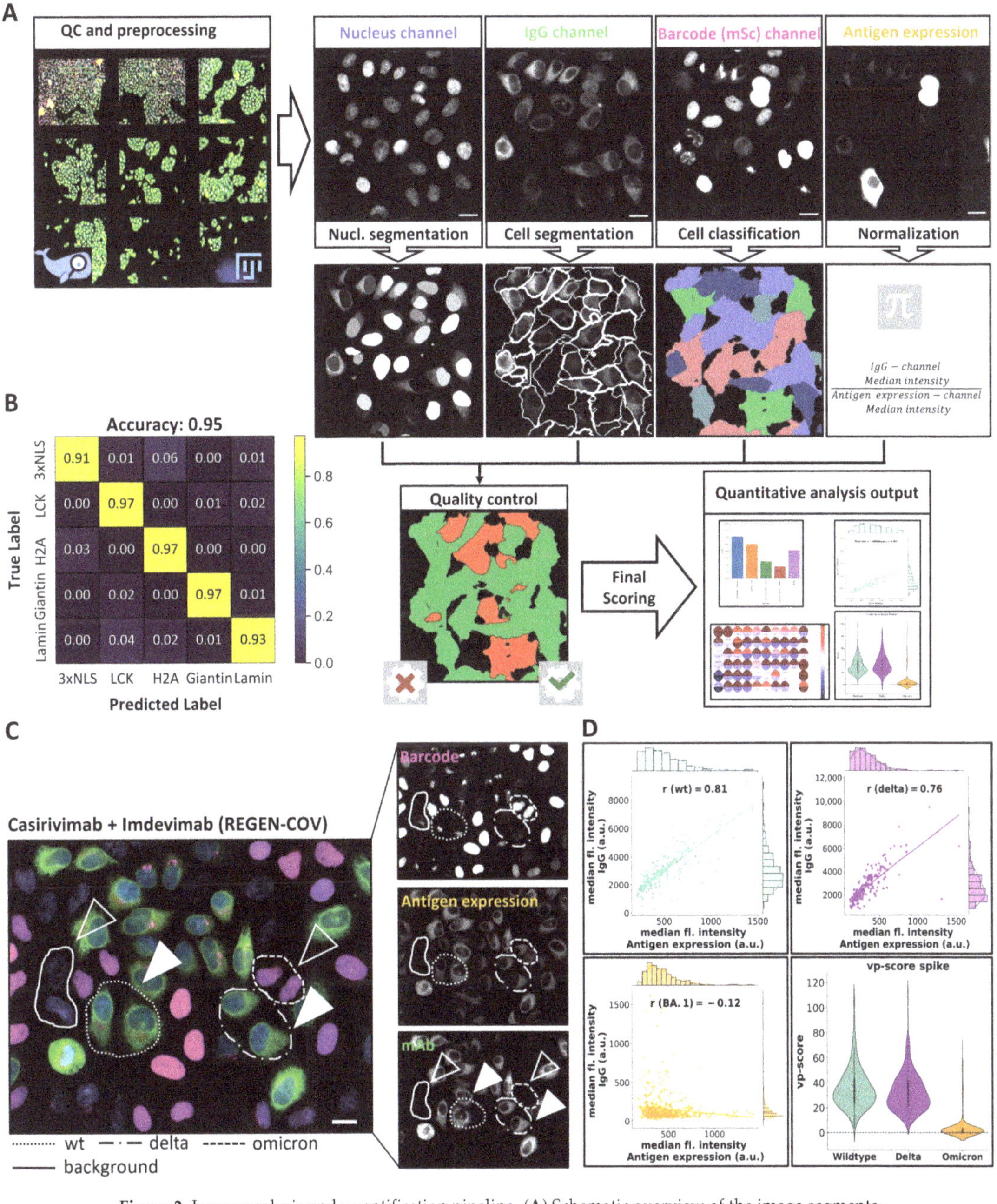

Figure 2. Image analysis and quantification pipeline. (**A**) Schematic overview of the image segmentation and classification workflow. A visual inspection of raw data (optional) and analysis results was conducted using FIJI [44] and MoBIE [33] software. Four channels were used for indicated purposes, and a final automated quality control was applied as described in the Materials and Methods section before computing the final scores. (**B**) Performance of the cell classification machine-learning model measured by confusion matrix showing the true cellular identity (True Label (manually classified), vertical) compared to predicted cellular identity (Predicted Label, horizontal). Color code (shown on

the right) represents the fraction of classified cells in an indicated category. (**C**) Images show results obtained using REGEN-COV therapeutic Ab on pooled cell lines. Individual channels ("IgG" (green), "Barcode (mScarlet)" (magenta) and "Antigen expression" (orange)) and overlay are shown. Cells expressing different variants are marked with different dashed line patterns (shown at the bottom of the image panel). Filled arrowheads indicate cells where REGEN-COV mAb binding can be detected, and empty arrowheads indicate cells lacking detectable mAb binding. Scale bar = 10 μm. (**D**) Quantification of REGEN-COV mAb binding using the pipeline shown in A. Graphs show a correlation analysis for wild-type (green), delta (magenta) and omicron BA.1 (orange) spike variants between mAb binding quantified by median fluorescence intensity in "IgG channel" per cell and antigen expression quantified by median fluorescence intensity in "Antigen expression channel" per cell. Histograms on the vertical and horizontal axis represent the frequency of cells with indicated intensity values. Pearson correlation coefficient (r) is indicated within the plot area. Violin plots on the bottom right illustrate vp-score distributions for wt, delta, and omicron (BA.1) variants, where the shape of each violin represents the distribution of vp-scores, and the white dot within each plot indicates median vp-score.

Cell segmentation was achieved by utilizing a machine learning model established in our previous work [25], employing a seeded watershed algorithm. It uses the nuclei segmented with StarDist [30], applied to the nucleus channel as seeds, and boundary predictions from a U-Net [31], applied to the IgG channel as a heightmap. This segmentation method routinely produced boundaries matching those observed visually in raw microscopy images in the IgG channel (Figure 2A).

Subsequently, we introduced a novel approach that automatically classified cells based on the expressed fluorescent markers with distinct subcellular localization. The approach relies on a deep neural network for supervised image classification (ResNet 34 [32]) (Figure 2A), applied to the image data of each segmented cell individually. The classification model was trained as described in the Materials and Methods section. To validate the effectiveness of our barcoding strategy, we applied the trained model to a new dataset that has not been used for training (unseen sample) comprising a mixture of cell lines expressing different barcodes. Remarkably, the model achieved an accuracy of 95.5% (Figure 2B). Notably, our model reliably distinguished the three different nuclear markers used in our setup (mSc-Lamin, mSc-H2A and 3xNLS-mSc) based on relatively subtle differences in nuclear staining patterns.

In summary, our approach combines machine learning-based segmentation and classification to automatically generate cell masks and categorize them into five classes based on the expressed viral protein. Potential unspecific antibody binding to cell surfaces was determined based on the IgG channel intensity of mSc-Lamin cells, which do not express any viral proteins. Antibody-binding was calculated as the ratio between median fluorescence intensity in IgG channels of viral protein-expressing cells and cells expressing solely mSc-Lamin (called viral protein (vp-score)), as described in the Materials and Methods section.

To further assess binding specificity, we obtained intensity measurements of viral protein expression from the "Antigen expression channel". It would be expected that specific antibody binding (IgG channel) is proportional to viral protein expression. Therefore, we calculated the Pearson's correlation coefficient (r) between these two channels.

To validate our analysis strategy, we applied the automated pipeline to the REGEN-COV mAb experiment described in Figure 1, which was previously only qualitatively evaluated. The derived quantitative data revealed negligible binding to the omicron variant (vp of 1.24 and r of −0.12) compared to binding to the wild type and delta spike variants, evidenced by high vp-score and Pearson's correlation coefficient (r) values (vp-score of 32.4 and 29.1, respectively; r of 0.81 and 0.76, respectively (Figure 2C,D)). To assess the sensitivity and dynamic range of the assay, we performed a titration experiment using the REGEN-COV mAb cocktail. mAb binding above the background was detectable for wild-type and delta variants at concentrations as low as 4 ng/mL of each mAb, reaching a plateau at around 200 ng/mL (Figure S4A). Notably, REGEN-COV comprises two distinct therapeutic mAbs—Casirivimab and Imdevimab. In individual dilution series experiments, Casirivimab exhibited stronger binding to the wild-type and delta spike variants compared to Imdevimab, with neither antibody demonstrating significant binding to the omicron variant (Figure S4B). Since Casirivimab and Imdevimab target different spike epitopes and can bind non-competitively [39], we investigated whether this was reflected in our assay. Cells were exposed to saturating concentrations of Casirivimab (1.8 μg/mL), Imdevimab (1.8 μg/mL), and a combination of both mAbs (where each mAb is present at half the concentration compared to the individual samples), and spike protein binding was assessed. The combination of both antibodies resulted in higher IgG channel fluorescence intensity measurements compared to individual mAb treatments (Figure S4C). This highlights the assay's capability to detect binding to distinct epitopes on the spike protein and the potential synergistic effects.

In summary, our proof-of-concept experiments using the REGEN-COV therapeutic mAb cocktail successfully validated the assay's capacity to quantitatively assess differential antibody binding to distinct viral proteins with high sensitivity as well as antibody binding to different epitopes on the same viral protein.

3.3. *Testing of Monoclonal Antibodies against SARS-CoV-2 Spike Protein*

Building on the results described above, we tested the capacity of our assay to characterize a comprehensive set of mAbs. For this, we obtained a panel of 14 mAbs with known specificity for wild-type, delta and omicron spike variants [29]. We selected mAbs, which exhibited differential binding patterns to the respective spike variants, as shown in Figure 3A (binders with EC50 < 1000 ng/mL are indicated with variant-specific colors).

The multiplex microscopy assay was employed (Figure 1A) together with the automated machine learning-based analysis pipeline (Figure 2A) to quantify the binding of each of the 14 mAbs at three different concentrations to the wt, delta and omicron spike variants. This yielded a total of 42 antibody–epitope combinations that were evaluated and compared to published data from Wang et al. [29]. Our analyses revealed differential binding of 14 mAbs across different spike mutants in a concentration-dependent manner (Figure 3A–C and Figure S5A). The specificity of the spike epitope binding determined here closely matched that previously reported by Wang et al. in almost all cases (39/42, Figure 3A). In two cases (mAb 6 and 12), the assay used here revealed, in addition to wild type, binding to a delta or omicron variant, respectively, which was not detected by ELISA measurements in Wang et al. (Figure 3A).

In the case of mAb 3, our assay failed to detect the binding to the omicron BA.1 variant (Figure 3A and Figure S5A) found by ELISA in Wang et al. To test whether PFA fixation prior to epitope staining had compromised the detection of certain epitopes, we performed antibody staining before PFA fixation. Indeed, this live cell labeling approach rescued the binding of mAb 3 to the omicron BA. 1 variant (Figure S5B).

In conclusion, the multiplex microscopy serology assay demonstrated the capacity to detect the differential binding of mAbs across various spike variants in very good agreement with the ELISA-based measurements, verifying the assay's potential to be used for mAb screening approaches.

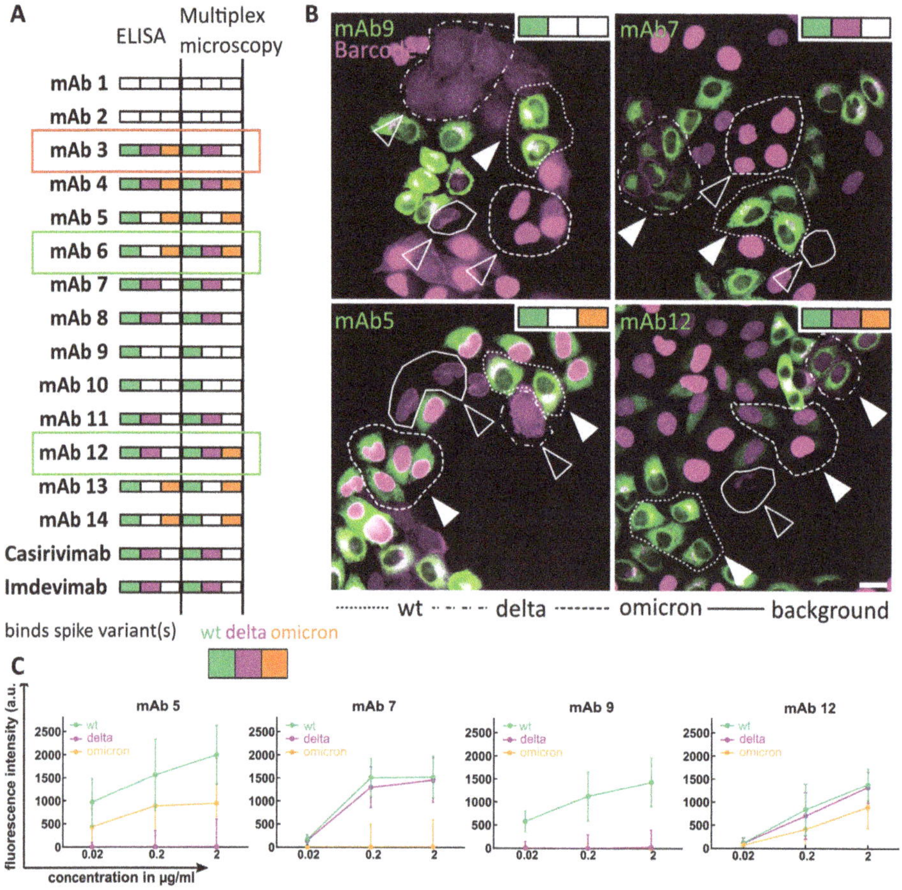

Figure 3. Assessment of mAb binding to different spike variants. (**A**) Qualitative comparison of differential mAb binding to different spike variants between ELISA (data from Wang et al. [29]) and the multiplex microscopy approach. Colored boxes (green (wt), magenta (delta) and orange (omicron)) indicate detected binding; white box: no binding. mAbs exhibiting some discrepancy between methods are marked with red (microscopy assay not detecting variant binding) and green (those where additional variant binding was detected by microscopy) boxes. Binding was evaluated as positive if median fluorescence intensity in the "IgG channel" was above 150 and more than three times higher than Casirivimab and Imdevimab binding to the omicron variant (negative control). (**B**) Representative images showing overlay of the indicated mAb and Barcode channels, illustrating differential spike binding specificity. Color bar in the top left corner indicates binding to the respective variant, and the gray box indicates no binding. Filled arrowheads: mAb binding detected; empty arrowheads: cells lacking detectable mAb binding. Scale bar = 10 µm. (**C**) Examples of quantitative data derived from results shown in B. Background-subtracted median fluorescence intensities are plotted against mAb concentration. Error bars show the standard deviation of median intensity values.

3.4. Screening of Human Sera

Next, we investigated the diagnostic utility of the assay for serological characterization of human samples. Therefore, we added a cell line expressing the SARS-CoV-2 nucleocapsid protein to our panel. The presence of the anti-nucleocapsid antibodies is indicative of exposure to SARS-CoV-2 and serves as a differentiator between vaccinated and acutely infected or convalescent individuals. To determine the specificity and sensitivity of the assay, we employed the positive and negative control cohorts established in our previous work [25]. The negative control cohort (n = 233) consisted of sera samples collected from healthy individuals, from patients infected with common cold coronaviruses (ccCoV), as well as patients infected with Mycoplasma pneumonia, Epstein–Barr virus (EBV) and Cytomegalovirus (CMV), all collected before 2019. The SARS-CoV-2 (positive control) cohort comprised 65 serum samples from 41 acutely infected patients confirmed positive for SARS-CoV-2 by RT-PCR in 2020 and collected between day 2 and 29 post-infection [25]. All samples were analyzed by our multiplex microscopy assay, the vp-score was calculated for spike (represented as an average vp-score across all three spike variants tested) and nucleocapsid proteins and compared to results obtained by a commercially available semi-quantitative SARS-CoV-2 ELISA (Euroimmun, Lübeck, Germany) approved for diagnostic use for the presence of SARS-CoV-2 spike-specific IgG.

In the negative cohort, 3 out of 233 samples (1.3%) had a false positive vp-score (vp > 1.3) for anti-spike, while 3 had a false positive vp-score (vp > 1.5) for anti-nucleocapsid without any overlap between the spike- and nucleocapsid-positive samples. A similar proportion (n = 7, 3%) scored false positive in ELISA measurements (Table 1, Figure 4A), but there was no overlap of false positive samples between our assay and ELISA. To evaluate the results from the positive cohort, we stratified the samples into three groups based on the day of sample collection post-infection (<11 days (n = 26); 11–14 days (n = 24); >14 days (n = 15)). The stratification was based on previous research on SARS-CoV-2 IgG seroconversion after first infection, where seroconversion for IgG targeting the spike protein was detected between day 10 and 14 post-infection with ELISA and CLIA [45–47]. All (15/15) of the samples collected >14 days post-infection exhibited a positive vp-score for both spike and nucleocapsid, compared to 14/15 (93.3%) in the ELISA assay. In the "11–14 days" and "<11 days" groups, 19/24 (79.1%) and 14/26 (54%) scored positive in our assay compared to 16/24 (66.7%) and 6/26 (23.1%) determined by ELISA (Table 1, Figure 4A). In general, the vp-score values correlated well with the ELISA scores (Pearson correlation r = 0.76) (Figure 4B). In summary, the results demonstrate that the multiplex microscopy serology assay can provide semi-quantitative measurements of spike and nucleocapsid reactive antibodies in human sera with specificity and sensitivity similar or better than an ELISA assay approved for clinical diagnostic use.

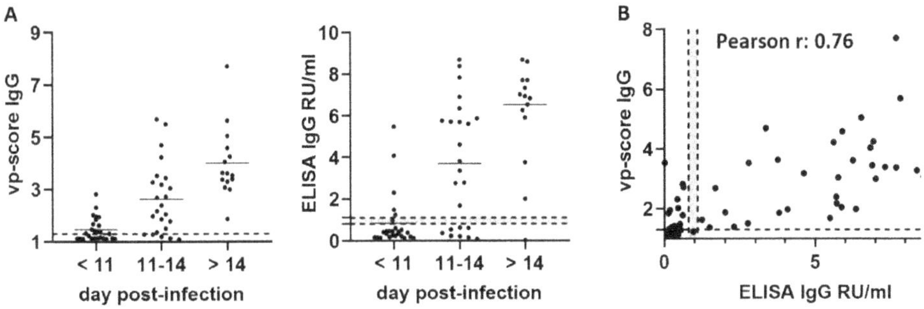

Figure 4. Comparison of multiplex microscopy assay and ELISA in the detection of SARS-CoV-2 spike antibodies in patient sera. (**A**) Samples from the positive control cohort have been stratified into three groups based on the day post-infection and analyzed by multiplex microscopy assay (**left**) and ELISA (**right**). Each dot represents one serum sample. Line = mean value; (**B**) correlation between vp-score obtained by multiplex microscopy assay and ELISA measurements; dotted line at 1.3 for vp-score = empirically determined cut-off value used to classify sera as negative or positive; dotted lines with gray area between 0.8 and 1.1 RU/mL in ELISA plots represent borderline situations where sera cannot be classified as positive or negative according to the manufacturer's protocol.

Table 1. Summary of positive results for control samples obtained by ELISA and multiplex microscopy assay.

Cohort	Microscopy Assay Spike IgG	ELISA Spike IgG
Negative cohort n = 233	3/233 (1.3%)	7/233 (3%)
SARS-CoV-2 cohort < 11 days post-infection n = 26	14/26 (53.8%)	6/26 (23.1%)
SARS-CoV-2 cohort 11–14 days post-infection n = 24	19/24 (79.1%)	16/24 (66.7%)
SARS-CoV-2 cohort > 14 days post-infection n = 15	15/15 (100%)	14/15 (93.3%)

Finally, we employed the assay to analyze the serological status of vaccinated individuals. This is particularly relevant in the context of a potential need for recurring vaccinations. For this, we examined sera from a cohort of n = 50 individuals who had received two doses of the Pfizer-BioNTech BNT162b2 vaccine in early 2021 at least 14 days before sample collection (Figure 5A). Vaccinees had not experienced a prior SARS-CoV-2 infection according to their own statements. Of the 50 tested sera, 47 scored positive for spike antibodies in our assay (Figure 5A), indicating that two doses of vaccine had resulted in the generation of reactive antibodies. Notably, the three samples (wells C3, F7 and F9, Figure 5A) that scored negative in this assay were also negative when tested with ELISA measuring anti-spike IgG (scoring 0.98, 0 and 0.03 RU/mL, respectively, with a cutoff for positivity being > 1.1 RU/mL), indicating low response to the vaccine. The vp-score for nucleocapsid was negative for 47 of 50 samples. Only three sera exhibited positive reactivity to the nucleocapsid protein in both IF assay and ELISA (wells B3, C4 and E11), indicating prior infection with SARS-CoV-2.

We also examined differences in antibody binding to the different spike variants in the vaccinee cohort (Figure 5A). Reactivity of the sera to all three spike variants was detected, indicating the presence of antibodies targeting conserved epitopes. However, in the majority of vaccinee samples, we detected more antibody binding to wild-type and delta spike proteins, while binding to the omicron variant was often weaker. This pattern of differential binding is not surprising considering that the BNT162b2 vaccine is based on the wild-type spike with the delta variant being relatively similar (8 mutations in the spike protein) and omicron considerably more divergent (33 mutations in the spike protein) [16,48].

To further investigate the effectiveness of a semi-quantitative assessment of antibody binding affinity across various spike variants and apply it in a more complex clinical situation, we analyzed a small cohort of six patients, where sequencing confirmed acute SARS-CoV-2 infection with the omicron variant (Figure 5B). The samples were collected at least 15 days post-infection except for sample 3 (8 days post-infection). Five of these individuals were unvaccinated before the omicron infection (samples 1–5), while one was vaccinated with the BNT162b2 vaccine prior to the omicron breakthrough infection (sample 6).

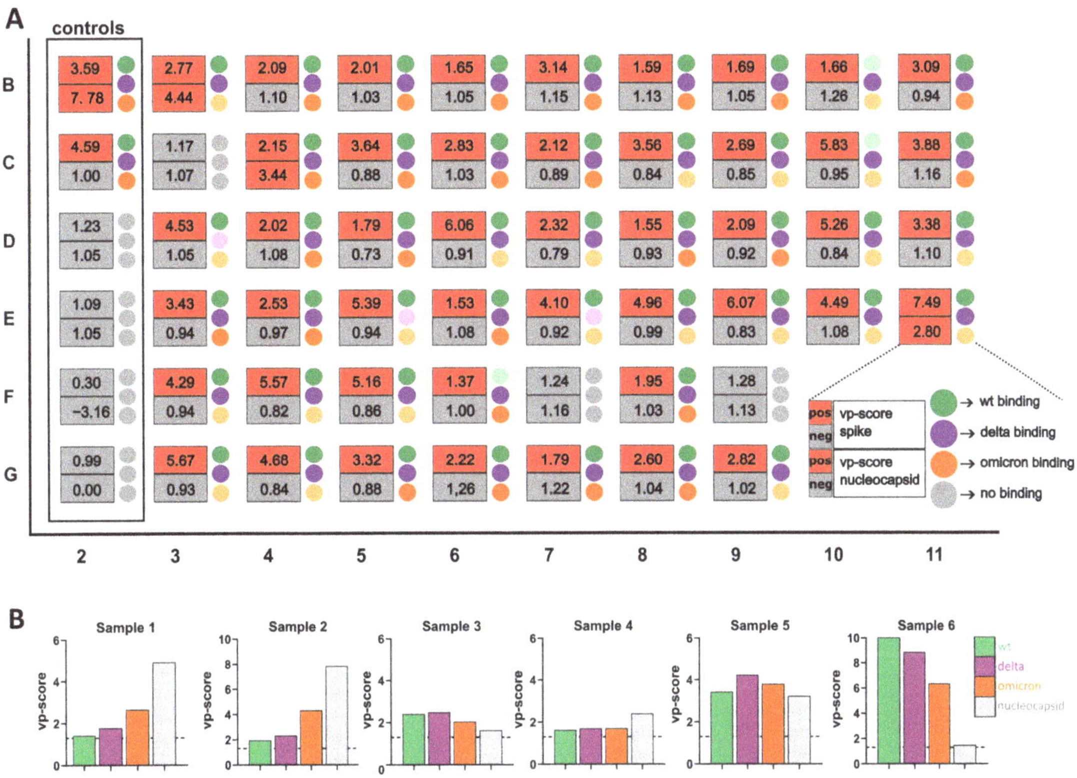

Figure 5. Analysis of the serological status of vaccinated and omicron-infected individuals. (**A**) Exemplary data showing the multiplex microscopy results of sera obtained from 50 vaccinees in the multi-well plate. The cells in the first column (2) are used as controls and are incubated with sera obtained from SARS-CoV-2-infected and -vaccinated individuals (wells B2 and C2, positive controls), pre-pandemic sera (wells D2 and E2, negative controls) and no sera (wells F2 and G2, background and bleed through control wells). The numbers in each box show the vp-score for spike binding (upper number, red = positive (vp > 1.3), gray = negative) and nucleocapsid binding (lower number, red = positive (vp > 1.5), gray = negative. The colored circles on the side of each box represent binding to different spike variants (green (wt), magenta (delta), orange (omicron) and gray (no binding). Dim color represents at least 15% weaker binding to the indicated variant compared to the variant where the strongest binding was measured. (**B**) Bar graphs show the vp-scores for different spike variants (green (wt), magenta (delta), orange (omicron)) and the nucleocapsid (gray) of omicron-infected individuals. Dotted line at 1.3 for vp-score = empirically determined cut-off value used to classify sera as negative or positive.

For two out of the five individuals who were unvaccinated, we detected antibodies that preferentially targeted the omicron variant while three sera displayed similar binding across all variants (Figure 5B). For the vaccinee who experienced an omicron breakthrough infection, we measured the highest vp-score for the wild-type variant, which is in line with reports indicating reactivation of preexisting cross-reactive B cells generated following breakthrough infection with a different strain [11,29,49,50] (Figure 5B, sample 5).

In summary, the multiplex microscopy serology assay was successfully employed in the semi-quantitative characterization of human samples in the context of determining the effectiveness of the vaccination procedure, previous (unknown) exposure to SARS-CoV-2 and disentangling preferential antibody binding across different spike variants, providing information relevant in situations of continued emergence of new virus variants as well as in clinical settings.

4. Discussion

We developed and validated a microscopy-based, multiplex serology assay, for the simultaneous detection and characterization of pathogen-specific antibodies. The assay integrates automated high-resolution microscopy in multi-well plates (Figure 1) and a robust machine learning analysis pipeline enabling a fully automated workflow (Figure 2). Serological assays like ELISA and (E)CLIA are a gold standard in antibody detection following viral infections [51]. However, the microscopy-based approach presented here has several unique aspects that may prove advantageous under certain conditions. Standard serological assays mainly employ purified recombinant proteins as antigens [20,52]. Potential complications may arise when proteins are expressed and purified under conditions that do not represent their native environment, compromising the attainment of biochemical properties essential for viral tropism and antigen specificity, such as folding and glycosylation patterns [20]. In addition, some proteins, in particular membrane proteins, are challenging to express and purify in large quantities *in vitro*. In the cell-based assay presented here, the protein of interest is expressed, processed and transported in a near-native cellular environment. This should facilitate the detection of antibodies that recognize the native conformation. The assay could be performed either with or without permeabilization. Here, we employed permeabilization since we were aiming for a parallel detection of intracellular and extracellular epitopes as well as to describe conditions that are broadly applicable for other viral antigens, independent of intracellular localization. Differential fixation can also reveal epitopes with different levels of glycosylation for comparison. The assay developed here, together with the panel of barcoded cell lines, represents a streamlined, rapidly adaptable, biosafety level 1-compatible toolbox that primarily requires the sequence of the protein of interest for cloning into a lentiviral vector. While some optimization of expression might be necessary when adapting the assay to other viral antigens, this renders the assay useful in the context of emerging pathogens since it can be rapidly developed as soon as the genomic sequence of the pathogen and relevant open reading frames are known.

A unique feature of the assay, distinguishing it from other microscopy-based serological assays described recently [25,53,54], is the multiplexing approach that uses fluorescent protein localization rather than its spectral characteristics as a barcoding strategy (Figures 1 and S2). Together with machine learning-driven cell classification and analysis (Figure 2), our assay is able to identify and characterize the differential binding of antibodies to various viral epitopes and variants within one experiment. Pietiäinen et al. previously demonstrated multiplex serology assay testing for various viral SARS-CoV-2 proteins [53]. Although the assay by Pietiäinen et al. investigated binding to S, N and M proteins of SARS-CoV-2, the multiplex approach faced constraints due to limitations in the number of available spectral channels and had to employ separate wells for each viral antigen. Our multiplex strategy not only simplifies the experimental process but also enhances its efficiency by allowing the simultaneous analysis of multiple targets. The approach can be extended to a larger number of viral proteins by employing the additional barcoding markers identified in this study (e.g., those localizing to lysosomes, ER or mitochondria). Although the system is limited by the number of markers with unique subcellular localization, machine learning-based pattern recognition and classification can recognize even subtle differences in localization patterns and are constantly improving. This is demonstrated by the fact that three different nuclear markers (H2A, lamin and NLS) were accurately classified by machine learning based on relatively subtle staining pattern differences. In addition, combining two or more barcodes in one cell will dramatically increase the number of different subcellular localization patterns that can be used for cell classification. Thus, we believe that the number of barcode proteins suitable for this multiplexing approach can be substantially expanded. Since barcoding uses only one spectral channel, one can take advantage of other spectral channels to assess, e.g., the presence of IgM and IgA in parallel to IgG.

In this proof of principle study, we applied the assay to the most relevant use-case situations, e.g., the testing of mAbs with potential therapeutic application (Figure 3) and serological analysis of human sera (Figures 4 and 5). Therapeutic mAbs represent a crucial treatment pillar in the early phase of pandemics and form an essential component of the portfolio of treatment options against newly emerging pathogens [55,56]. Our analysis of therapeutic mAb REGEN-COV and 14 mAbs from Wang et al. with known spike-variant binding specificity was largely in agreement with published data [29] (Figures 3 and S5). The few cases where we detected a discrepancy between the multiplex microscopy assay and ELISA highlight the subtle differences between methods. Given these results, as well as the costs and risks involved in mAb development, we imagine the multiplex microscopy assay could be integrated into a large-scale, commercial mAb discovery pipeline as an alternative approach to validate positive hits obtained by ELISA. Here, a multiplex option and the capacity to detect synergistic binding could be exploited to generate cells expressing different segments of the immunodominant protein to test for specific antibody combinations.

The multiplexing microscopy approach is particularly relevant for serological analysis of human serum samples, which often contain a complex array of antibodies against various viral components, depending on the vaccination status and previous exposure to viral variants. The lack of overlap in false positives between ELISA and multiplex microscopy assay (Figure 4, Table 1) indicates the potential for leveraging a dual-readout strategy to achieve extremely accurate specificity in critical applications such as serological screening in low seroprevalence populations [26]. Our analysis of vaccinee and omicron-infected individuals revealed differential antibody binding to spike variants and nucleocapsids consistent with their vaccination status and medical history (Figure 5). The ability of the assay to provide a "serological fingerprint" of SARS-CoV-2 binding antibodies for each individual enables addressing the inherent variability in the antibody landscape. This is particularly relevant in the context of the extensive vaccination efforts and continuously evolving virus variants.

The main limitation of the approach is the requirement for expertise and infrastructure for cell culture, microscopy and image analysis. The assay is, therefore, not considered as an alternative to commercially available assays for routine diagnostics. However, it can present an important experimental alternative in the context of newly emerging viral pathogens when information about immuno-relevant epitopes has yet to be obtained and when well-established serological assays are not available or do not reflect rapid viral evolution in a pandemic scenario.

5. Conclusions

In conclusion, we developed a microscopy assay suitable for the detection of antibodies against multiple viral proteins simultaneously. For multiplexing, we used a barcoding strategy that relies on subcellular localization of mScarlet-tagged peptides/proteins enabling evaluation of antibody binding to a large number of targets by using only a single spectral channel. Together with automated microscopy and the developed machine learning-based image segmentation, classification and analysis, the assay is compatible with high throughput applications. We applied this approach for quantitation of antibody binding to different variants of SARS-CoV-2 spike and nucleocapsid proteins in sera of patients or vaccinees, as well as for the study of reactivity of monoclonal antibodies. Importantly, given the compatibility with biosafety level 1 and the potential for rapid adaptability to emerging pathogens, the assay represents an important resource in the context of pandemic preparedness and an alternative to standard serological assays based on purified proteins.

Supplementary Materials: The following supporting information can be downloaded at: https://www.mdpi.com/article/10.3390/v16091473/s1, Figure S1: Schematic representation of plasmids generated for this study; Figure S2: Evaluation of subcellular localization of mScarlet-tagged peptides and proteins in living and fixed cells; Figure S3: Sorting of the cell lines; Figure S4: Quantification of the sensitivity and specificity of multiplex microscopy assay in mAb evaluation; Figure S5: mAb binding quantification; Table S1: List of plasmids used and generated in the study; Table S2: List of cell lines used or generated in the study; Table S3: List of used mAbs from Wang et al. and their binding specificities.

Author Contributions: Conceptualization, V.L., C.P., B.M., A.K., T.G.M. and H.-G.K.; methodology, N.S., V.S., M.K. and T.G.M.; software, C.P. and A.K.; formal analysis, N.S., V.S., F.M. and V.L.; investigation, N.S., V.S. and M.K.; resources, F.M., P.S. and U.M.; writing—original draft preparation, N.S. and V.L.; writing—review and editing, C.P., F.M. and B.M.; supervision, V.L. and B.M. All authors have read and agreed to the published version of the manuscript.

Funding: This research was funded by Deutsches Zentrum fuer Infektionsforschung (DZIF), grant number TTU 04.710 (V.L. and H.-G.K.). The research of C.P. was supported by the Deutsche Forschungsgemeinschaft (DFG, German Research Foundation) under Germany's Excellence Strategy—EXC 2067/1-390729940. The study was funded in part by the Ministry of Science, Research and the Arts Baden-Württemberg, Germany, within the framework of the special funding line for COVID-19 research.

Institutional Review Board Statement: The study was conducted in accordance with the Declaration of Helsinki and approved by the Institutional Ethics Committee of University Hospital Heidelberg (ethics votum No. S-148/2020).

Informed Consent Statement: Informed consent was obtained from all subjects involved in the study.

Data Availability Statement: The original contributions presented in the study are included in the article. Requests for reagents developed in this work such as cell lines and plasmids as well as raw images can be directed to the corresponding author. The image analysis pipeline is available as an open-source software at https://github.com/kreshuklab/covid-if-2 (accessed on 4 August 2024).

Acknowledgments: We would like to acknowledge the microscopy support from the Infectious Diseases Imaging Platform (IDIP) at the Center for Integrative Infectious Disease Research, Heidelberg, Germany. We would like to thank Michel C. Nussenzweig (The Rockefeller University, New York, NY, USA) for providing the mAbs for testing. We would like to thank the EMBL IT Services Department for providing computational infrastructure and support. We thank the patients who participated in this study. We acknowledge Addgene for providing the following plasmids: 3xnls-mScarlet-I (plasmid #98816), mScarlet-I-LaminB (plasmid #98831), pmScarlet-i_Giantin_C1 (plasmid #85050), pmScarlet-i_H2A_C1 (plasmid #85053), pLifeAct_mScarlet-i_N1 (plasmid #85056), Lck-mScarlet-I (plasmid #98821) were gifts from Dorus Gadella (Bindels et al. 2017); pcDNA3.3-SARS2-B.1.617.2 was a gift from David Nemazee (plasmid #172320); pCAGGS-S-B.1.617.2-FLAG was a gift from Daniel Conway (plasmid #180375). Individual images used in the graphical abstract courtesy of the medical illustrations database—https://smart.servier.com/ (accessed on 4 August 2024).

Conflicts of Interest: The authors declare no conflicts of interest. The funders had no role in the design of the study; in the collection, analyses, or interpretation of data; in the writing of the manuscript; or in the decision to publish the results.

References

1. Zhou, P.; Yang, X.L.; Wang, X.G.; Hu, B.; Zhang, L.; Zhang, W.; Si, H.R.; Zhu, Y.; Li, B.; Huang, C.L.; et al. A pneumonia outbreak associated with a new coronavirus of probable bat origin. *Nature* **2020**, *579*, 270–273. [CrossRef] [PubMed]
2. Zhu, N.; Zhang, D.; Wang, W.; Li, X.; Yang, B.; Song, J.; Zhao, X.; Huang, B.; Shi, W.; Lu, R.; et al. A Novel Coronavirus from Patients with Pneumonia in China, 2019. *N. Engl. J. Med.* **2020**, *382*, 727–733. [CrossRef] [PubMed]
3. Clemente-Suarez, V.J.; Dalamitros, A.A.; Beltran-Velasco, A.I.; Mielgo-Ayuso, J.; Tornero-Aguilera, J.F. Social and Psychophysiological Consequences of the COVID-19 Pandemic: An Extensive Literature Review. *Front. Psychol.* **2020**, *11*, 580225. [CrossRef] [PubMed]
4. Ahmed, F.; Shafer, L.; Malla, P.; Hopkins, R.; Moreland, S.; Zviedrite, N.; Uzicanin, A. Systematic review of empiric studies on lockdowns, workplace closures, and other non-pharmaceutical interventions in non-healthcare workplaces during the initial year of the COVID-19 pandemic: Benefits and selected unintended consequences. *BMC Public Health* **2024**, *24*, 884. [CrossRef]

5. Alizadeh, H.; Sharifi, A.; Damanbagh, S.; Nazarnia, H.; Nazarnia, M. Impacts of the COVID-19 pandemic on the social sphere and lessons for crisis management: A literature review. *Nat. Hazards* **2023**, *117*, 2139–2164. [CrossRef] [PubMed]
6. Ledford, H. US authorization of first COVID vaccine marks new phase in safety monitoring. *Nature* **2020**, *588*, 377–378. [CrossRef]
7. Sahin, U.; Muik, A.; Derhovanessian, E.; Vogler, I.; Kranz, L.M.; Vormehr, M.; Baum, A.; Pascal, K.; Quandt, J.; Maurus, D.; et al. COVID-19 vaccine BNT162b1 elicits human antibody and T(H)1 T cell responses. *Nature* **2020**, *586*, 594–599. [CrossRef]
8. World Health Organization. WHO Coronavirus (COVID-19) Dashboard. Available online: https://data.who.int/dashboards/covid19/vaccines?n=c (accessed on 15 July 2024).
9. Mlcochova, P.; Kemp, S.A.; Dhar, M.S.; Papa, G.; Meng, B.; Ferreira, I.; Datir, R.; Collier, D.A.; Albecka, A.; Singh, S.; et al. SARS-CoV-2 B.1.617.2 Delta variant replication and immune evasion. *Nature* **2021**, *599*, 114–119. [CrossRef]
10. Cele, S.; Jackson, L.; Khoury, D.S.; Khan, K.; Moyo-Gwete, T.; Tegally, H.; San, J.E.; Cromer, D.; Scheepers, C.; Amoako, D.G.; et al. Omicron extensively but incompletely escapes Pfizer BNT162b2 neutralization. *Nature* **2022**, *602*, 654–656. [CrossRef]
11. Servellita, V.; Syed, A.M.; Morris, M.K.; Brazer, N.; Saldhi, P.; Garcia-Knight, M.; Sreekumar, B.; Khalid, M.M.; Ciling, A.; Chen, P.Y.; et al. Neutralizing immunity in vaccine breakthrough infections from the SARS-CoV-2 Omicron and Delta variants. *Cell* **2022**, *185*, 1539–1548.e5. [CrossRef]
12. Schäfer, A.; Muecksch, F.; Lorenzi, J.C.C.; Leist, S.R.; Cipolla, M.; Bournazos, S.; Schmidt, F.; Gazumyan, A.; Baric, R.S.; Robbiani, D.F.; et al. Antibody potency, effector function and combinations in protection from SARS-CoV-2 infection in vivo. *bioRxiv* **2020**, preprints.
13. Tuccori, M.; Ferraro, S.; Convertino, I.; Cappello, E.; Valdiserra, G.; Blandizzi, C.; Maggi, F.; Focosi, D. Anti-SARS-CoV-2 neutralizing monoclonal antibodies: Clinical pipeline. *MAbs* **2020**, *12*, 1854149. [CrossRef] [PubMed]
14. Cao, Y.; Wang, J.; Jian, F.; Xiao, T.; Song, W.; Yisimayi, A.; Huang, W.; Li, Q.; Wang, P.; An, R.; et al. Omicron escapes the majority of existing SARS-CoV-2 neutralizing antibodies. *Nature* **2022**, *602*, 657–663. [CrossRef] [PubMed]
15. Planas, D.; Bruel, T.; Staropoli, I.; Guivel-Benhassine, F.; Porrot, F.; Maes, P.; Grzelak, L.; Prot, M.; Mougari, S.; Planchais, C.; et al. Resistance of Omicron subvariants BA.2.75.2, BA.4.6, and BQ.1.1 to neutralizing antibodies. *Nat. Commun.* **2023**, *14*, 824. [CrossRef] [PubMed]
16. Planas, D.; Veyer, D.; Baidaliuk, A.; Staropoli, I.; Guivel-Benhassine, F.; Rajah, M.M.; Planchais, C.; Porrot, F.; Robillard, N.; Puech, J.; et al. Reduced sensitivity of SARS-CoV-2 variant Delta to antibody neutralization. *Nature* **2021**, *596*, 276–280. [CrossRef]
17. Pollan, M.; Perez-Gomez, B.; Pastor-Barriuso, R.; Oteo, J.; Hernan, M.A.; Perez-Olmeda, M.; Sanmartin, J.L.; Fernandez-Garcia, A.; Cruz, I.; Fernandez de Larrea, N.; et al. Prevalence of SARS-CoV-2 in Spain (ENE-COVID): A nationwide, population-based seroepidemiological study. *Lancet* **2020**, *396*, 535–544. [CrossRef]
18. Wiegand, R.E.; Deng, Y.; Deng, X.; Lee, A.; Meyer, W.A., 3rd; Letovsky, S.; Charles, M.D.; Gundlapalli, A.V.; MacNeil, A.; Hall, A.J.; et al. Estimated SARS-CoV-2 antibody seroprevalence trends and relationship to reported case prevalence from a repeated, cross-sectional study in the 50 states and the District of Columbia, United States-25 October 2020–26 February 2022. *Lancet Reg. Health Am.* **2023**, *18*, 100403. [CrossRef]
19. Ernst, E.; Wolfe, P.; Stahura, C.; Edwards, K.A. Technical considerations to development of serological tests for SARS-CoV-2. *Talanta* **2021**, *224*, 121883. [CrossRef]
20. Spencer, K.A.; Osorio, F.A.; Hiscox, J.A. Recombinant viral proteins for use in diagnostic ELISAs to detect virus infection. *Vaccine* **2007**, *25*, 5653–5659. [CrossRef]
21. Zhang, P.; Woen, S.; Wang, T.; Liau, B.; Zhao, S.; Chen, C.; Yang, Y.; Song, Z.; Wormald, M.R.; Yu, C.; et al. Challenges of glycosylation analysis and control: An integrated approach to producing optimal and consistent therapeutic drugs. *Drug Discov. Today* **2016**, *21*, 740–765. [CrossRef]
22. Zheng, X.; Duan, R.H.; Gong, F.; Wei, X.; Dong, Y.; Chen, R.; Yue Liang, M.; Tang, C.; Lu, L. Accuracy of serological tests for COVID-19: A systematic review and meta-analysis. *Front. Public Health* **2022**, *10*, 923525. [CrossRef] [PubMed]
23. van Tol, S.; Mogling, R.; Li, W.; Godeke, G.J.; Swart, A.; Bergmans, B.; Brandenburg, A.; Kremer, K.; Murk, J.L.; van Beek, J.; et al. Accurate serology for SARS-CoV-2 and common human coronaviruses using a multiplex approach. *Emerg. Microbes Infect.* **2020**, *9*, 1965–1973. [CrossRef] [PubMed]
24. Butt, J.; Murugan, R.; Hippchen, T.; Olberg, S.; van Straaten, M.; Wardemann, H.; Stebbins, E.; Krausslich, H.G.; Bartenschlager, R.; Brenner, H.; et al. From Multiplex Serology to Serolomics-A Novel Approach to the Antibody Response against the SARS-CoV-2 Proteome. *Viruses* **2021**, *13*, 749. [CrossRef] [PubMed]
25. Pape, C.; Remme, R.; Wolny, A.; Olberg, S.; Wolf, S.; Cerrone, L.; Cortese, M.; Klaus, S.; Lucic, B.; Ullrich, S.; et al. Microscopy-based assay for semi-quantitative detection of SARS-CoV-2 specific antibodies in human sera: A semi-quantitative, high throughput, microscopy-based assay expands existing approaches to measure SARS-CoV-2 specific antibody levels in human sera. *Bioessays* **2021**, *43*, e2000257. [CrossRef]
26. Tonshoff, B.; Muller, B.; Elling, R.; Renk, H.; Meissner, P.; Hengel, H.; Garbade, S.F.; Kieser, M.; Jeltsch, K.; Grulich-Henn, J.; et al. Prevalence of SARS-CoV-2 Infection in Children and Their Parents in Southwest Germany. *JAMA Pediatr.* **2021**, *175*, 586–593. [CrossRef] [PubMed]
27. Trotard, M.; Tsopoulidis, N.; Tibroni, N.; Willemsen, J.; Binder, M.; Ruggieri, A.; Fackler, O.T. Sensing of HIV-1 Infection in Tzm-bl Cells with Reconstituted Expression of STING. *J. Virol.* **2016**, *90*, 2064–2076. [CrossRef]
28. Trinh, T.; Jessee, J.; Bloom, F.; Hirsch, V. STBL2: An Escherichia coli strain for the stable propagation of retroviral clones and direct repeat sequences. *Focus* **1994**, *16*, 78–80.

29. Wang, Z.; Zhou, P.; Muecksch, F.; Cho, A.; Ben Tanfous, T.; Canis, M.; Witte, L.; Johnson, B.; Raspe, R.; Schmidt, F.; et al. Memory B cell responses to Omicron subvariants after SARS-CoV-2 mRNA breakthrough infection in humans. *J. Exp. Med.* **2022**, *219*, e20221006. [CrossRef] [PubMed]
30. Schmidt, U.; Weigert, M.; Broaddus, C.; Myers, G. *Cell Detection with Star-Convex Polygons*; Springer International Publishing: Cham, Switzerland, 2018; pp. 265–273.
31. Ronneberger, O.; Fischer, P.; Brox, T. *U-Net: Convolutional Networks for Biomedical Image Segmentation*; Springer International Publishing: Cham, Switzerland, 2015; pp. 234–241.
32. He, K.; Zhang, X.; Ren, S.; Sun, J. Deep residual learning for image recognition. In Proceedings of the IEEE Conference on Computer Vision and Pattern Recognition, Las Vegas, NV, USA, 27–30 June 2016; pp. 770–778.
33. Pape, C.; Meechan, K.; Moreva, E.; Schorb, M.; Chiaruttini, N.; Zinchenko, V.; Martinez Vergara, H.; Mizzon, G.; Moore, J.; Arendt, D.; et al. MoBIE: A Fiji plugin for sharing and exploration of multi-modal cloud-hosted big image data. *Nat. Methods* **2023**, *20*, 475–476. [CrossRef]
34. Chertkova, A.O.; Mastop, M.; Postma, M.; van Bommel, N.; van der Niet, S.; Batenburg, K.L.; Joosen, L.; Gadella, T.W.J.; Okada, Y.; Goedhart, J. Robust and Bright Genetically Encoded Fluorescent Markers for Highlighting Structures and Compartments in Mammalian Cells. *bioRxiv* **2020**, preprints.
35. Schmidt, F.; Weisblum, Y.; Muecksch, F.; Hoffmann, H.H.; Michailidis, E.; Lorenzi, J.C.C.; Mendoza, P.; Rutkowska, M.; Bednarski, E.; Gaebler, C.; et al. Measuring SARS-CoV-2 neutralizing antibody activity using pseudotyped and chimeric viruses. *J. Exp. Med.* **2020**, *217*, e20201181. [CrossRef]
36. Chen, H.Y.; Huang, C.; Tian, L.; Huang, X.; Zhang, C.; Llewellyn, G.N.; Rogers, G.L.; Andresen, K.; O'Gorman, M.R.G.; Chen, Y.W.; et al. Cytoplasmic Tail Truncation of SARS-CoV-2 Spike Protein Enhances Titer of Pseudotyped Vectors but Masks the Effect of the D614G Mutation. *J. Virol.* **2021**, *95*, e0096621. [CrossRef] [PubMed]
37. Gaeta, A.; Angeloni, A.; Napoli, A.; Pucci, B.; Cinti, L.; Roberto, P.; Colaiacovo, F.; Berardelli, E.; Farina, A.; Antonelli, G.; et al. Anti-N SARS-CoV-2 assays for evaluation of natural viral infection. *J. Immunol. Methods* **2023**, *518*, 113486. [CrossRef] [PubMed]
38. Wu, W.; Cheng, Y.; Zhou, H.; Sun, C.; Zhang, S. The SARS-CoV-2 nucleocapsid protein: Its role in the viral life cycle, structure and functions, and use as a potential target in the development of vaccines and diagnostics. *Virol. J.* **2023**, *20*, 6. [CrossRef] [PubMed]
39. Baum, A.; Ajithdoss, D.; Copin, R.; Zhou, A.; Lanza, K.; Negron, N.; Ni, M.; Wei, Y.; Mohammadi, K.; Musser, B.; et al. REGN-COV2 antibodies prevent and treat SARS-CoV-2 infection in rhesus macaques and hamsters. *Science* **2020**, *370*, 1110–1115. [CrossRef] [PubMed]
40. Baum, A.; Fulton, B.O.; Wloga, E.; Copin, R.; Pascal, K.E.; Russo, V.; Giordano, S.; Lanza, K.; Negron, N.; Ni, M.; et al. Antibody cocktail to SARS-CoV-2 spike protein prevents rapid mutational escape seen with individual antibodies. *Science* **2020**, *369*, 1014–1018. [CrossRef]
41. Brehm, T.T.; Pfefferle, S.; von Possel, R.; Karolyi, M.; Zoufaly, A.; Wichmann, D.; Kobbe, R.; Emmerich, P.; Norz, D.; Aepfelbacher, M.; et al. Clinical efficacy and in vitro neutralization capacity of monoclonal antibodies for severe acute respiratory syndrome coronavirus 2 delta and omicron variants. *J. Med. Virol.* **2022**, *94*, 5038–5043. [CrossRef]
42. Bruel, T.; Hadjadj, J.; Maes, P.; Planas, D.; Seve, A.; Staropoli, I.; Guivel-Benhassine, F.; Porrot, F.; Bolland, W.H.; Nguyen, Y.; et al. Serum neutralization of SARS-CoV-2 Omicron sublineages BA.1 and BA.2 in patients receiving monoclonal antibodies. *Nat. Med.* **2022**, *28*, 1297–1302. [CrossRef]
43. Cao, Y.; Yisimayi, A.; Jian, F.; Song, W.; Xiao, T.; Wang, L.; Du, S.; Wang, J.; Li, Q.; Chen, X.; et al. BA.2.12.1, BA.4 and BA.5 escape antibodies elicited by Omicron infection. *Nature* **2022**, *608*, 593–602. [CrossRef]
44. Schindelin, J.; Arganda-Carreras, I.; Frise, E.; Kaynig, V.; Longair, M.; Pietzsch, T.; Preibisch, S.; Rueden, C.; Saalfeld, S.; Schmid, B.; et al. Fiji: An open-source platform for biological-image analysis. *Nat. Methods* **2012**, *9*, 676–682. [CrossRef]
45. Lou, B.; Li, T.D.; Zheng, S.F.; Su, Y.Y.; Li, Z.Y.; Liu, W.; Yu, F.; Ge, S.X.; Zou, Q.D.; Yuan, Q.; et al. Serology characteristics of SARS-CoV-2 infection after exposure and post-symptom onset. *Eur. Respir. J.* **2020**, *56*. [CrossRef]
46. Nair, S.; Chen, X. Biology of severe acute respiratory syndrome coronavirus 2 (SARS-CoV-2) and the humoral immunoresponse: A systematic review of evidence to support global policy-level actions and research. *Glob. Health J.* **2022**, *6*, 38–43. [CrossRef] [PubMed]
47. Zheng, J.; Deng, Y.; Zhao, Z.; Mao, B.; Lu, M.; Lin, Y.; Huang, A. Characterization of SARS-CoV-2-specific humoral immunity and its potential applications and therapeutic prospects. *Cell Mol. Immunol.* **2022**, *19*, 150–157. [CrossRef] [PubMed]
48. Suzuki, R.; Yamasoba, D.; Kimura, I.; Wang, L.; Kishimoto, M.; Ito, J.; Morioka, Y.; Nao, N.; Nasser, H.; Uriu, K.; et al. Attenuated fusogenicity and pathogenicity of SARS-CoV-2 Omicron variant. *Nature* **2022**, *603*, 700–705. [CrossRef] [PubMed]
49. Kaku, C.I.; Bergeron, A.J.; Ahlm, C.; Normark, J.; Sakharkar, M.; Forsell, M.N.E.; Walker, L.M. Recall of preexisting cross-reactive B cell memory after Omicron BA.1 breakthrough infection. *Sci. Immunol.* **2022**, *7*, eabq3511. [CrossRef] [PubMed]
50. Quandt, J.; Muik, A.; Salisch, N.; Lui, B.G.; Lutz, S.; Kruger, K.; Wallisch, A.K.; Adams-Quack, P.; Bacher, M.; Finlayson, A.; et al. Omicron BA.1 breakthrough infection drives cross-variant neutralization and memory B cell formation against conserved epitopes. *Sci. Immunol.* **2022**, *7*, eabq2427. [CrossRef] [PubMed]
51. Rogier, E.W.; Giorgi, E.; Tetteh, K.; Sepulveda, N. Editorial: Current research on serological analyses of infectious diseases. *Front. Med.* **2023**, *10*, 1154584. [CrossRef]
52. Mongkonwattanaporn, T.; Lertwatcharasarakul, P.; Rukkwamsuk, T. Development of in-house ELISA based on recombinant gag proteins of small ruminant lentiviruses isolated from goats in Thailand. *Sci. Rep.* **2024**, *14*, 3636. [CrossRef]

53. Pietiainen, V.; Polso, M.; Migh, E.; Guckelsberger, C.; Harmati, M.; Diosdi, A.; Turunen, L.; Hassinen, A.; Potdar, S.; Koponen, A.; et al. Image-based and machine learning-guided multiplexed serology test for SARS-CoV-2. *Cell Rep. Methods* **2023**, *3*, 100565. [CrossRef]
54. Williams, D.M.; Hornsby, H.R.; Shehata, O.M.; Brown, R.; Gallis, M.; Meardon, N.; Newman, T.A.H.; Plowright, M.; Zafred, D.; Shun-Shion, A.S.M.; et al. Establishing SARS-CoV-2 membrane protein-specific antibodies as a valuable serological target via high-content microscopy. *iScience* **2023**, *26*, 107056. [CrossRef]
55. Focosi, D.; McConnell, S.; Casadevall, A.; Cappello, E.; Valdiserra, G.; Tuccori, M. Monoclonal antibody therapies against SARS-CoV-2. *Lancet Infect. Dis.* **2022**, *22*, e311–e326. [CrossRef] [PubMed]
56. Taylor, P.C.; Adams, A.C.; Hufford, M.M.; de la Torre, I.; Winthrop, K.; Gottlieb, R.L. Neutralizing monoclonal antibodies for treatment of COVID-19. *Nat. Rev. Immunol.* **2021**, *21*, 382–393. [CrossRef] [PubMed]

viruses

MDPI

Article

Discordant Antigenic Properties of Soluble and Virion SARS-CoV-2 Spike Proteins

Sameer Kumar [1], Souradip Dasgupta [1], Mohammad M. Sajadi [1,2,3], Greg A. Snyder [1,3], Anthony L. DeVico [1,3] and Krishanu Ray [1,4,*]

1 Division of Vaccine Research, Institute of Human Virology, University of Maryland School of Medicine, 725 West Lombard Street, Baltimore, MD 21201, USA
2 Division of Clinical Research, Institute of Human Virology, University of Maryland School of Medicine, 725 West Lombard Street, Baltimore, MD 21201, USA
3 Department of Medicine, University of Maryland School of Medicine, 725 West Lombard Street, Baltimore, MD 21201, USA
4 Department of Biochemistry and Molecular Biology, University of Maryland School of Medicine, 725 West Lombard Street, Baltimore, MD 21201, USA
* Correspondence: kray@som.umaryland.edu

Abstract: Efforts to develop vaccine and immunotherapeutic countermeasures against the COVID-19 pandemic focus on targeting the trimeric spike (S) proteins of SARS-CoV-2. Vaccines and therapeutic design strategies must impart the characteristics of virion S from historical and emerging variants onto practical constructs such as soluble, stabilized trimers. The virus spike is a heterotrimer of two subunits: S1, which includes the receptor binding domain (RBD) that binds the cell surface receptor ACE2, and S2, which mediates membrane fusion. Previous studies suggest that the antigenic, structural, and functional characteristics of virion S may differ from current soluble surrogates. For example, it was reported that certain anti-glycan, HIV-1 neutralizing monoclonal antibodies bind soluble SARS-CoV-2 S but do not neutralize SARS-CoV-2 virions. In this study, we used single-molecule fluorescence correlation spectroscopy (FCS) under physiologically relevant conditions to examine the reactivity of broadly neutralizing and non-neutralizing anti-S human monoclonal antibodies (mAbs) isolated in 2020. Binding efficiency was assessed by FCS with soluble S trimers, pseudoviruses and inactivated wild-type virions representing variants emerging from 2020 to date. Anti-glycan mAbs were tested and compared. We find that both anti-S specific and anti-glycan mAbs exhibit variable but efficient binding to a range of stabilized, soluble trimers. Across mAbs, the efficiencies of soluble S binding were positively correlated with reactivity against inactivated virions but not pseudoviruses. Binding efficiencies with pseudoviruses were generally lower than with soluble S or inactivated virions. Among neutralizing mAbs, potency did not correlate with binding efficiencies on any target. No neutralizing activity was detected with anti-glycan antibodies. Notably, the virion S released from membranes by detergent treatment gained more efficient reactivity with anti-glycan, HIV-neutralizing antibodies but lost reactivity with all anti-S mAbs. Collectively, the FCS binding data suggest that virion surfaces present appreciable amounts of both functional and nonfunctional trimers, with neutralizing anti-S favoring the former structures and non-neutralizing anti-glycan mAbs binding the latter. S released from solubilized virions represents a nonfunctional structure bound by anti-glycan mAbs, while engineered soluble trimers present a composite structure that is broadly reactive with both mAb types. The detection of disparate antigenicity and immunoreactivity profiles in engineered and virion-associated S highlight the value of single-virus analyses in designing future antiviral strategies against SARS-CoV-2.

Keywords: fluorescence correlation spectroscopy; SARS-CoV-2 spike; antigenicity; Covi-mAbs; anti-HIV glycan mAbs; in vitro binding assay

1. Introduction

The COVID-19 global pandemic has claimed at least 7 million lives worldwide, owing to the periodic emergence and circulation of new SARS-CoV-2 variants [1–3]. Newer variants seem more transmissible and partially resistant to early vaccines [4–7] and neutralizing antibodies (NAbs) [8–11] elicited by early strains. Although vaccines and drugs have reduced infections and deaths, it is highly likely that virus evolution and emergence will be a recurring, worldwide concern for the foreseeable future. Consequently, COVID-19 prevention and treatment must be an ongoing, evolving effort reliant on the ability to understand and counteract genotypic and phenotypic changes that alter SARS-CoV-2 infectivity, transmissibility, and immune escape.

SARS-CoV-2 envelope (Env) spike (S) trimers are critical targets for antiviral countermeasures. These structures mediate virus attachment to and fusion with host cells and are targets for neutralizing antibodies [12,13]. Evolutionary mutations constantly occurring in the human host population modulate both replication functions and immune recognition. On virion surfaces, S is a trimer of two interacting subunits [14], S1 and S2, that mediate attachment to host cells (e.g., human respiratory epithelial cells) via the angiotensin-converting enzyme 2 (ACE2) receptor [15]. S1 contains the N-terminal domain (NTD), the receptor binding domain (RBD), and S1–S2 furin cleavage site. S2 contains an S2 protease site that is cleaved (e.g., by TMPRSS2 or cathepsins) to release its fusion peptide. As with all coronaviruses, the receptor-liganded SARS-CoV-2 trimer transitions through distinct conformational intermediates that "open" the trimer apex and initiate fusogenic activity [16–19]. Unlike the unliganded S proteins of seasonal human coronaviruses, which are constitutively organized as "closed" trimeric structures [20,21], SARS-CoV-2 trimers variably toggle between closed (all RBDs down toward the virion surface) and "open" (one or more RBDs "up") conformations [17,18,22–24] depending on strain. When the S1 RBD of each S protomer flips to an "up" orientation (pointed away from the virion), the ACE2 receptor is captured [25]. The trimers then transition through additional conformational intermediates that fix the trimer apex in an "open" state with S protomers "up"; shed S1; and place the exposed S2 fusion domain toward the host cell membrane [16–19,26].

Several lines of evidence associate the emergence and dominance of new SARS-CoV-2 variants with certain mutations in S [27]. Such relationships have been observed since the first stage of the pandemic, when emerging variants with an S1 D614G mutation appeared to have greater transmissibility [28,29]. Subsequently, a host of mutations have been identified with the potential to influence attachment, membrane fusion, and entry functions [30]; confer replication/fitness advantages [31–33]; and/or evade immune responses [28,29,34]. The Alpha (B.1.1.7), Beta (B.1.351), and Gamma (P.1) variants express an N501Y mutation in RBD, which could enhance ACE2 binding. The B.1.351 and P.1 strains also include a E484K S mutation that reduces sensitivity to NAbs elicited by earlier variants [35]. Delta (B.1.617.2) variants harbor L452R and P681R mutations, in RBD and near the furin cleavage site, respectively, which also reduce sensitivity to antibody neutralization versus previous variants [31,36,37]. Compared to ancestral (Wuhan) variants, the more recent Omicron BA.1 [38,39] expresses 32 unique amino acid positions in S [40,41]. Notably, the neutralizing activity of serum antibodies from SARS-CoV-2 vaccinees to the ancestral (Wuhan) strain is lower against Omicron subvariants compared to the Wuhan variant [42,43]. As of 2023 [44]; the XBB.1.5 descendant of the BA.2 lineage expresses an F486P S mutation that enhances attachment to ACE2 [45,46]. In view of these findings, it is apparent that S evolution in the human population will be a key barometer for understanding the future course of the COVID-19 pandemic.

The structure–function relationships of SARS-CoV-2 trimer sequences have been extensively studied using indirect in vitro analyses (e.g., infectivity or neutralization) or the ensemble imaging/computational modeling of static structures [17–19,23,24,35,47,48] such as stabilized soluble trimers. Structural analyses of engineered soluble S trimers have been important for understanding how sequence changes alter protein–protein interactions [35,49]. However, such approaches may fail to fully capture the antigenic properties of natural S trimers on virion surfaces. Coarse-grain model simulations of interactions between membrane-bound ACE2 dimers and virion S trimers predicted the disparate structural and functional behaviors of membrane-bound versus soluble S trimers [50]. Further, Förster resonance energy transfer (FRET) imaging of particles pseudotyped with sequence-modified (for fluorophore attachment) SARS-CoV-2 trimers (from a 2020 variant) detected a dynamic structure, exhibiting constitutive fluctuations between four conformational states ranging from all RBD up versus all RBD down protomer positions. The favored state changes as receptor liganding ensues [51]. Caveats of this approach include the need for sequence modifications of S (for placing FRET dye pairs); the presentation of one or a few such modified trimers on a virion; and uncertainties that the modified trimer represents the disposition of all spikes across the entire virion. Nevertheless, evidence of structural dynamics must be considered in the context of virus–NAb or monoclonal antibody (mAb) interactions and immune evasion. Differences between soluble and virion-associated S might explain why certain anti-HIV NAbs are broadly reactive with glycan patches on both HIV gp120 and SARS-CoV-2 S, but neutralize only the former [52,53] and not the latter [52,53]. Such dichotomies warrant further investigation.

Here, we explore, for the first time, the antigenicity of SARS-CoV-2 S and virions in solution, using physiologically relevant mAb concentrations. Our analyses are enabled by fluorescence correlation spectroscopy (FCS), which we successfully applied in order to define the antigenicity of HIV-1 virions with respect to neutralizing and non-neutralizing antibody interactions [54–56]. This approach allows direct analyses of unadulterated virus surface proteins at the level of single particles. Immunoreactivity was explored using a panel of non-neutralizing and broadly neutralizing anti-S mAbs, including ones (designated "Covi-series") we isolated from an infected individual in year 2020 of the pandemic. The epitope targets for these mAbs have been reported [49,57]. These antibodies allowed us to comprehensively track changes in S antigenicity from 2020 forward. A collection of cross-reactive anti-glycan mAbs [58] which neutralize HIV and bind SARS-CoV-2 virions [59,60] were also examined for comparison. Collectively, our data reflect the evolving binding efficiencies of early anti-S mAbs with SARS-CoV-2 variants over time. Importantly, data for individual strains also indicate the antigenic, structural, and/or functional heterogeneity of virion-associated S, which mediates the binding of neutralizing and non-neutralizing mAbs. This heterogeneity is only partially captured by matching soluble S trimers or S-pseudotyped particles.

2. Materials and Methods

Production and labeling of anti-SARS-CoV-2 spike mAbs. All anti-spike mAbs labeled "Covi" were isolated from donor ID3, who was infected in South Asia in February 2020. The method of isolation, as well as individual mAbs Covi-9, Covi-11, and Covi-17, have been described before [49,57]. In previous publications, Covi-9, Covi-11, and Covi-17 were referred to as CoVIC-78, CoVIC-79, and CoVIC-96, respectively [49], and Covi-11 as Cov11 [57]. Covi-10, Covi-21, and Covi-24 have not been previously reported. CR3022 was obtained from BEI Resources, and 2G12 (#AB002) was purchased from Polymun scientific. PGT121 (ARP12343) and PGT126 (ARP12344) were obtained from HIV reagent program contributed by International AIDS Vaccine Initiative. Fragment of antigen binding (Fab) of 2G12 was generated and purified as described [55]. The nonspecific human IgG1 used as negative control was purchased from Calbiochem, San Diego, CA, USA. All mAbs were fluorescently labeled using Alexa Fluor 647 dye (A647) and purified as described [54]. Briefly, succinimidyl ester moiety of A647 dye reacts with primary amines of the antibody in slightly alkaline condition to form a stable amide bond, producing the dye labeled antibody. The reaction was incubated for 2 h at room temperature. The labeled antibody was purified from free dye by centrifuging in spin column at 1100× *g* for 5 min. The labeled antibodies were characterized by a UV–visible spectrometer (Nanodrop 2000, ThermoFisher Scientific, Wilmington, DE, USA).

SARS-CoV-2 recombinant spike proteins and virions. The following SARS-CoV-2 reagents were obtained through BEI Resources, NIAID, NIH; gamma-irradiated SARS-CoV-2 viruses of ancestral (Wuhan) variant (USA/MD-HP01542/2021) and Lineage B.1.617.2; Delta variant (hCoV-19/USA/MD-HP05285/2021) and full-length SARS-CoV-2 spike protein variants; Alpha (B.1.1.7), Beta (B.1.351), Gamma (P.1), Delta (B.1.617.2), and Omicron (B.1.1.529, BA.2, BA.2.75, BQ.1.1, and BA.4); spike trimer (S1+S2) his-tag (D614G); SARS-CoV-2 perfusion spike ectodomain Hexapro; and recombinant SARS-CoV-2 soluble stabilized spike (Wuhan). The recombinant Wuhan spike protein lacks the signal sequence and contains 1196 residues (ectodomain) of the SARS-CoV-2 spike glycoprotein. It is also modified to remove the polybasic S1/S2 cleavage site (RRAR to A; residues 682 to 685), stabilized with a pair of mutations (K986P and V987P), and includes a thrombin cleavage site, T4 fold on trimerization domain, and C-terminal hexa-histidine tag. Both Wuhan isolate USA/MD-HP01542/2021 and delta isolate hCoV-19/USA/MD-HP05285/2021 virions were gamma-irradiated (5×10^6 RADs) on dry ice, followed by sonication.

RBD sequences were aligned using Clustal Omega [61]. RBD mutations were mapped onto a surface representation of PDB 7A94 using ChimeraX [62].

Cell culture and production of pseudoviruses. HEK293T and TZM-bl Hela cells were obtained from the American Type Culture Collection (ATCC) and maintained in complete Dulbecco's modified Eagle's medium (DMEM) without phenol red. HEK293T cells with stable expressing ACE2 receptor (HEK293T-ACE2) was maintained in complete DMEM. To produce the SARS-CoV-2 pseudovirus, the envelope plasmid (PSF361.2) was constructed first using the spike gene from 2019NCoV S gene (GenScript, Piscataway, NJ, USA) and cloned into pCAGGS expression vector (GenScript, USA). The primers used are as follows: sf1147, CTCTGAATTCGCCATGCCACCATGTTCGTCTTCC; and sf1148, CTCTACC-CGGGATCCGATTTAGGTGTAATGC. Furthermore, for better expression, an additional Kozak sequence was added to the N-terminal end. SARS-CoV-2 pseudovirus was generated by co-transfecting the envelope-deficient HIV-1 backbone plasmid, pNL4-3-ΔE-luc (obtained through AIDS Research and Reference Reagent Program, NIAID, NIH, Bethesda, MD, USA), and SARS-CoV-2 spike plasmid (PSF361.2) in HEK293T cells using the FuGENE (Promega, USA) transfection reagents. Cell supernatant was isolated and concentrated using PEG-it (System Biosciences, USA). Quantification of p24 concentration was performed using ELISA kit from R&D system. Control particles lacking envelope (delE) were produced by transfection with pNL4-3-ΔE-EGFP backbone alone as reported earlier [56].

Fluorescence correlation spectroscopy (FCS). FCS has been proven to be a valuable method for evaluating virion–mAb interactions at the single-molecule level for several years [54–56,63–65]. Briefly, a 15 μL reaction containing 3 μg/mL p24 equivalent of SARS-CoV-2 pseudovirus was treated with nonspecific IgG1 (100 μg/mL) for 90 min at 37 °C to block the nonspecific interaction. Subsequently, 1 nM Alexa Fluor 647-conjugated mAb (A647–mAb) was added to the reaction mixture and incubated for 90 min at 37 °C to form immune complexes. Approximately 11 μL of the reaction mixture was loaded onto a glass coverslip sample chamber and sealed before FCS measurements were performed. A 3 μg/mL p24 level of pseudovirus contained approximately 3×10^{10} virions/mL. SARS-CoV-2 gamma-irradiated virions (BEI) contained approximately 1×10^9 genome copies/mL, which was further concentrated by ultracentrifugation to 1×10^{10} genome copies/mL. FCS reaction conditions rely on limiting protein concentrations that tend to be more physiologically relevant than other reaction systems. Our main goal in configuring reaction conditions was to achieve optimal sensitivity to detect differences in binding efficiency (% antibody bound to antigen) among mAbs. Such conditions occur with limiting amounts of mAb. The reactions we employed (ex: SARS-CoV-2 Wuhan variant) reflect a condition of 3×10^{10} particles/mL (pseudovirus) or 1×10^{10} particles/mL (inactivated virus) of fluid. Assuming 30 spikes per virion, the overall spike concentration should be ~1.5 nM. Accordingly, we used roughly equimolar (1 nM) concentrations of mAb in the reactions. In the case of purified recombinant SARS-CoV-2 spike protein, a reaction mixture containing 10 nM of spike protein and 5 nM of A647–mAbs was incubated for 30 min at room temperature to form the spike–mAb complexes. As above, 11 μL of this reaction mixture was used for the experiments each time. Different stoichiometric ratios of mAbs to spike protein were used to determine the optimal condition for binding in solution. A molar ratio of 1:2 of mAb to spike protein was selected as it exhibited the maximum percent of binding. FCS measurements were performed in a customized ISS Q2 confocal microscope. The excitation source was Fianium SC-400 super-continuum laser. NKT super-select AOTF filter was used to select the excitation wavelength of 635 nm, which was reflected by a dichroic mirror to an Olympus high-numerical-aperture (NA) water objective (60×; NA 1.2) and focused onto the solution sample. The fluorescence was collected by avalanche photodiodes through a dichroic beam splitter and a band-pass (650–720 nm; Chroma) filter, thus eliminating the scattered excitation light and collecting the fluorescence from the A647-labeled probes in the spectral region of interest. The data acquisition was enabled by a B&H SPC-150 card operated in a photon time-tag time-resolved (TTTR) mode. The recorded data were analyzed using the ISS VistaVision software V4.2 to assess the in vitro binding of A647-labeled mAbs to recombinant spike proteins or SARS-CoV-2 virions. The autocorrelation plots were generated by the VistaVision software and fitted with one-species or two-species 3D-Gaussian diffusion model. The fitting processes yielded the translational diffusion coefficients of A647-labeled mAbs, A647-labeled mAb-bound spike proteins of different variants, and mAb-bound virions. We also determined the fractions of free mAbs, mAbs-bound to SARS-CoV-2 recombinant spike protein, or virions in the reaction mixture.

Neutralization assay. The neutralization capability of the Covi-mAbs were assayed using the ACE2 expressing HEK293T cells (HEK293T-ACE2). The SARS-CoV-2 pseudovirus has an HIV-1 backbone containing luciferase gene which, upon infection, can be measured as relative luminescence units (RLUs). For the HIV-1 bnAbs, the neutralizing activity was determined using the TZM-bl target cells [66], which measure Tat-driven luciferase expression after single round of infection. Briefly, 50 μL of SARS-CoV-2 pseudovirus of a $TCID_{50}$ value 20,000 was incubated with 50 μL of 2–7-fold serially diluted test antibodies (starting at 10–50 μg/mL) for 1 h at 37 °C in a CO_2 incubator. Next, the HEK293T-ACE2 cells (10,000 cells/well) were added to the virus–mAb solution and incubated at 37 °C in 5% CO_2 incubator. A cell-only control without the virus and a virus-only control without the antibody were included in this assay. After 72 h, Steady-Glo reagent (Promega, Madison, WI, USA) was added and RLU was measured using Victor 3 plate reader (PerkinElmer, Hopkinton, MA, USA) and the percentage of neutralization was calculated.

Statistical analyses. Using the GraphPad prism software (Version 10), data were analyzed using either Mann–Whitney's T-test or the analysis of variance (ANOVA). The statistical significance was defined as $p < 0.05$.

3. Results

3.1. Variable Solution Binding of Anti-S mAbs to Recombinant Soluble S from Different SARS-CoV-2 Strains

MAb interactions with soluble S were determined by FCS, adopting procedures we previously used to study the antigenicity of HIV Env. FCS is a sensitive analytical technique that measures fluctuations in the signal intensity of a fluorophore as it passes in and out of a defined focal volume. Such signal fluctuations reflect the fluorophore diffusion kinetic (coefficient), which is a function of mass. Thus, when a fluorophore-labeled molecule binds to a larger target, its diffusion coefficient decreases, owing to the larger mass of the resulting complex. The fraction of labeled molecules assuming a slower diffusion coefficient quantitatively reflects the target binding efficiency. As diffusion coefficients are proportional to mass, FCS measures also validate that fluorescent signals emanate from the expected reaction complexes. For example, a 150 kD free IgG molecule displays a diffusion coefficient of ~55 $\mu m^2/s$; one bound to a 140 kD soluble S protein should diffuse at ~35 $\mu m^2/s$.

To produce a perspective of antigenic changes in variants arising over time, we tested mAbs (e.g., "Covi" series) derived from an individual infected with the earliest known cases of SARS-CoV-2. The panel (Table 1) included neutralizing mAbs Covi-9, Covi-10, Covi-11, Covi-17, and Covi-24, as well as non-neutralizing mAbs such as CR3022 and Covi-21. CR3022 binds to the relatively conserved conformation of the spike protein (amino acids 318–510) in the S1 domain of the SARS-CoV, as well as SARS-CoV-2 strains [67,68]. Members of the panel have been characterized structurally [49,57].

We first tested the mAbs (conjugated to A647 dye) for reactivity with S from the ancestral (Wuhan) variant discovered in Wuhan, China in 2019 [69]. Sequence analysis of the Wuhan variant showed a distinctive polybasic amino acid incorporation in the spike protein that appeared to make it more virulent than other related coronaviruses [70]. MAb CR3022 showed a clear shift in the autocorrelation curve, which fit a two-species diffusion model in which one species exhibited the high diffusion coefficient (55 $\mu m^2/s$) of free mAb and the other showed a lower value of approximately 35 $\mu m^2/s$ corresponding to the mAb–S complex (Figure 1a). Applying the FCS-fitting routine (see Methods), we calculated the fractional component of the 35 $\mu m^2/s$ species, defining the percentage of the mAb-bound spike in the reaction mixture, i.e., binding efficiency. CR3022 showed a high binding efficiency (approximately 80% bound to S). An irrelevant IgG1 negative control demonstrated only a one-species diffusion rate at 55 $\mu m^2/s$ (Figure 1b), indicating no S binding.

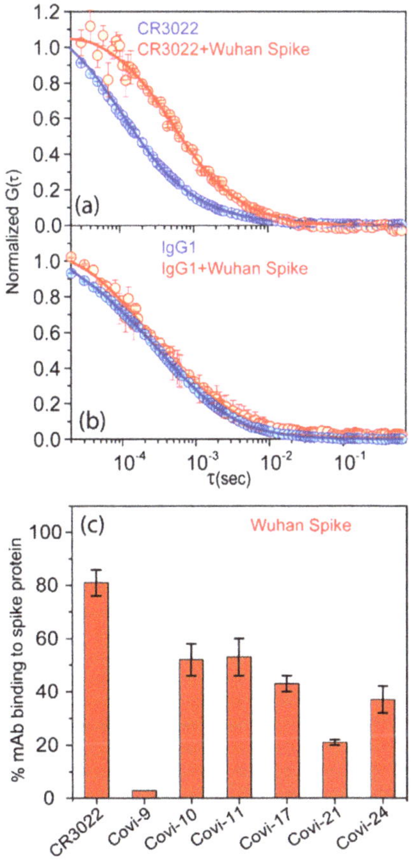

Figure 1. A647-dye-labeled anti-spike mAbs exhibit variable binding efficiencies in reactions with the SARS-CoV-2 spike protein of the Wuhan variant. Representative FCS autocorrelation plots used

to calculate binding efficiencies (see text) are shown in panels (**a**,**b**). Antibodies were analyzed with or without exposure to recombinant spike protein. (**a**) mAb CR3022; broadly reactive positive control. (**b**) Non-specific human IgG1; negative control. (**c**) FCS autocorrelation plots such as in panels (**a**,**b**) were used to determine the binding efficiencies (% antibody bound) of various labeled mAbs with the SARS-CoV-2 spike protein. All experiments were performed three times. Data are presented as the mean of three experiments ± sem.

Table 1. mAbs selected for the current study.

Antibody	Specificity
CR3022	anti-spike S1 (RBD 7)—non-neutralizing
Covi-9	anti-RBD—neutralizing (RBD 5)
Covi-10	anti-RBD—neutralizing
Covi-11	anti-RBD—neutralizing (RBD 2)
Covi-17	anti-spike—neutralizing (RBD 5)
Covi-21	non-neutralizing
Covi-24	anti-spike—neutralizing
2G12	anti-HIV Env, targets glycan patch
PGT-121	anti-HIV Env, targets V3 glycan
PGT-126	anti-HIV Env, targets V3 glycan
Non-specific IgG1	negative control

When analyzed alone, all other test mAbs displayed auto-correlation plots with single-diffusion coefficients of ~55 $\mu m^2/s$ (Supplemental Figure S1). In the presence of soluble Wuhan S, autocorrelation curves consistently fit the two-species diffusion model (one species at 55 $\mu m^2/s$, the other at 35 $\mu m^2/s$) with no other species present. The Covi-series mAbs exhibited a range of binding efficiencies to Wuhan S, ranging from 2% (Covi-9) to 53% (Covi-11) (Figure 1c). Structural studies indicated that Covi-11 Fab binds to the SARS-CoV-2 RBD in a unique orientation, possibly contributing to its neutralization potency. As described earlier, Covi-11 employs a heavy chain VH 3–53 and a light chain VK 3–20 germline sequence towards attaching to the RBD. Most residues of the Covi-11 epitope footprint match with the ACE2-binding site on the RBD [57]. Covi-17 also showed a higher binding efficiency of 43%. Notably, this mAb (coded as CoVIC-096 in [49]) was evaluated by the Coronavirus Immunotherapy Consortium (CoVIC) [71] and determined to have particularly broad and potent neutralizing activity. Negative-stain electron microscopy (NS-EM) characterization suggested that mAb Covi-17 crosslinks two recombinant S proteins [49], possibly explaining the neutralization potency. However, in our reactions, we did not observe species with diffusion coefficients of ~20 $\mu m^2/sec$, indicative of complexes formed by two trimers and two mAbs [49]. The discordance between the NS-EM and FCS data might be explained by differences in reaction conditions. The FCS conditions comprised concentrations of reactants (0.15 µg/mL) mixed in solution at 1:2 molar ratios, whereas, for the NS-EM experiments [49], the complex was formed by incubating the spike and antibody at the same concentrations of 0.25 µg/µL which was equal to a concentration of ~1.6 µM. The FCS experiments were performed at a 1 nM (0.15 µg/mL) concentration of mAbs, which is a ~1600-fold less concentration than the NS-EM experiments. FCS is a sensitive single-molecule-based microscopic technique in which the correlation curve is generated from the fluctuations in fluorescence intensity within the confocal volume in a certain time frame. As FCS is a single (or few)-molecule-based detection technique, based on observations in femtoliter confocal volumes, it cannot be performed at protein concentrations approaching 1 µM. FCS is most sensitive and best performed in the limiting concentration range of 1 to 10 nM. Should a large, cross-linked

S protein complex form, FCS would detect it, provided such an object produces a few percent of the total signal events. As we do not observe such a species, we conclude that crosslinking is a phenomenon that mainly occurs in reactions with high concentrations of antigen and antibody. We do not exclude that these interactions are possible.

SARS-CoV-2 variants that emerged post-(Alpha, Beta, Gamma, Delta, and Omicron) acquired S mutations that alter neutralizing epitopes and/or modulate the constitutive structure of the trimer as an "open" or "closed" structure [72–74]. The cross-reactive mAb CR3022 demonstrated substantial immunoreactivity with all these variants (40 to 80% in S-bound form), whereas the negative control IgG1 showed negligible to no evidence of S binding (Figure 2).

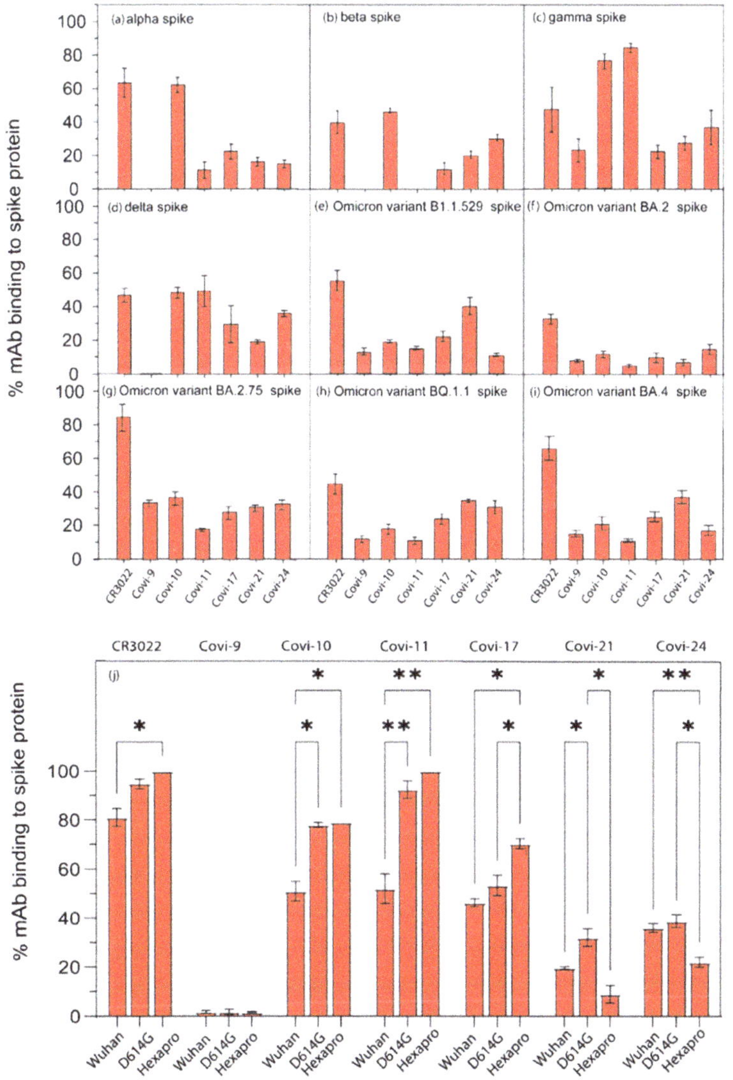

Figure 2. Anti-spike mAbs exhibit variable reactivity with diverse SARS-CoV-2 spike proteins. FCS was used to analyze mAb-binding efficiencies in reactions with the spike proteins of (**a**) Alpha, (**b**) Beta,

(**c**) Gamma, (**d**) Delta, (**e**) Omicron (Lineage B.1.1.529), (**f**) Omicron (Lineage BA.2), (**g**) Omicron (Lineage BA.2.75), (**h**) Omicron (Lineage BQ.1.1), and (**i**) Omicron (Lineage BA.4). (**j**) mAb-binding efficiencies in reactions with the spike proteins of Wuhan, D614G mutant, and Hexapro mutant were compared. Data for mAb-binding efficiencies in reactions with the spike proteins of Wuhan variant from Figure 1c was repeated for comparison. Non-specific IgG1 and CR3022 mAb were used as negative and positive controls, respectively. Data are presented as the mean of triplicate measurements ± SEM. Each experiment was performed three times with similar results. * and ** represent p values of <0.05 and <0.01 respectively.

The Alpha variants, first identified in the United Kingdom in 2020 [75], contain an N501Y mutation in S. The Alpha variant tested here showed efficient binding to Covi-10 (62% of the antibody bound to S) but low or no reactivity with any other mAb (Figure 2a). The S protein of Beta variants, first identified in South Africa in 2020, contain a E484K mutation that enables escape from convalescent antisera [75,76]. In this case, only Covi-10 showed substantial immunoreactivity (42% mAb-bound; Figure 2b). Notably, the S trimer of the Gamma variant (Lineage P.1), identified in Brazil in 2020, exhibited broader and more efficient immunoreactivity ranging from 84% (Covi-11) to 23% (Covi-17) mAb-bound (Figure 2c). The S of the Delta variant, first identified in India in 2020, presents several mutations including L452R that reduces sensitivity to neutralizing antibodies elicited by previous strains [75]. Although the Covi mAbs were derived from a subject infected early in the pandemic, four of them (Covi-11, Covi-10, Covi-17, and Covi-24) exhibited binding efficiencies of $\geq$ 30% mAb attached to S (Figure 2d). The Covi-series mAbs displayed a range of binding efficiencies (15–40 percent complexed to S) across Omicron subvariants (B.1.1.529, BA.2, BA.2.75, BQ.1.1, and BA.4), depending on the reaction mixture (Figure 2e–i).

We further explored mAb binding to S variants encoding sequence changes that impact structural behavior. The D614G mutation near the S1/S2 boundary of SARS-CoV-2 was reported to allow more open S trimer structures recognized by multiple neutralizing antibodies [29]. Compared to Wuhan S, the D614G mutant was indeed more reactive with Covi-10, 11, 17, 21, and 24, whereas Covi-9 was not bound (Figure 2j). The S1 subunit of the SARS-CoV-2 spike protein is prone to proteases, which results in degradation [77]. The Hexapro spike, based on a soluble Wuhan S, contained six proline substitutions (F817P, A892P, A899P, A942P, K986P, and V987P) that counteract such thermal instability and degradation [78]. Compared to an unmodified soluble Wuhan S, Covi-10, -11, and -17 showed more efficient binding to the Hexapro spike. Nearly all Covi-11 was detected in S-bound form (Figure 2j). In contrast, Covi-9 and 21 showed very weak binding to Hexapro.

The occurrence of mutations in the RBD domain is a conventional explanation for cross-variant differences in mAb-binding efficiencies. This possibility was explored with anti-RBD mAbs Covi-10, -11, and -17, for which the defined epitope and/or structural information is available [49,57]. Figure 3 shows RBD sequences for the listed test variants as they emerged chronologically (Figure 3a), with residue changes mapped onto the respective inner, outer, and top RBD epitope surfaces (Figure 3b) as previously defined by Callaway et al. [79]. The generally lower binding efficiencies with Omicron versus earlier variants is consistent with extensive changes in RBD residue positions, many of which are predicted to establish mAb–RBD contacts [49,57,74,75,79]. In earlier variants (Alpha and Beta), the negligible Covi-11- and Covi-17-binding efficiencies (versus Wuhan) paired with the appearance of mutations at only a single RBD position (417), a predicted mAb contact site. Binding efficiency was higher when mutations occurred elsewhere in the RBD, for example, with Covi-11 and the Beta versus Gamma or Delta strains.

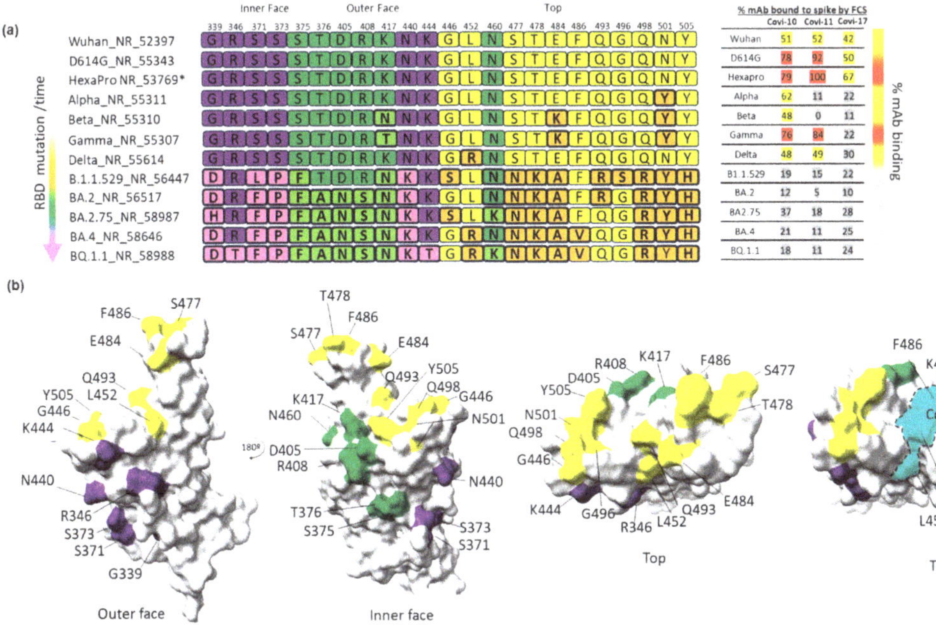

Figure 3. Sequence alignment of variants of concern (VOC) RBD's mutations and percent binding of mAbs (Covi-10, 11, and 17) to soluble spike protein of different variants. (**a**) Sequence alignment of RBD mutations in VOCs (**left**) compared with the percent mAb binding as measured in solution by FCS (**right**). * Hexapro is a vaccine reagent and is not a VOC. Shown are RBD mutations found on the outer face (purple), inner face (green), and top surface (yellow). VOC RBD mutations deviating from consensus are shown as light purple, light green, and orange. The mAb-binding efficiencies in reactions with the spike proteins of different variants as determined by FCS are color-coded as follows: 75–100% (red), 42–74% (yellow), 0–41% (grey–white). (**b**) Surface representation of RBD mutations. VOC RBD mutations are mapped onto model (PDB:7A94) and colored corresponding to outer (purple), inner (green), and top (yellow) RBD surface views (as previously defined by Callaway et al. [79]). Covi-11-binding epitope [57] on RBD is outlined and shown in cyan (**bottom right**).

3.2. Variable Solution Binding of Anti-Spike mAbs to SARS-CoV-2 Virions

The above experiments demonstrating variable binding to soluble S trimers prompted us to examine whether similar patterns occur with virus-membrane-associated S. Since FCS is not amenable to the biocontainment analyses of live SARS-CoV-2, the analyses comprised HIV virion-based particles pseudotyped with S trimers, as well as gamma-irradiated (5×10^6 RADs) wild-type virions which retain intact surface structures. We tested two strains, Wuhan (hCoV-19/USA/MD-HP05285/2021) and Delta (hCoV-19/USA/MD-HP05285/2021), for which soluble S trimers, and inactivated viruses with identical RBD sequences are available. In virion reactions, any bound mAb should demonstrate a "slow" diffusion coefficient of ~6 $\mu m^2/s$, corresponding to the ~100 nm size of SARS-CoV-2 virion targets. In accordance with expectations, all virus–mAb reactions produced autocorrelation curves fitting a two-species model; free mAb (55 $\mu m^2/s$) and virion-bound mAb (6 $\mu m^2/s$). As shown in Figure 4a,b, the mAbs again showed a range of binding efficiencies, with both Wuhan pseudoviruses and matched irradiated wild-type virions. In general, the binding efficiencies (percentages of mAbs bound) with virions was lower than what was measured with soluble S (Figure 1). This difference may be explained by higher molar amounts of the target in reactions with the latter. However, there was only a nonsignificant trend toward a direct relationship between mAb-binding efficiencies (percentages of mAbs bound) with the Wuhan pseudovirus versus soluble S trimers (Figure 4c). In comparison, there was a positive and statistically significant ($p = 0.03$) direct relationship between mAb-binding efficiencies with the inactivated virus versus soluble S (Figure 4d). Moreover, there was a significant direct relationship between mAb-binding efficiencies with pseudoviruses versus inactivated wild-type viruses ($p = 0.01$), although the percent mAb-bound was consistently higher with the latter target. Similar results were obtained with the Delta variant targets (Supplemental Figure S2). As shown in Figure S2b, as with the Wuhan variant, there was a statistically significant direct relationship between mAb-binding efficiencies with the inactivated virus versus soluble S ($p = 0.01$) for Delta variants. In control experiments (Supplemental Figure S3), we verified that the test mAbs had negligible to undetectable non-specific interactions with virions by testing delE pseudoviruses, which were produced without an envelope expression construct [56] (see Methods). When mixed with these particles, all mAbs fit only a single-species diffusion model with the signature diffusion coefficient (55 $\mu m^2/s$) of free antibody.

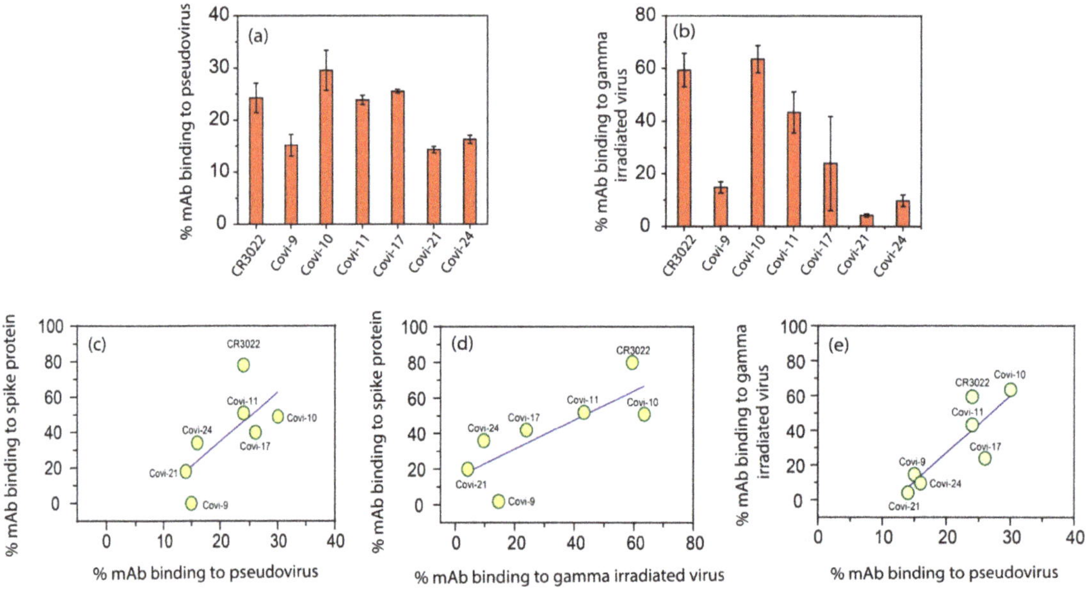

Figure 4. Binding of anti-spike mAbs to SARS-CoV-2 varies between pseudovirions versus gamma-irradiated inactivated virions of Wuhan variant. Autocorrelation plots were used to determine the binding efficiencies (% antibody bound) of a panel of A647-labeled anti-spike mAbs reacted with (**a**) SARS-CoV-2 pseudovirions or (**b**) gamma-irradiated inactivated SARS-CoV-2 virions. Relationships between mAb-binding efficiencies with recombinant Wuhan SARS-CoV-2 spike trimers vs. Wuhan (**c**) pseudovirions ($p = 0.08$ (ns), $R^2 = 0.48$) or (**d**) gamma-irradiated virion ($p = 0.03$ (*), $R^2 = 0.62$). (**e**) Correlation between mAb-binding efficiencies with pseudovirion vs. gamma-irradiated virion ($p = 0.01$ (*), $R^2 = 0.73$). All experiments were performed three times. Data are presented as the mean of three experiments ± SEM.

3.3. Neutralizing Capacity of Covi-mAbs against SARS-CoV-2 Pseudoviruses and Correlation of Covi-mAbs Binding Profiles by FCS

To examine relationships between virion binding and neutralization, the panel of anti-S mAbs were tested in infectivity assays (see Methods) with the Wuhan strain pseudoviruses and HEK293T-ACE2 target cells. As shown in Figure 5a, the mAbs showed a range of neutralization potencies. Covi-17 was the most potent, with an IC50 of 0.01 μg/mL. In agreement with previous reports [80], mAb CR3022 was not neutralizing. We then compared percent virion (Figure 5b) or S binding (Figure 5c) detected by FCS versus percent neutralization at the mAb concentration (0.15 μg/mL) used in the fluorescent binding reaction.

Although trends were apparent, there were no statistically significant relationships between binding efficiencies and neutralization potency regardless of whether non-neutralizing mAbs CR3022 and mAb Covi-21 were included in the data (Figure 5b, Supplemental Figure S4). Similarly, there was no significant relationship between neutralization potency and inactivated virion binding (Supplemental Figure S4). For the neutralizing mAbs, we detected IC50 values of 0.01 to 0.21 µg/mL, which equate with ~0.1 to 1.4 nM, which is within the optimum sensitivity range of FCS. Accordingly, FCS binding experiments were performed (Supplementary Figure S5) where mAb concentrations were matched with the IC50 value of the neutralizing antibodies. As in the previous analyses, there were no statistically significant relationships between binding efficiencies and neutralization potency regardless of whether non-neutralizing mAbs CR3022 and Covi-21 were included in the data. The inclusion of mAbs at high molar excess over target would result in a lower % of total antibody in the bound form across the board. This is shown in Supplementary Figure S6, which compares 0.1, 1, and 5 nM of antibodies. Also shown is that, as expected, a low antibody concentration versus target (0.1 nM) reduces signal detection but does not significantly decrease the % antibody bound, i.e., reaction efficiency.

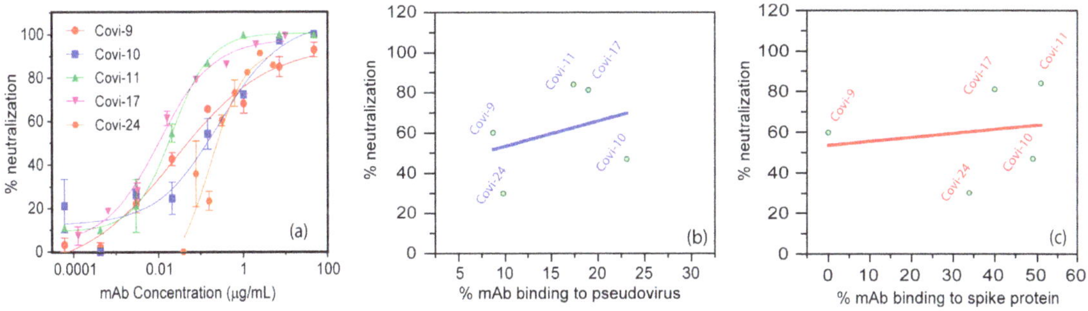

Figure 5. Relationships between mAb-binding efficiency and in vitro neutralization activity against SARS-CoV-2 pseudovirus of Wuhan variant. (**a**) Dose–effect neutralization curves of Covi-series mAbs in assays with the SARS-CoV-2 pseudovirus of Wuhan variant and HEK293T-ACE2 cells (see Methods). Data are presented as the mean of triplicate measurements ± SEM. Relationships between mAb-binding efficiency measured with pseudoviruses (**b**) or recombinant spike protein (**c**) were compared with their neutralization effect at 0.15 µg/mL. Each dot in (**b**) or (**c**) represents a single mAb.

3.4. Discordant Binding of Cross-Reactive Anti-Glycan Antibodies to SARS-CoV-2 S Proteins

Both HIV-1 and SARS-CoV-2 Env proteins present glycosylated domains and epitopes [81,82]. Antibodies against such epitopes on HIV-1 Env (e.g., 2G12, PGT121, and PGT126) are broadly HIV-neutralizing [83], accordingly termed bnAbs. It has been reported that anti-glycan bnAbs bind soluble S but fail to neutralize SARS-CoV-2 [52,53,84]. In agreement, we failed to detect SARS-CoV-2 neutralization with the antibodies even when tested at up to 10 µg/mL. To elucidate the nature of this dichotomy, FCS was performed using bnAbs 2G12, PGT121, and PGT126 in reactions with S proteins or SARS-CoV-2 pseudoviruses. Similar to the anti-S Covi mAbs, binding efficiencies varied according to virus strain and antibody combinations (Figure 6a). However, in some cases, the bnAbs bound soluble S with efficiencies equal or better than mAb CR3022 (e.g., 2G12, PGT 121, and PGT 126 with Omicron S). Conversely, binding efficiencies with Hexapro were overall lower with bnAbs versus mAb CR3022. To probe whether steric constraints impacted anti-glycan bnAb binding to soluble trimers, the binding efficiency of a smaller Fab fragment of 2G12 was compared with whole IgG. Higher binding efficiencies of Fab vs. whole IgG were apparent and statistically significant for reactions with Alpha ($p < 0.001$), and Gamma ($p < 0.01$) S proteins but not with other S variants (Figure 6a). The Wuhan strain pseudovirus was used as a test case to examine anti-glycan bnAb binding to virions. All bnAbs showed low binding efficiencies (Supplemental Figure S7), equivalent to those of Covi-neutralizing and non-neutralizing mAbs and CR3022. Notably, 2G12 Fab was not significantly more reactive with virions compared to whole IgG, suggesting more extreme steric constraints at the virion surface than occurs with soluble trimers.

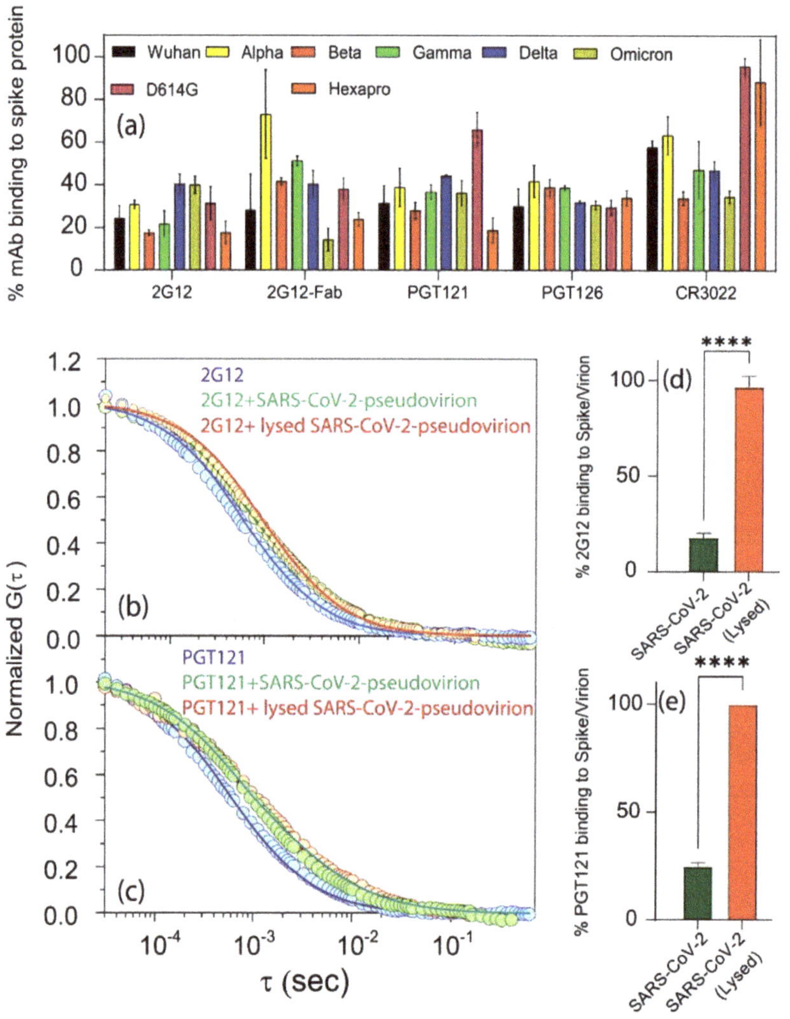

Figure 6. SARS-CoV-2 reactivity with anti-HIV-1 glycan mAbs varies according to the strain and context of spike proteins. In all experiments, CR3022 is included as a positive control and nonspecific human IgG1 as negative control. (**a**) Comparative binding efficiencies of A647-labeled mAbs 2G12, PGT121, PGT126, and Fab 2G12 with soluble spike proteins. mAb 2G12 and PGT121 reactivity was further tested against spike proteins extant on whole pseudovirions or released from particles by detergent lysis. Diffusion coefficients were used to determine the spike protein disposition (virion-bound or released into solution; see text) reactive with the mAb. Fluorescence autocorrelation plots of (**b**) mAb 2G12; or (**c**) mAb PGT121 under the indicated reaction conditions. Autocorrelation plots and diffusion coefficients of signals were used to determine (see text) the binding efficiencies of mAb 2G12 (**d**) or PGT121 (**e**). All experiments were performed three times. Data are presented as the mean of three experiments ± SEM. **** represents a p value of <0.0001.

We next examined whether anti-glycan bnAb binding to virions is limited by the nature of the S trimer presentation in the virus membrane. Accordingly, we used anti-glycan bnAbs 2G12 and PGT121 to probe the Wuhan pseudovirions, treated or untreated by detergent to release free S trimers (see Methods). MAb CR3022 was tested as a control. As shown in Figure 6b,c, with untreated/intact virions, the anti-glycan bnAbs showed low binding efficiencies consistent with the previous experiments. In comparison, detergent treatment produced a remarkably different binding profile. The autocorrelation curves fit a two-species diffusion model, but one comprising only free IgG and a mAb-bound species with a diffusion coefficient (35 $\mu m^2/s$) matching the predicted size of a mAb-free trimer complex. No residual virion binding was evident. Complexes of antibody and dissociated S protomers, which would display diffusion coefficients of ~45 $\mu m^2/sec$, were not observed. Most importantly, both 2G12 and PGT121 demonstrated extremely high binding efficiencies with the released S trimers (97% and 100%, respectively; Figure 6d,e), which were otherwise poorly or not reactive with the anti-S mAbs (Supplemental Figure S8).

4. Discussion

Soluble SARS-CoV-2 S proteins will be a key point of focus in COVID-19 vaccine design for the foreseeable future. Non-pathogenic S-pseudotyped viruses will continue to support assessments of humoral neutralizing antibody activity and other antiviral immune functions. Therefore, the elucidation of antigenic differences between these reagents versus native virions will inform the refinement of immunotherapeutic and/or vaccine countermeasures against COVID-19. Towards this end, here, we utilized a single-molecule-based analytical approach to define antigenic differences in unalloyed soluble SARS-CoV-2 spike proteins, pseudovirions, or inactivated virions representing virus variants emerging over the course of the COVID-19 pandemic.

We used test mAbs isolated early in the pandemic to elucidate any trends in antigenic changes in variants emerging over time. All test mAbs except Covi-9 demonstrated binding to soluble S trimers from the contemporaneous Wuhan strain, but with variable efficiency (Figure 1c and Supplemental Figure S1). Notably, the non-neutralizing mAb CR3022 [80] demonstrated the most efficient binding. As expected, each mAb demonstrated variable anti-S binding efficiencies across variants (Figure 2) that emerged after the Wuhan strain. Except for CR3022, mAb-binding efficiencies trended lower for more recently emerging (e.g., Omicron) strains. A variant of the Wuhan strain with the D614G mutation, which increases transmissibility and cell entry [27], showed a higher binding efficiency to Covi-10 and Covi-11 compared to the parental virus (Figure 2j). The Hexapro spike, which has six proline substitutions in the Wuhan strain that enable higher expression and stability, and a prefusion S conformation [78], also demonstrated more efficient binding to Covi-10 and Covi-11 but showed less efficient binding to Covi-21 and Covi-24. The latter data indicate that structural flexibility might be an important factor in mAb binding.

The appearance of evolutionary changes in RBD target epitopes were associated with changes in mAb-binding efficiency. For the three mAbs (Covi-10, 11, and 17) evaluated in this context, binding efficiencies trended downward with more recently occurring strains in concert with the accumulation of point mutations in the RBD (Figure 3). Covi-11 was an interesting case where the binding efficiency was severely reduced in the Alpha and Beta strains due to mutations at contact position 417, but was retained for Gamma in concert with other mutations at positions 493 and 498 in the RBD (Figure 3b) [57]. Thus, immune evasion from some mAbs may not be fixed in time, and instead appear and disappear according to other selective pressures on S that drive sequence changes.

Notably, binding patterns and efficiencies across the panel of mAbs also varied according to whether S was presented on a soluble trimer, pseudovirus, or inactivated wild-type virion (Figures 1, 2, and 4) with the same epitope sequences. There were trends but no statistically significant relationships between various mAb-binding efficiencies on soluble trimers versus pseudoviruses (Figure 4c) but significant direct relationships between binding efficiencies on inactivated virions versus pseudovirions or soluble S (Figure 4e,d).

For individual mAbs the comparatively low percent binding on pseudoviruses was likely due to lower levels of surface S trimer versus wild-type virions. Notably, no significant relationships appeared between the neutralization potency versus binding to any target. These data indicate that S antigenicity varies according to its mode of expression and that biochemical variables other than the overall virion binding efficiency impact the anti-S neutralization capacity (see below).

Insights into the lack of a correlation between binding efficiency and neutralizing activity were provided by analyses of HIV-neutralizing bnAbs that cross-react with SARS-CoV-2 S via their specificity for S glycan clusters [52,59,85–89]. Using our in vitro single-molecule fluorescence-based solution binding approach, we observed variable cross-strain binding of such bnAbs (2G12, PGT121, and PGT126) to soluble S proteins and pseudoviruses (Figure 6) but no neutralizing activity, in accordance with previous reports [52,53]. The relatively lower binding efficiencies with pseudoviruses (Figure 6b–e, Supplemental Figure S7) was not readily tracked to steric hindrance on virion surfaces, as the binding efficiency of 2G12 Fab was not significantly higher than whole IgG (Supplemental Figure S7). However, the detergent release of S trimers from virion membranes efficiently bound bnAbs (Figure 6d,e) but poorly reacted with Covi mAbs (Supplemental Figure S8), indicating that they assume a nonfunctional structure that, nevertheless, retains and, perhaps, optimizes the glycan presentation patterns recognized by the bnAbs.

Collectively, the FCS data are consistent with a scenario in which mAb binding to SARS-CoV-2 virions depends not only on the epitope sequence but also factors that modulate the known structural flexibility of S trimers [49,74,90,91]. We consider one heuristic model where virion trimers asynchronously swing between two end states, one functional and the other nonfunctional. The former is recognized by neutralizing antibodies (thus allowing them to neutralize) and the latter by anti-glycan mAbs, which are not antiviral by virtue of targeting a nonfunctional structure. It is possible that a certain fraction of spikes become locked in the nonfunctional form, depending on the virus or method of virion production. Thus, virions of a given strain can be antigenic for both neutralizing (e.g., Covi) and non-neutralizing antibodies (e.g., anti-glycan), with moderate binding efficiencies. It is likely that certain neutralizing mAbs also recognize structural intermediates between the two end states. Such fine preferences, perhaps combined with differences in S-binding affinities, could further explain why virion-neutralizing activity does not correlate with mAb-binding efficiency. In any case, our data suggest that the release of native S from virions causes a broad conversion to the nonfunctional state, improving anti-glycan mAb binding and abrogating reactivity with neutralizing anti-S antibodies. Stabilized soluble S trimers may represent a fixed intermediate in the virion S structural repertoire, being recognized by neutralizing mAbs, non-neutralizing anti-S mAbs, and anti-glycan Abs. This property would explain why soluble-S-binding- versus pseudovirion-binding efficiencies are not correlated.

5. Conclusions

Three main conclusions can be drawn from the FCS data, which reflect interactions between unperturbed reactants at physiologically relevant concentrations. First, FCS is sufficiently sensitive for differentiating anti-S mAb-binding efficiencies according to variants with altered target epitope sequences. Importantly, FCS measures reflected that mAbs elicited by an early SARS-CoV-2 strain demonstrate reduced FCS binding efficiencies with more recently emerging strains. Second, S trimers for a given variant display discernable differences in antigenic properties and immunoreactivity according to whether they are expressed as soluble or virion-associated molecules. Thus, immunochemical analyses of mAbs with soluble S might over- or underestimate reactivity with virions. Soluble S trimers are undoubtedly valuable for studies of anti-SARS-CoV-2 humoral immunity and vaccine development. Continued efforts to develop versions that more closely mirror the characteristics of virion trimers should be useful in refining vaccine and immunotherapeutic designs. Analytical approaches such as FCS can aid such efforts. Third, the structural flexibility and consequent heterogeneity of S trimers likely allow virion immunoreactivity with both neutralizing and non-neutralizing anti-S or anti-glycan antibodies. The possibility that non-neutralizing mechanisms of humoral immunity (e.g., ADCC) against SARS-CoV-2 occur via the antibody recognition of nonfunctional S structures merits further investigation.

Supplementary Materials: The following supporting information can be downloaded at: https://www.mdpi.com/article/10.3390/v16030407/s1, Figure S1: Anti-spike mAbs binding to SARS-CoV-2 spike protein of the Wuhan variant; Figure S2: Binding of anti-spike mAbs to gamma irradiated inactivated virions of Delta variant.; Figure S3: Non-specific binding of anti-spike mAbs to virions of lacking any Env protein; Figure S4: Relationships between Covi-series mAbs binding efficiency and neutralization; Figure S5: Binding of anti-spike mAbs to SARS-CoV-2 pseudovirions of Wuhan variant; Figure S6: Binding of anti-spike mAbs to SARS-CoV-2 pseudovirions of Wuhan variant at varied concentration range; Figure S7: Comparative binding efficiencies of mAbs 2G12, PGT121, PGT126, CR3022 and Fab 2G12 with SARS-CoV-2 pseudovirion of Wuhan variant; Figure S8: MAb CR3022 and Covi- mAbs reactivity against spike proteins extant on whole pseudovirions or released from particles by detergent lysis.

Author Contributions: Conceptualization, K.R.; methodology, S.K., S.D., G.A.S., M.M.S., A.L.D. and K.R.; investigation, S.K., S.D., G.A.S., M.M.S., A.L.D. and K.R.; resources, M.M.S., A.L.D. and K.R.; data curation, S.K., S.D., G.A.S., M.M.S., A.L.D. and K.R.; writing—original draft preparation, S.K. and K.R.; writing—review and editing, S.K., S.D., G.A.S., M.M.S., A.L.D. and K.R.; visualization, G.A.S.; supervision, K.R.; project administration, K.R.; funding acquisition, M.M.S. and K.R. All authors have read and agreed to the published version of the manuscript.

Funding: The research reported in this paper was supported by the National Institute of Allergy and Infectious Diseases of the National Institutes of Health under Award Number R01-AI150447 (KR), R01-AI172487 (KR), R61-AI176561 (KR), and R01-AI147870 (MMS). The research was also supported by the US Department of Veterans Affairs Biomedical Laboratory Research and Development Service I01BX005469 (MMS). Molecular graphics and analyses were performed with UCSF ChimeraX, developed by the Resource for Biocomputing, Visualization, and Informatics at the University of California, San Francisco, with support from National Institutes of Health R01-GM129325 and the Office of Cyber Infrastructure and Computational Biology, National Institute of Allergy and Infectious Diseases. The content is solely the authors' responsibility and does not necessarily represent the official views of the National Institutes of Health or US government.

Institutional Review Board Statement: The study (HP-00090821) was approved by the University of Maryland Baltimore IRB on 27 March 2020.

Informed Consent Statement: Informed consent was obtained from the subject in this study.

Data Availability Statement: All data are included in the main manuscript and supplementary figures.

Acknowledgments: The following SARS-CoV-2-related reagents were obtained through BEI Resources, NIAID, NIH: NR-52397—soluble stabilized spike protein (Wuhan), NR-55343—spike protein (D614G), NR-53769—spike protein (Hexapro), NR-55311—spike protein B.1.1.7 (Alpha), NR-55310—spike protein B.1.351 (Beta), NR-55307—spike protein P.1 (Gamma), NR-55614—spike protein B.1.617.2 (Delta), NR-56447—spike protein B.1.1.529 (Omicron), NR-56517—spike protein BA.2 (Omicron), NR-58646—spike protein BA.4 (Omicron), NR-58987—spike protein BA.2.75 (Omicron), NR-58988—spike protein BQ.1.1 (Omicron), NR-52481—mAb CR3022, NR-52287—gamma-irradiated virion (2019-nCoV/USA-WA1/2020, Wuhan), and NR-56129—gamma-irradiated virion (hCoV-19/USA/MD-HP05285/2021, Delta). The authors thank James Foulke for technical assistance in preparing plasmids and pseudovirions.

Conflicts of Interest: The authors declare no conflict of interest.

References

1. Oude Munnink, B.B.; Sikkema, R.S.; Nieuwenhuijse, D.F.; Molenaar, R.J.; Munger, E.; Molenkamp, R.; van der Spek, A.; Tolsma, P.; Rietveld, A.; Brouwer, M.; et al. Transmission of SARS-CoV-2 on mink farms between humans and mink and back to humans. *Science* **2021**, *371*, 172–177. [CrossRef]
2. Banerjee, A.; Mossman, K.; Baker, M.L. Zooanthroponotic potential of SARS-CoV-2 and implications of reintroduction into human populations. *Cell Host Microbe* **2021**, *29*, 160–164. [CrossRef]

3. Fenollar, F.; Mediannikov, O.; Maurin, M.; Devaux, C.; Colson, P.; Levasseur, A.; Fournier, P.E.; Raoult, D. Mink, SARS-CoV-2, and the Human-Animal Interface. *Front. Microbiol.* **2021**, *12*, 663815. [CrossRef]
4. Anderson, E.J.; Rouphael, N.G.; Widge, A.T.; Jackson, L.A.; Roberts, P.C.; Makhene, M.; Chappell, J.D.; Denison, M.R.; Stevens, L.J.; Pruijssers, A.J.; et al. Safety and Immunogenicity of SARS-CoV-2 mRNA-1273 Vaccine in Older Adults. *N. Engl. J. Med.* **2020**, *383*, 2427–2438. [CrossRef]
5. Polack, F.P.; Thomas, S.J.; Kitchin, N.; Absalon, J.; Gurtman, A.; Lockhart, S.; Perez, J.L.; Perez Marc, G.; Moreira, E.D.; Zerbini, C.; et al. Safety and Efficacy of the BNT162b2 mRNA COVID-19 Vaccine. *N. Engl. J. Med.* **2020**, *383*, 2603–2615. [CrossRef]
6. Pegu, A.; O'Connell, S.E.; Schmidt, S.D.; O'Dell, S.; Talana, C.A.; Lai, L.; Albert, J.; Anderson, E.; Bennett, H.; Corbett, K.S.; et al. Durability of mRNA-1273 vaccine-induced antibodies against SARS-CoV-2 variants. *Science* **2021**, *373*, 1372–1377. [CrossRef]
7. Abu-Raddad, L.J.; Chemaitelly, H.; Butt, A.A.; National Study Group for COVID-19 Vaccination. Effectiveness of the BNT162b2 COVID-19 Vaccine against the B.1.1.7 and B.1.351 Variants. *N. Engl. J. Med.* **2021**, *385*, 187–189. [CrossRef] [PubMed]
8. Chen, P.; Nirula, A.; Heller, B.; Gottlieb, R.L.; Boscia, J.; Morris, J.; Huhn, G.; Cardona, J.; Mocherla, B.; Stosor, V.; et al. SARS-CoV-2 Neutralizing Antibody LY-CoV555 in Outpatients with COVID-19. *N. Engl. J. Med.* **2021**, *384*, 229–237. [CrossRef] [PubMed]
9. Hansen, J.; Baum, A.; Pascal, K.E.; Russo, V.; Giordano, S.; Wloga, E.; Fulton, B.O.; Yan, Y.; Koon, K.; Patel, K.; et al. Studies in humanized mice and convalescent humans yield a SARS-CoV-2 antibody cocktail. *Science* **2020**, *369*, 1010–1014. [CrossRef] [PubMed]
10. Rappazzo, C.G.; Tse, L.V.; Kaku, C.I.; Wrapp, D.; Sakharkar, M.; Huang, D.; Deveau, L.M.; Yockachonis, T.J.; Herbert, A.S.; Battles, M.B.; et al. Broad and potent activity against SARS-like viruses by an engineered human monoclonal antibody. *Science* **2021**, *371*, 823–829. [CrossRef] [PubMed]
11. Shanmugaraj, B.; Siriwattananon, K.; Wangkanont, K.; Phoolcharoen, W. Perspectives on monoclonal antibody therapy as potential therapeutic intervention for Coronavirus disease-19 (COVID-19). *Asian Pac. J. Allergy Immunol.* **2020**, *38*, 10–18. [CrossRef] [PubMed]
12. Jiang, S.; Hillyer, C.; Du, L. Neutralizing Antibodies against SARS-CoV-2 and Other Human Coronaviruses. *Trends Immunol.* **2020**, *41*, 355–359. [CrossRef] [PubMed]
13. Gobeil, S.M.; Janowska, K.; McDowell, S.; Mansouri, K.; Parks, R.; Stalls, V.; Kopp, M.F.; Manne, K.; Li, D.; Wiehe, K.; et al. Effect of natural mutations of SARS-CoV-2 on spike structure, conformation, and antigenicity. *Science* **2021**, *373*, eabi6226. [CrossRef] [PubMed]
14. Li, F. Structure, Function, and Evolution of Coronavirus Spike Proteins. *Annu. Rev. Virol.* **2016**, *3*, 237–261. [CrossRef]
15. Hoffmann, M.; Kleine-Weber, H.; Schroeder, S.; Kruger, N.; Herrler, T.; Erichsen, S.; Schiergens, T.S.; Herrler, G.; Wu, N.H.; Nitsche, A.; et al. SARS-CoV-2 Cell Entry Depends on ACE2 and TMPRSS2 and Is Blocked by a Clinically Proven Protease Inhibitor. *Cell* **2020**, *181*, 271–280.e8. [CrossRef]
16. Cai, Y.; Zhang, J.; Xiao, T.; Peng, H.; Sterling, S.M.; Walsh, R.M., Jr.; Rawson, S.; Rits-Volloch, S.; Chen, B. Distinct conformational states of SARS-CoV-2 spike protein. *Science* **2020**, *369*, 1586–1592. [CrossRef]
17. Ke, Z.; Oton, J.; Qu, K.; Cortese, M.; Zila, V.; McKeane, L.; Nakane, T.; Zivanov, J.; Neufeldt, C.J.; Cerikan, B.; et al. Structures and distributions of SARS-CoV-2 spike proteins on intact virions. *Nature* **2020**, *588*, 498–502. [CrossRef]
18. Walls, A.C.; Park, Y.J.; Tortorici, M.A.; Wall, A.; McGuire, A.T.; Veesler, D. Structure, Function, and Antigenicity of the SARS-CoV-2 Spike Glycoprotein. *Cell* **2020**, *181*, 281–292.e6. [CrossRef]
19. Wrapp, D.; Wang, N.; Corbett, K.S.; Goldsmith, J.A.; Hsieh, C.L.; Abiona, O.; Graham, B.S.; McLellan, J.S. Cryo-EM structure of the 2019-nCoV spike in the prefusion conformation. *Science* **2020**, *367*, 1260–1263. [CrossRef]
20. Li, Z.; Tomlinson, A.C.; Wong, A.H.; Zhou, D.; Desforges, M.; Talbot, P.J.; Benlekbir, S.; Rubinstein, J.L.; Rini, J.M. The human coronavirus HCoV-229E S-protein structure and receptor binding. *Elife* **2019**, *8*, e51230. [CrossRef]
21. Walls, A.C.; Tortorici, M.A.; Frenz, B.; Snijder, J.; Li, W.; Rey, F.A.; DiMaio, F.; Bosch, B.J.; Veesler, D. Glycan shield and epitope masking of a coronavirus spike protein observed by cryo-electron microscopy. *Nat. Struct. Mol. Biol.* **2016**, *23*, 899–905. [CrossRef] [PubMed]
22. Mansbach, R.A.; Chakraborty, S.; Nguyen, K.; Montefiori, D.C.; Korber, B.; Gnanakaran, S. The SARS-CoV-2 Spike variant D614G favors an open conformational state. *Sci. Adv.* **2021**, *7*, eabf3671. [CrossRef] [PubMed]
23. Xu, C.; Wang, Y.; Liu, C.; Zhang, C.; Han, W.; Hong, X.; Wang, Y.; Hong, Q.; Wang, S.; Zhao, Q.; et al. Conformational dynamics of SARS-CoV-2 trimeric spike glycoprotein in complex with receptor ACE2 revealed by cryo-EM. *Sci. Adv.* **2021**, *7*, eabe5575. [CrossRef] [PubMed]
24. Yuan, Y.; Cao, D.; Zhang, Y.; Ma, J.; Qi, J.; Wang, Q.; Lu, G.; Wu, Y.; Yan, J.; Shi, Y.; et al. Cryo-EM structures of MERS-CoV and SARS-CoV spike glycoproteins reveal the dynamic receptor binding domains. *Nat. Commun.* **2017**, *8*, 15092. [CrossRef] [PubMed]
25. Song, W.; Gui, M.; Wang, X.; Xiang, Y. Cryo-EM structure of the SARS coronavirus spike glycoprotein in complex with its host cell receptor ACE2. *PLoS Pathog.* **2018**, *14*, e1007236. [CrossRef] [PubMed]
26. Shang, J.; Wan, Y.; Luo, C.; Ye, G.; Geng, Q.; Auerbach, A.; Li, F. Cell entry mechanisms of SARS-CoV-2. *Proc. Natl. Acad. Sci. USA* **2020**, *117*, 11727–11734. [CrossRef] [PubMed]
27. Plante, J.A.; Liu, Y.; Liu, J.; Xia, H.; Johnson, B.A.; Lokugamage, K.G.; Zhang, X.; Muruato, A.E.; Zou, J.; Fontes-Garfias, C.R.; et al. Spike mutation D614G alters SARS-CoV-2 fitness. *Nature* **2021**, *592*, 116–121. [CrossRef]

28. Hou, Y.J.; Chiba, S.; Halfmann, P.; Ehre, C.; Kuroda, M.; Dinnon, K.H., 3rd; Leist, S.R.; Schafer, A.; Nakajima, N.; Takahashi, K.; et al. SARS-CoV-2 D614G variant exhibits efficient replication ex vivo and transmission in vivo. *Science* **2020**, *370*, 1464–1468. [CrossRef]
29. Korber, B.; Fischer, W.M.; Gnanakaran, S.; Yoon, H.; Theiler, J.; Abfalterer, W.; Hengartner, N.; Giorgi, E.E.; Bhattacharya, T.; Foley, B.; et al. Tracking Changes in SARS-CoV-2 Spike: Evidence that D614G Increases Infectivity of the COVID-19 Virus. *Cell* **2020**, *182*, 812–827.e19. [CrossRef]
30. Zhu, C.; He, G.; Yin, Q.; Zeng, L.; Ye, X.; Shi, Y.; Xu, W. Molecular biology of the SARS-CoV-2 spike protein: A review of current knowledge. *J. Med. Virol.* **2021**, *93*, 5729–5741. [CrossRef]
31. Liu, C.; Ginn, H.M.; Dejnirattisai, W.; Supasa, P.; Wang, B.; Tuekprakhon, A.; Nutalai, R.; Zhou, D.; Mentzer, A.J.; Zhao, Y.; et al. Reduced neutralization of SARS-CoV-2 B.1.617 by vaccine and convalescent serum. *Cell* **2021**, *184*, 4220–4236.e13. [CrossRef]
32. Liu, Y.; Liu, J.; Johnson, B.A.; Xia, H.; Ku, Z.; Schindewolf, C.; Widen, S.G.; An, Z.; Weaver, S.C.; Menachery, V.D.; et al. Delta spike P681R mutation enhances SARS-CoV-2 fitness over Alpha variant. *Cell Rep.* **2022**, *39*, 110829. [CrossRef] [PubMed]
33. Scudellari, M. How the coronavirus infects cells—And why Delta is so dangerous. *Nature* **2021**, *595*, 640–644. [CrossRef] [PubMed]
34. Dejnirattisai, W.; Huo, J.; Zhou, D.; Zahradník, J.; Supasa, P.; Liu, C.; Duyvesteyn, H.M.E.; Ginn, H.M.; Mentzer, A.J.; Tuekprakhon, A.; et al. SARS-CoV-2 Omicron-B.1.1.529 leads to widespread escape from neutralizing antibody responses. *Cell* **2022**, *185*, 467–484.e15. [CrossRef] [PubMed]
35. Wang, P.; Nair, M.S.; Liu, L.; Iketani, S.; Luo, Y.; Guo, Y.; Wang, M.; Yu, J.; Zhang, B.; Kwong, P.D.; et al. Antibody resistance of SARS-CoV-2 variants B.1.351 and B.1.1.7. *Nature* **2021**, *593*, 130–135. [CrossRef] [PubMed]
36. Planas, D.; Veyer, D.; Baidaliuk, A.; Staropoli, I.; Guivel-Benhassine, F.; Rajah, M.M.; Planchais, C.; Porrot, F.; Robillard, N.; Puech, J.; et al. Reduced sensitivity of SARS-CoV-2 variant Delta to antibody neutralization. *Nature* **2021**, *596*, 276–280. [CrossRef]
37. Zhang, J.; Xiao, T.; Cai, Y.; Lavine, C.L.; Peng, H.; Zhu, H.; Anand, K.; Tong, P.; Gautam, A.; Mayer, M.L.; et al. Membrane fusion and immune evasion by the spike protein of SARS-CoV-2 Delta variant. *Science* **2021**, *374*, 1353–1360. [CrossRef]
38. Syed, A.M.; Ciling, A.; Taha, T.Y.; Chen, I.P.; Khalid, M.M.; Sreekumar, B.; Chen, P.Y.; Kumar, G.R.; Suryawanshi, R.; Silva, I.; et al. Omicron mutations enhance infectivity and reduce antibody neutralization of SARS-CoV-2 virus-like particles. *Proc. Natl. Acad. Sci. USA* **2022**, *119*, e2200592119. [CrossRef]
39. Viana, R.; Moyo, S.; Amoako, D.G.; Tegally, H.; Scheepers, C.; Althaus, C.L.; Anyaneji, U.J.; Bester, P.A.; Boni, M.F.; Chand, M.; et al. Rapid epidemic expansion of the SARS-CoV-2 Omicron variant in southern Africa. *Nature* **2022**, *603*, 679–686. [CrossRef]
40. Chen, J.; Wei, G.W. Omicron BA.2 (B.1.1.529.2): High Potential for Becoming the Next Dominant Variant. *J. Phys. Chem. Lett.* **2022**, *13*, 3840–3849. [CrossRef]
41. Kumar, S.; Karuppanan, K.; Subramaniam, G. Omicron (BA.1) and sub-variants (BA.1.1, BA.2, and BA.3) of SARS-CoV-2 spike infectivity and pathogenicity: A comparative sequence and structural-based computational assessment. *J. Med. Virol.* **2022**, *94*, 4780–4791. [CrossRef]
42. Liu, L.; Iketani, S.; Guo, Y.; Chan, J.F.; Wang, M.; Liu, L.; Luo, Y.; Chu, H.; Huang, Y.; Nair, M.S.; et al. Striking antibody evasion manifested by the Omicron variant of SARS-CoV-2. *Nature* **2022**, *602*, 676–681. [CrossRef]
43. Iketani, S.; Liu, L.; Guo, Y.; Liu, L.; Chan, J.F.; Huang, Y.; Wang, M.; Luo, Y.; Yu, J.; Chu, H.; et al. Antibody evasion properties of SARS-CoV-2 Omicron sublineages. *Nature* **2022**, *604*, 553–556. [CrossRef]
44. Callaway, E. Coronavirus variant XBB.1.5 rises in the United States—Is it a global threat? *Nature* **2023**, *613*, 222–223. [CrossRef]
45. Qu, P.; Faraone, J.N.; Evans, J.P.; Zheng, Y.M.; Carlin, C.; Anghelina, M.; Stevens, P.; Fernandez, S.; Jones, D.; Panchal, A.R.; et al. Enhanced evasion of neutralizing antibody response by Omicron XBB.1.5, CH.1.1, and CA.3.1 variants. *Cell Rep.* **2023**, *42*, 112443. [CrossRef] [PubMed]
46. Parums, D.V. Editorial: The XBB.1.5 ('Kraken') Subvariant of Omicron SARS-CoV-2 and its Rapid Global Spread. *Med. Sci. Monit.* **2023**, *29*, e939580. [CrossRef] [PubMed]
47. Benton, D.J.; Wrobel, A.G.; Xu, P.; Roustan, C.; Martin, S.R.; Rosenthal, P.B.; Skehel, J.J.; Gamblin, S.J. Receptor binding and priming of the spike protein of SARS-CoV-2 for membrane fusion. *Nature* **2020**, *588*, 327–330. [CrossRef] [PubMed]
48. Zhou, T.; Tsybovsky, Y.; Gorman, J.; Rapp, M.; Cerutti, G.; Chuang, G.Y.; Katsamba, P.S.; Sampson, J.M.; Schon, A.; Bimela, J.; et al. Cryo-EM Structures of SARS-CoV-2 Spike without and with ACE2 Reveal a pH-Dependent Switch to Mediate Endosomal Positioning of Receptor-Binding Domains. *Cell Host Microbe* **2020**, *28*, 867–879.e5. [CrossRef] [PubMed]
49. Hastie, K.M.; Li, H.; Bedinger, D.; Schendel, S.L.; Dennison, S.M.; Li, K.; Rayaprolu, V.; Yu, X.; Mann, C.; Zandonatti, M.; et al. Defining variant-resistant epitopes targeted by SARS-CoV-2 antibodies: A global consortium study. *Science* **2021**, *374*, 472–478. [CrossRef] [PubMed]
50. Pak, A.J.; Yu, A.; Ke, Z.; Briggs, J.A.G.; Voth, G.A. Cooperative multivalent receptor binding promotes exposure of the SARS-CoV-2 fusion machinery core. *Nat. Commun.* **2022**, *13*, 1002. [CrossRef]
51. Lu, M.; Uchil, P.D.; Li, W.; Zheng, D.; Terry, D.S.; Gorman, J.; Shi, W.; Zhang, B.; Zhou, T.; Ding, S.; et al. Real-Time Conformational Dynamics of SARS-CoV-2 Spikes on Virus Particles. *Cell Host Microbe* **2020**, *28*, 880–891.e8. [CrossRef]
52. Mannar, D.; Leopold, K.; Subramaniam, S. Glycan reactive anti-HIV-1 antibodies bind the SARS-CoV-2 spike protein but do not block viral entry. *Sci. Rep.* **2021**, *11*, 12448. [CrossRef]
53. Williams, W.B.; Meyerhoff, R.R.; Edwards, R.J.; Li, H.; Manne, K.; Nicely, N.I.; Henderson, R.; Zhou, Y.; Janowska, K.; Mansouri, K.; et al. Fab-dimerized glycan-reactive antibodies are a structural category of natural antibodies. *Cell* **2021**, *184*, 2955–2972.e25. [CrossRef] [PubMed]

54. Agrawal, P.; DeVico, A.L.; Foulke, J.S., Jr.; Lewis, G.K.; Pazgier, M.; Ray, K. Stoichiometric Analyses of Soluble CD4 to Native-like HIV-1 Envelope by Single-Molecule Fluorescence Spectroscopy. *Cell Rep.* **2019**, *29*, 176–186.e4. [CrossRef] [PubMed]
55. Ray, K.; Mengistu, M.; Orlandi, C.; Pazgier, M.; Lewis, G.K.; DeVico, A.L. Concurrent Exposure of Neutralizing and Non-neutralizing Epitopes on a Single HIV-1 Envelope Structure. *Front. Immunol.* **2019**, *10*, 1512. [CrossRef] [PubMed]
56. Ray, K.; Mengistu, M.; Yu, L.; Lewis, G.K.; Lakowicz, J.R.; DeVico, A.L. Antigenic properties of the HIV envelope on virions in solution. *J. Virol.* **2014**, *88*, 1795–1808. [CrossRef] [PubMed]
57. Tolbert, W.D.; Chen, Y.; Sun, L.; Benlarbi, M.; Ding, S.; Manickam, R.; Pangaro, E.; Nguyen, D.N.; Gottumukkala, S.; Cote, M.; et al. The molecular basis of the neutralization breadth of the RBD-specific antibody CoV11. *Front. Immunol.* **2023**, *14*, 1178355. [CrossRef] [PubMed]
58. Stephenson, K.E.; Julg, B.; Tan, C.S.; Zash, R.; Walsh, S.R.; Rolle, C.P.; Monczor, A.N.; Lupo, S.; Gelderblom, H.C.; Ansel, J.L.; et al. Safety, pharmacokinetics and antiviral activity of PGT121, a broadly neutralizing monoclonal antibody against HIV-1: A randomized, placebo-controlled, phase 1 clinical trial. *Nat. Med.* **2021**, *27*, 1718–1724. [CrossRef] [PubMed]
59. Moldt, B.; Le, K.M.; Carnathan, D.G.; Whitney, J.B.; Schultz, N.; Lewis, M.G.; Borducchi, E.N.; Smith, K.M.; Mackel, J.J.; Sweat, S.L.; et al. Neutralizing antibody affords comparable protection against vaginal and rectal simian/human immunodeficiency virus challenge in macaques. *AIDS* **2016**, *30*, 1543–1551. [CrossRef]
60. Hessell, A.J.; Rakasz, E.G.; Poignard, P.; Hangartner, L.; Landucci, G.; Forthal, D.N.; Koff, W.C.; Watkins, D.I.; Burton, D.R. Broadly neutralizing human anti-HIV antibody 2G12 is effective in protection against mucosal SHIV challenge even at low serum neutralizing titers. *PLoS Pathog.* **2009**, *5*, e1000433. [CrossRef]
61. Madeira, F.; Pearce, M.; Tivey, A.R.N.; Basutkar, P.; Lee, J.; Edbali, O.; Madhusoodanan, N.; Kolesnikov, A.; Lopez, R. Search and sequence analysis tools services from EMBL-EBI in 2022. *Nucleic Acids Res.* **2022**, *50*, W276–W279. [CrossRef] [PubMed]
62. Meng, E.C.; Goddard, T.D.; Pettersen, E.F.; Couch, G.S.; Pearson, Z.J.; Morris, J.H.; Ferrin, T.E. UCSF ChimeraX: Tools for structure building and analysis. *Protein Sci.* **2023**, *32*, e4792. [CrossRef] [PubMed]
63. Gohain, N.; Tolbert, W.D.; Orlandi, C.; Richard, J.; Ding, S.; Chen, X.; Bonsor, D.A.; Sundberg, E.J.; Lu, W.; Ray, K.; et al. Molecular basis for epitope recognition by non-neutralizing anti-gp41 antibody F240. *Sci. Rep.* **2016**, *6*, 36685. [CrossRef]
64. Mengistu, M.; Ray, K.; Lewis, G.K.; DeVico, A.L. Antigenic properties of the human immunodeficiency virus envelope glycoprotein gp120 on virions bound to target cells. *PLoS Pathog.* **2015**, *11*, e1004772. [CrossRef] [PubMed]
65. Tolbert, W.D.; Sherburn, R.; Gohain, N.; Ding, S.; Flinko, R.; Orlandi, C.; Ray, K.; Finzi, A.; Lewis, G.K.; Pazgier, M. Defining rules governing recognition and Fc-mediated effector functions to the HIV-1 co-receptor binding site. *BMC Biol.* **2020**, *18*, 91. [CrossRef]
66. Sarzotti-Kelsoe, M.; Bailer, R.T.; Turk, E.; Lin, C.L.; Bilska, M.; Greene, K.M.; Gao, H.; Todd, C.A.; Ozaki, D.A.; Seaman, M.S.; et al. Optimization and validation of the TZM-bl assay for standardized assessments of neutralizing antibodies against HIV-1. *J. Immunol. Methods* **2014**, *409*, 131–146. [CrossRef]
67. Sankhala, R.S.; Dussupt, V.; Chen, W.H.; Bai, H.; Martinez, E.J.; Jensen, J.L.; Rees, P.A.; Hajduczki, A.; Chang, W.C.; Choe, M.; et al. Antibody targeting of conserved sites of vulnerability on the SARS-CoV-2 spike receptor-binding domain. *Structure* **2024**, *32*, 131–147.e7. [CrossRef] [PubMed]
68. Yuan, M.; Wu, N.C.; Zhu, X.; Lee, C.D.; So, R.T.Y.; Lv, H.; Mok, C.K.P.; Wilson, I.A. A highly conserved cryptic epitope in the receptor binding domains of SARS-CoV-2 and SARS-CoV. *Science* **2020**, *368*, 630–633. [CrossRef]
69. Zhou, P.; Yang, X.L.; Wang, X.G.; Hu, B.; Zhang, L.; Zhang, W.; Si, H.R.; Zhu, Y.; Li, B.; Huang, C.L.; et al. A pneumonia outbreak associated with a new coronavirus of probable bat origin. *Nature* **2020**, *579*, 270–273. [CrossRef]
70. Hu, B.; Guo, H.; Zhou, P.; Shi, Z.L. Characteristics of SARS-CoV-2 and COVID-19. *Nat. Rev. Microbiol.* **2021**, *19*, 141–154. [CrossRef]
71. Coronavirus Immunotherapy Consortium (COVIC). Available online: https://covic.lji.org/ (accessed on 21 January 2024).
72. Ding, C.; He, J.; Zhang, X.; Jiang, C.; Sun, Y.; Zhang, Y.; Chen, Q.; He, H.; Li, W.; Xie, J.; et al. Crucial Mutations of Spike Protein on SARS-CoV-2 Evolved to Variant Strains Escaping Neutralization of Convalescent Plasmas and RBD-Specific Monoclonal Antibodies. *Front. Immunol.* **2021**, *12*, 693775. [CrossRef] [PubMed]
73. Lan, J.; Ge, J.; Yu, J.; Shan, S.; Zhou, H.; Fan, S.; Zhang, Q.; Shi, X.; Wang, Q.; Zhang, L.; et al. Structure of the SARS-CoV-2 spike receptor-binding domain bound to the ACE2 receptor. *Nature* **2020**, *581*, 215–220. [CrossRef]
74. Singh, J.; Vashishtha, S.; Kundu, B. Spike Protein Mutation-Induced Changes in the Kinetic and Thermodynamic Behavior of Its Receptor Binding Domains Explain Their Higher Propensity to Attain Open States in SARS-CoV-2 Variants of Concern. *ACS Cent. Sci.* **2023**, *9*, 1894–1904. [CrossRef] [PubMed]
75. Carabelli, A.M.; Peacock, T.P.; Thorne, L.G.; Harvey, W.T.; Hughes, J.; Consortium, C.-G.U.; Peacock, S.J.; Barclay, W.S.; de Silva, T.I.; Towers, G.J.; et al. SARS-CoV-2 variant biology: Immune escape, transmission and fitness. *Nat. Rev. Microbiol.* **2023**, *21*, 162–177. [CrossRef] [PubMed]
76. Cox, M.; Peacock, T.P.; Harvey, W.T.; Hughes, J.; Wright, D.W.; Consortium, C.-G.U.; Willett, B.J.; Thomson, E.; Gupta, R.K.; Peacock, S.J.; et al. SARS-CoV-2 variant evasion of monoclonal antibodies based on in vitro studies. *Nat. Rev. Microbiol.* **2023**, *21*, 112–124. [CrossRef]
77. Liu, J.; Kan, M.; Zhang, L.; Yue, Y.; Wang, S.; Hong, M.; Hong, X. Rapid Degradation of SARS-CoV-2 Spike S Protein by A Specific Serine Protease. *Molecules* **2022**, *27*, 1882. [CrossRef]
78. Hsieh, C.L.; Goldsmith, J.A.; Schaub, J.M.; DiVenere, A.M.; Kuo, H.C.; Javanmardi, K.; Le, K.C.; Wrapp, D.; Lee, A.G.; Liu, Y.; et al. Structure-based design of prefusion-stabilized SARS-CoV-2 spikes. *Science* **2020**, *369*, 1501–1505. [CrossRef]

79. Callaway, H.M.; Hastie, K.M.; Schendel, S.L.; Li, H.; Yu, X.; Shek, J.; Buck, T.; Hui, S.; Bedinger, D.; Troup, C.; et al. Bivalent intra-spike binding provides durability against emergent Omicron lineages: Results from a global consortium. *Cell Rep.* **2023**, *42*, 112014. [CrossRef]
80. Wu, N.C.; Yuan, M.; Bangaru, S.; Huang, D.; Zhu, X.; Lee, C.D.; Turner, H.L.; Peng, L.; Yang, L.; Burton, D.R.; et al. A natural mutation between SARS-CoV-2 and SARS-CoV determines neutralization by a cross-reactive antibody. *PLoS Pathog.* **2020**, *16*, e1009089. [CrossRef]
81. Zhao, X.; Chen, H.; Wang, H. Glycans of SARS-CoV-2 Spike Protein in Virus Infection and Antibody Production. *Front. Mol. Biosci.* **2021**, *8*, 629873. [CrossRef]
82. Rathore, U.; Saha, P.; Kesavardhana, S.; Kumar, A.A.; Datta, R.; Devanarayanan, S.; Das, R.; Mascola, J.R.; Varadarajan, R. Glycosylation of the core of the HIV-1 envelope subunit protein gp120 is not required for native trimer formation or viral infectivity. *J. Biol. Chem.* **2017**, *292*, 10197–10219. [CrossRef] [PubMed]
83. Sok, D.; Burton, D.R. Recent progress in broadly neutralizing antibodies to HIV. *Nat. Immunol.* **2018**, *19*, 1179–1188. [CrossRef]
84. Mishra, N.; Kumar, S.; Singh, S.; Bansal, T.; Jain, N.; Saluja, S.; Kumar, R.; Bhattacharyya, S.; Palanichamy, J.K.; Mir, R.A.; et al. Cross-neutralization of SARS-CoV-2 by HIV-1 specific broadly neutralizing antibodies and polyclonal plasma. *PLoS Pathog.* **2021**, *17*, e1009958. [CrossRef] [PubMed]
85. Julien, J.P.; Sok, D.; Khayat, R.; Lee, J.H.; Doores, K.J.; Walker, L.M.; Ramos, A.; Diwanji, D.C.; Pejchal, R.; Cupo, A.; et al. Broadly neutralizing antibody PGT121 allosterically modulates CD4 binding via recognition of the HIV-1 gp120 V3 base and multiple surrounding glycans. *PLoS Pathog.* **2013**, *9*, e1003342. [CrossRef] [PubMed]
86. Saunders, K.O.; Nicely, N.I.; Wiehe, K.; Bonsignori, M.; Meyerhoff, R.R.; Parks, R.; Walkowicz, W.E.; Aussedat, B.; Wu, N.R.; Cai, F.; et al. Vaccine Elicitation of High Mannose-Dependent Neutralizing Antibodies against the V3-Glycan Broadly Neutralizing Epitope in Nonhuman Primates. *Cell Rep.* **2017**, *18*, 2175–2188. [CrossRef] [PubMed]
87. Stewart-Jones, G.B.; Soto, C.; Lemmin, T.; Chuang, G.Y.; Druz, A.; Kong, R.; Thomas, P.V.; Wagh, K.; Zhou, T.; Behrens, A.J.; et al. Trimeric HIV-1-Env Structures Define Glycan Shields from Clades A, B, and G. *Cell* **2016**, *165*, 813–826. [CrossRef]
88. Trkola, A.; Purtscher, M.; Muster, T.; Ballaun, C.; Buchacher, A.; Sullivan, N.; Srinivasan, K.; Sodroski, J.; Moore, J.P.; Katinger, H. Human monoclonal antibody 2G12 defines a distinctive neutralization epitope on the gp120 glycoprotein of human immunodeficiency virus type 1. *J. Virol.* **1996**, *70*, 1100–1108. [CrossRef]
89. Watanabe, Y.; Allen, J.D.; Wrapp, D.; McLellan, J.S.; Crispin, M. Site-specific glycan analysis of the SARS-CoV-2 spike. *Science* **2020**, *369*, 330–333. [CrossRef]
90. Gowthaman, R.; Guest, J.D.; Yin, R.; Adolf-Bryfogle, J.; Schief, W.R.; Pierce, B.G. CoV3D: A database of high resolution coronavirus protein structures. *Nucleic Acids Res.* **2021**, *49*, D282–D287. [CrossRef]
91. Yin, R.; Guest, J.D.; Taherzadeh, G.; Gowthaman, R.; Mittra, I.; Quackenbush, J.; Pierce, B.G. Structural and energetic profiling of SARS-CoV-2 receptor binding domain antibody recognition and the impact of circulating variants. *PLoS Comput. Biol.* **2021**, *17*, e1009380. [CrossRef] [PubMed]

viruses

MDPI

Article

An Improved Workflow for the Quantification of Orthohantavirus Infection Using Automated Imaging and Flow Cytometry

Laura Menke [1] **and Christian Sieben** [1,2,*]

[1] Nanoscale Infection Biology, Helmholtz Centre for Infection Research, Inhoffenstr. 7, 38124 Braunschweig, Germany; laura.menke@helmholtz-hzi.de
[2] Institute of Genetics, Technische Universität Braunschweig, Spielmannstr. 7, 38106 Braunschweig, Germany
* Correspondence: christian.sieben@helmholtz-hzi.de; Tel.: +49-531-6181-2950

Abstract: Determination of the infectious titer is a central requirement when working with pathogenic viruses. The plaque or focus assay is a commonly used but labor- and time-consuming approach for determining the infectious titer of orthohantavirus samples. We have developed an optimized virus quantification approach that relies on the fluorescence-based detection of the orthohantavirus nucleocapsid protein (N) in infected cells with high sensitivity. We present the use of flow cytometry but highlight fluorescence microscopy in combination with automated data analysis as an attractive alternative to increase the information retrieved from an infection experiment. Additionally, we offer open-source software equipped with a user-friendly graphical interface, eliminating the necessity for advanced programming skills.

Keywords: orthohantavirus; virus quantification; imaging; bunyavirales

1. Introduction

Orthohantaviruses are enveloped viruses that contain three circular single-stranded RNA genome segments of negative sense, namely, the small (S-segment), medium (M-segment) and large segment (L-segment) [1,2]. Encoded on the S-segment is the nucleocapsid (N) protein [3], which is often used for virus quantification due to its early and abundant production [4–6]. Research on orthohantaviruses started with the discovery of the Hantaan virus (HTNV) in 1978 [7], and morphological and genetic insights on HTNV followed as soon as virus propagation and titration could be established in cell culture systems [8,9]. Here especially, the viral adaptation to Vero E6 cells was an important milestone [10]. Since then, several protocols for orthohantavirus titration in cell culture have been developed. The widespread foci assay titrates the number of infectious particles by detecting the viral N protein in infected cells [9,11,12]. The foci assay is closely related to the plaque assay where plaques of dead cells are visualized as empty spots following a general cell staining procedure (i.e., with methylene blue). Since orthohantaviruses are non-lytic, infected cells are detected via the expression of viral proteins. Specifically, a semisolid overlay medium ensures that after infection, viral spread can only occur as far as neighboring cells, which leads to a patch of infected cells that can be counted as a focus [9,11,12]. A derivative of the fluorescence-based foci quantification protocol is viral antigen detection via chemiluminescence, which potentially offers improved sensitivity [13]. One drawback of the foci assay is its labor- and time-consuming protocol, which takes 5–11 days depending on the viral strain [13]. Recently, a flow cytometry-based titration assay was established using Andes orthohantavirus (ANDV). This assay has been shown to reduce the time required for the detection of viral replication during infection [14]. The authors state that the tested foci assays can be compared with their improved protocol, in which ANDV detection was possible as early as 6 h post-infection.

Here, we provide an improved virus quantification protocol based on fluorescence imaging and flow cytometry that is compatible with other orthohantavirus strains such as Puumala virus (PUUV) and Tula virus (TULV). While earlier time points can also be detected via fluorescence microscopy, our data suggest that PUUV and TULV should best be quantified after 48 h. We provide an automated imaging workflow together with a data analysis package which allows for the routine quantification of large amounts of single-cell data (5000–10,000 cells) in a time-efficient way. An additional advantage of titration through imaging is the simultaneous extraction of structural information, eliminating the need for an extra experimental setting. This represents a significant benefit, enabling discrimination between the early and late stages of infection.

2. Material and Methods

2.1. Cells, Viruses and Reagents

Vero E6 cells were purchased from ATCC and grown in DMEM (Thermo Fisher Scientific, Waltham, MA, USA: DMEM (1x) + GlutaMAX^{TM}-I, [+] 2.5 g/L D-Glucose, [-] Pyruvat) containing 10% fetal calf serum (FCS) (Merck, Darmstadt, Germany) and 1% penicillin/streptomycin (Pen/Strep) (Thermo Fisher Scientific, Waltham, MA, USA). Deionized water processed by an EMD Millipore^{TM} Milli-Q^{TM} advantage A10 ultrapure water purification system served as the basis for all buffers and solutions. All disposable cell culture materials were purchased from Sarstedt (Nümbrecht, Germany), TPP (Trasadingen, Switzerland), Thermo Fisher Scientific (Waltham, MA, USA) and Greiner bio-one (Kremsmünster, Austria). Nucleocapsid (N) proteins were detected using primary antibodies 5E11 (Abcam, Cambridge, UK; cat. ab34757) [15] or monoclonal antibody anti-TULV-1, generously provided by Jens Peter Teifke and Sven Reiche (Friedrich-Loeffler-Institut, Greifswald, Germany). We used Alexa Fluor 568 (cat. A11031) and Alexa Fluor 488 secondary antibodies (cat. A11029) (Thermo Fisher Scientific, Waltham, MA, USA).

2.2. Orthohantavirus Propagation

Hantaviruses were propagated on Vero E6 cells (ATCC: CCL-1586, Manassas, Virginia) under strict standard operating procedures using biosafety level 2 (BSL2) facilities and practice. For virus propagation, T75 flasks with Vero E6 cells were seeded to reach a density of 80–90% confluence after overnight incubation at 37 °C in 5% carbon dioxide (CO_2). The cells were infected using a multiplicity of infection (MOI) of 0.01–0.1 in DMEM + 2% FCS + 1% Pen/Strep. After 1 h at 37 °C (+5% CO_2), the inoculum was removed and the medium replaced with fresh DMEM + 2% FCS + 1% Pen/Strep. The infected cells were further incubated for 7–10 days, whereby after day 4, the supernatant could be collected on a daily basis and, later, either quantified passage per passage or pooled using concentration methods. Collected supernatants were centrifuged for 30 min at 1200–2000 g at 4 °C to remove cell debris then transferred into 50–100 kDa filtration units (Amicon) (Sartorius AG, Göttingen, Germany) and centrifuged at 4000 g until volume decreased to 1–3 mL. The retentates were mixed, resuspended and frozen at −80 °C until further use.

2.3. Preparation of Glass Coverslips for Microscopy

A total of 200 mL washing solution containing 60% hydrochloric acid (HCl) and 40% ethanol (EtOH) were mixed, and 120 mL of HCl was added and heated to 37 °C. The solution was filtered through a pleated filter. Coverslips were washed carefully in a 1 L glass bottle on a roll shaker for at least 30 min. Afterwards the HCl/EtOH mix was discarded. Coverslips were washed 3 times with water and then dried on Whatman paper. Finally, the slides were sterilized at 180 °C.

2.4. Preparation of a Fluorophore-Conjugated Primary Antibodies

A total of 20 µL of the antibody 5E11 (Abcam, Cambridge, UK (cat. ab34757)) [15], together with 20 µL phosphate-buffered saline (PBS; filtered) (Thermo Fisher Scientific, Waltham, MA, USA), 10 µL $NaHCO_3$ (pH 8–8.5; 0.5 M) and 0.5 µL dye conjugate (CF-488A Dye; Biotium, Fremont, CA, USA) (4 µg; 64 µM final; dye overload ca. 1:50) were incubated for 30 min at room temperature in the dark. Amersham NAP-5 Sephadex columns (Cytiva, Marlborough, USA) were rinsed with phosphate-buffered saline (PBS) before the sample was applied and eluted with PBS. Run through was collected drop-wise in a 96-well plate (well after well), and the dye/protein content of each well was measured using a NanoDrop (Thermo Fisher Scientific, Waltham, MA, USA). The wells with the highest concentration were pooled and measured again to obtain the final concentration (protocol is based on [16]).

2.5. Immunofluorescence Assay (IFA) for Flow Cytometry

A total of 100,000 Vero E6 cells were seeded in a 24-well plate (Corning Incorporated, New York, NY, USA) and incubated with DMEM + 10% FCS + 1% Pen/Strep for 24 h at 37 °C in 5% CO_2. Dependent on the type of experiment, either an MOI of 1 was used or a fixed volume (5 µL or 50 µL). For infections, the cell layer should be 80% confluent. The used virus volume was filled to 300 µL per well with DMEM + 2% FCS + 1% Pen/Strep and incubated for 1 h at 37 °C. The cells were washed twice with PBS and further incubated for 48 h with DMEM + 2% FCS + 1% Pen/Strep. After washing with PBS, the cells were treated with Accutase (Merck, Darmstadt, Germany) for 5 min at 37 °C. The detached cells were transferred into a tube, which was filled with PBS and then centrifuged (all following centrifugation steps: 2000× *g* for 5 min at 4 °C). Near-IR fluorescent reactive dye (Thermo Fisher Scientific, Waltham, MA, USA) was used 1:1000 in PBS and added for 30 min at room temperature (RT). As a washing step, the tubes were filled with PBS and centrifuged. We used 4% paraformaldehyde (PFA) for 30 min at RT to fix the samples, followed by a dilution/washing step and centrifugation. The cells were permeabilized for 15 min using 0.5% saponin in PBS at RT, followed by a dilution/washing step and centrifugation. Antibodies were added in a buffer consisting of PBS, 2% FCS, 0.4% 0.5 M ethylenediaminetetraacetic acid (EDTA) and 0.5% saponin and incubated for 1 h at RT. After each antibody-staining step, two washing steps were performed. The samples were analyzed using a BD LSR-II flow cytometer using the software BD FACSDiva (version 9.0). Data were further processed using FlowJo_V10.9.0.

2.6. Immunofluorescence Assay (IFA) for Microscopy

A total of 100,000 Vero E6 cells were seeded in a 24-well plate (Corning Incorporated, New York, USA)) containing 12 mm coverslips (Epredia, Kalamazoo, MI, USA). Alternatively, 50,000 cells were seeded in a 48-well plate containing 7 mm coverslips. The cells were incubated in DMEM + 10% FCS + 1% Pen/Strep for 24 h at 37 °C in 5% CO_2. Dependent on the type of experiment, either an MOI of 1 was used or a fixed volume (5 µL or 50 µL). For infections, the cell layer should be 80% confluent. The used virus volume was filled up to 300 µL per well with DMEM + 2% FCS + 1% Pen/Strep and incubated for 1 h at 37 °C. The cells were washed with PBS and further incubated for 48 h with DMEM + 2%FCS + 1% Pen/Strep. We used 4% PFA for 15 min at RT to fix the cells (Thermo Fisher Scientific, Waltham, MA, USA). As a quenching solution, we used 300 mM glycine (Serva, Heidelberg, Germany) in PBS and then permeabilized the cells in 0.1% Triton X-100 (Thermo Fisher Scientific, Waltham, MA, USA) in PBS for 30 min at RT. As a blocking solution, 1% bovine serum albumin (BSA) (Thermo Fisher Scientific, Waltham, MA, USA) was mixed with 0.1% Tween20 (Thermo Fisher Scientific, Waltham, MA, USA) in PBS and added for 30 min at RT. All following antibodies or dyes were diluted in the BSA-Tween-PBS-blocking solution. The first antibody, as well as the secondary antibody was incubated for 1 h at RT, followed by 20 min incubation with Phalloidin (1:400 dilution pf a 66 µM stock solution; manufacturer's specification; Thermo Fisher Scientific, Waltham, MA, USA; cat. A22287) and 10 min of Hoechst (1 µg/µL; Thermo Fisher Scientific, Waltham, MA, USA; cat. H3570) staining. Between all steps, the samples were washed three times with PBS and once with water before the samples were mounted in ProLong TM Gold antifade reagent (Thermo Fisher Scientific, Waltham, MA, USA). The coverslips were dried for at least 24 h. As a sealant, CoverGripTM (Biotium, Fremont, CA, USA) was used and dried for at least 24 h.

2.7. Fluorescence Microscopy

Fluorescence microscopy was performed on a Nikon Ti2-Eclipse microscope equipped with the spinning disk confocal module CSU-W1 (Yokogawa) as well as an LED widefield illumination light source (pE-4000, CoolLED). Images were acquired with a S Plan Fluor 20× and Plan Fluor 60× oil objective (Nikon, Minator, Tokyo, Japan), a PCOedge 4.2LT or a Zyla 4.2 sCMOS camera (Andor, Belfast, UK) and 405/488/561/638 nm laser lines (Omicron, Rodgau-Dudenhofen, Germany) controlled by NIS-Elements software (version 5.11.01). Z-stack images were acquired with a step size of 300 nm.

2.8. Data Processing and Statistical Analyses

Image analysis was carried out using CellProfiler (version 4.2.1) and NIS-Elements (version 5.11.01) (Nikon, Minator, Tokyo, Japan). Further post-processing was conducted using our custom software Datalyze (Supplementary Software). Further data processing steps and statistical analyses were carried out in NIS-Elements, Inkscape, Excel 2010 and Graphpad Prism 9. Graphical illustrations were created with https://app.biorender.com (accessed on 4 July 2023).

3. Results

3.1. PUUV and TULV Grow to Low Titers, Requiring Time-Consuming Quantifications

Recently, imaging-based virus quantification has become increasingly popular due to its ease of use and the additional amount of information contained in microscopy images [17,18]. Pseudotyped or recombinant viruses encoding a cytosolic fluorescent reporter further enable the quick detection of infected cells. However, some orthohantavirus strains grow to relatively low titers compared with other enveloped RNA viruses and a recombinant system is not yet available. In our hands, PUUV reached a 100-fold lower titer compared to influenza virus A/Puerto Rico/8/34 and La Crosse encephalitis virus (LACV), another bunyavirus. Compared to VSV, the titer was even 10^4-fold lower (Supplementary Figure S1). The typical proliferation time is also considerably longer for orthohantaviruses, which typically require on average 10 days of incubation. Consequently, the small amount of virus material available is typically used up quickly, leading to a constant amplifications process, followed by quantifications of viral titers again for up to 10 days. We thus set out to optimize the orthohantavirus quantification methods to reduce the amount of infectious material and save time compared to the standard method.

3.2. We Present an Improved Flow Cytometry Workflow to Quantify PUUV and TULV with an Optimized Time Point and Reduced Material Requirement

Quantifying orthohantavirus infection using flow cytometry has been described as early as 6 h post-infection for ANDV infection [14]. We thus first tested this early time point after PUUV and TULV infection using the described protocol but could not detect any infected cells (Supplementary Figure S2). Instead, 48 h post-infection enabled a robust detection of infected cells (Figure 1B–E), suggesting that the time point requires optimization depending on the virus strain used. During our optimization, we could further reduce the required cell number by 80%. While the previous protocol describes the use of a 6-well plate format (well surface area: 9.6 cm^2), we could adjust the protocol to a 24-well plate format (well surface area: 1.9 cm^2). For the small amount of required starting material which still allows for the measurement of 10,000 cells, we want to highlight several critical steps in our protocol (Figure 1A). First, the cells were detached from the well plate using Accutase, which we found to be superior compared to trypsin or cell scrapers, which are often used in flow cytometry assays. The cells were collected in a buffer containing EDTA to prevent the cells from sticking together. Then, a live–dead marker was included, which enables the selection of only intact cells for the final analysis. Following this staining step, the cells were fixed in 4% PFA. The subsequent antibody staining requires cell permeabilization, for which we recommend using aponin rather than detergents such as Triton X-100. Orthohantavirus infection is then detected using primary antibodies against the viral nucleocapsid protein N diluted in a buffer containing saponin to retain the cells' permeability for the required intracellular staining. This step is followed by two washing steps and incubation with a secondary antibody, again followed by another two washing steps. Here, the use of a secondary antibody amplifies the signal since one primary antibody can bind more than one secondary antibody. However, due to the many washing steps, we observed a strong loss of cells: on average, 3–10% per washing step (Supplementary Figure S3). The use of a direct fluorophore-conjugated antibody can help overcome this limitation while retaining the use of small cell culture volume (24 well). As we describe in Section 2, the anti-N primary antibody was then conjugated to a selected bright dye in a separate process. Using a pre-conjugated primary antibody saved about 75 min during each of the quantification experiments, as well as at least two washing steps and the associated loss in cell number. As an additional benefit, the background signal was also reduced when compared to the use of a combined primary/secondary antibody in a mock infected control sample (Supplementary Figure S4). Finally, following a simple gating strategy to select only live single cells (Supplementary Figure S5), our optimized protocol enabled the counting of cells infected with PUUV and TULV after 48 h using a small number of input cells.

Microscopy in combination with automated image analysis is a highly sensitive approach for quantifying orthohantavirus infection phenotypes. The mere counting of infected cells based on the presence of a viral protein leaves out the opportunity to use the positional and structural information of the antibody-labeled proteins within the cellular context. Hence, we wondered if we could also use an imaging-based approach to quantify infection with PUUV and TULV in cell culture.

In general, quantification via microscopy requires even lower input material due to the specific and exhaustive imaging of whole cell populations and the low loss during sample preparation. We routinely use either 12 mm coverslips in a 24-well plate or 7 mm coverslips in a 48-well plate (well surface area: 1.1 cm^2). Compared to a 6-well format, this results in a reduction of 89% with respect to the required reagents. Vero E6 cells were infected for 48 h, as described above, then fixed with 4% PFA directly on the glass slides. No detachment and centrifugation steps were required, thus minimizing the cell loss compared to the flow cytometry workflow (Supplementary Figure S5). Nevertheless, on a 7 mm coverslip, it is still possible to acquire 15–25 non-overlapping images using a 21× objective, resulting in a cell count of 5000–10,000 cells for the evaluation, which is comparable with the flow cytometry workflow described above (Figure 2). Following fixation, the cells were permeabilized with Triton X-100 and blocked with a BSA-containing buffer, which was also used for the antibody dilutions. Since there is no risk of losing cell material during the following washing steps, we recommend using primary/secondary antibodies to increase the signal to noise. To enable single cell segmentation later on, we then stained the cell body using phalloidin as well as the nuclear DNA using Hoechst (Figure 2D–K). Images were taken at 20× magnification, whereby one image corresponds to approximately 200–500 Vero E6 cells (Figure 2B) that could still be segmented adequately as shown by the corresponding software-based mask in Figure 2C. For image processing, we used the licensed program NIS-Elements AR, as well as the open-source tool CellProfiler [19]. Both programs first segment the cell nuclei (Figure 2F), followed by detection of the cell boundaries (Figure 2G,J), whereby not only epithelial cells can be recognized but also fibroblast-like ones (Supplementary Figure S6). Following single-cell and nuclear recognition, segmentation of the N protein (Figure 2H,K) and association of the different masks can be used to distinguish infected from non-infected cells. Due to the mask association (per single-cell), virus particles non-specifically bound to the glass slide can easily be ignored. We provide a detailed walkthrough of the image analysis in our short tutorial (Supplementary Software; sample datasets can be downloaded from https://zenodo.org/records/10417810 (accessed on 7 January 2024)). Using this protocol, we then analyzed cells infected with PUUV and TULV at different time points post-infection (6 h, 12 h, 24 h and 48 h). Interestingly, a positive infection with PUUV, as well as with TULV, could already be detected at 6 h post-infection (Figure 3C), which was not possible using flow cytometry (Supplementary Figure S2). After 48 h, we found similar infection rates as measured by flow cytometry when using the same input virus (Supplementary Figure S7). Over the course of infection, we observed a correlation between the number of infected cells and the intensity of the segmented signal (Figure 3A,B). This observation implies that within the initial 48 h of infection, the two parameters are linked, and the quantification mainly monitors the primary infection. To test this conclusion, we used the supernatant of all time points to infect new Vero E6 cells for a subsequent 48 h period. Interestingly, we observed infection rates at background levels for all supernatant samples except the 48 h time point (Supplementary Figure S8), supporting our conclusion that we are effectively detecting the ongoing primary infection process (Figure 3). Based on these findings, we recommend the 48 h time point post-infection, characterized by the highest signal intensity, as the optimal time frame for quantifying PUUV and TULV stocks.

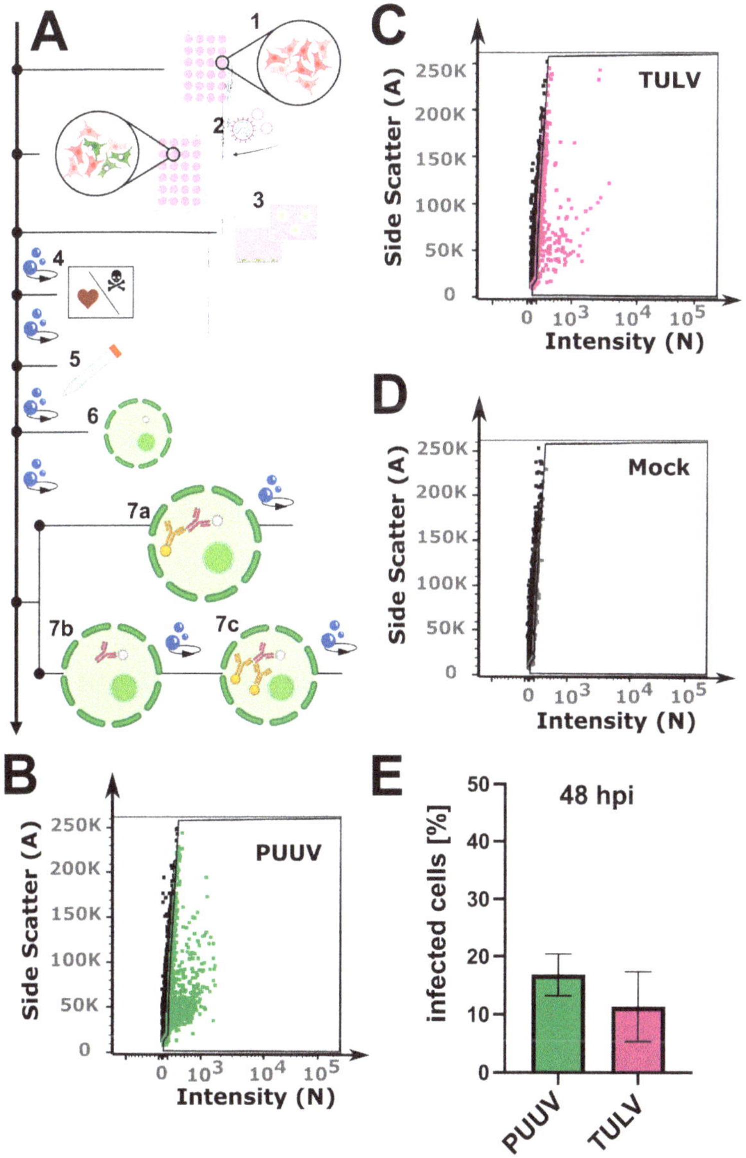

Figure 1. Flow cytometry workflow for quantifying orthohantavirus infection in cell culture. (**A**) Schematic overview of the protocol: (1) cell seeding; (2) infection with orthohantavirus for 1 h; (3) cell

detachment with Accutase; (4) live–dead staining; (5) fixation with paraformaldehyde (PFA); (6) cell permeabilization with saponin; (7) staining with antibodies: either with a conjugated antibody (7a) or with a primary (7b) and a secondary (7c) antibody. (**B–D**) Exemplary gate settings for quantification of newly propagated virus; their analysis in (**E**) is a bar graph displaying the percentage of infected cells (%). Shown are mean ± SD, N = 3.

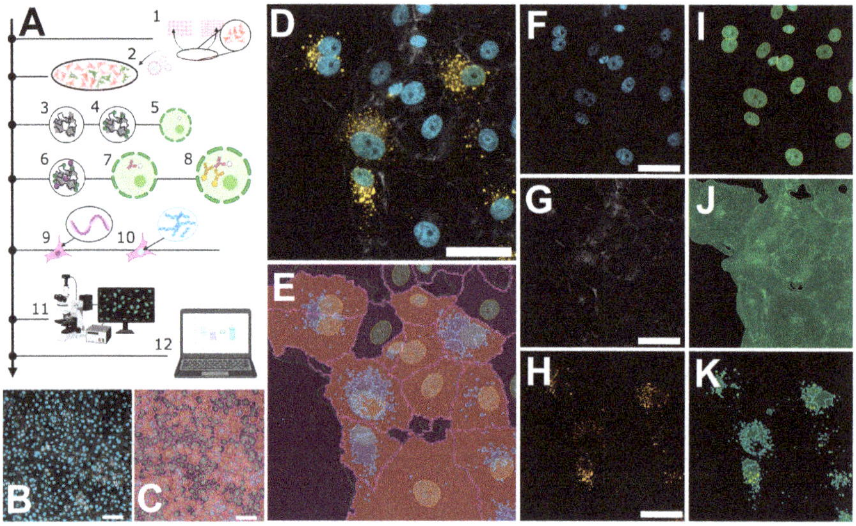

Figure 2. Microscopic workflow for quantifying orthohantavirus infection in cell culture. (**A**) Schematic overview of the protocol: cell seeding (1); infection with orthohantavirus for 1 h (2); cell fixation with paraformaldehyde (PFA) (3); blocking of unreacted aldehydes with glycine solution (4); permeabilization with Triton X-100 (5); blocking of unspecific binding sites using a BSA-containing buffer (6); staining with primary antibody against viral nucleocapsid protein (N) (7) and staining with a secondary antibody (8); phalloidin staining to label filamentous actin (9) and Hoechst for labelling dsDNA (10); and imaging via fluorescence microscopy (11) and analysis via custom software tools (12). (**B**,**C**) Exemplary images recorded with 20× magnification, showing ~500 cells. (**D–K**) Pairwise original images together with their masks resulting from software-based segmentation. (**D**) A merged image of three individual channels in (**F–H**): Hoechst (**F**) and its mask for nuclei recognition (**I**); phalloidin staining (**G**) for recognizing cell bodies or cell boundaries; and Anti-N staining (**H**) to reveal viral particles (**K**). The overlay of (**I–K**) is shown in (**E**). Scale bars = 50 μm.

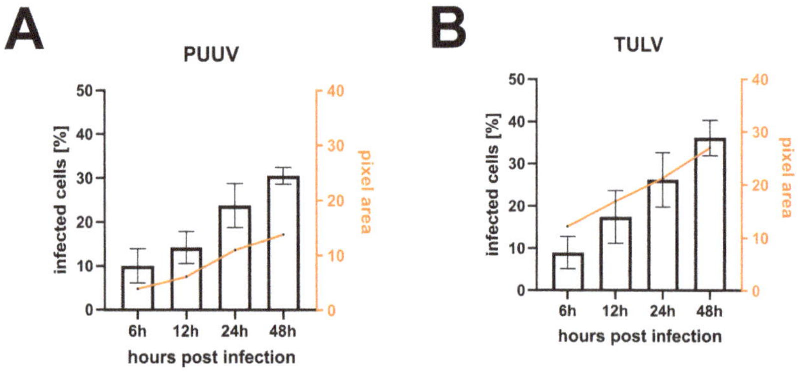

Figure 3. *Cont.*

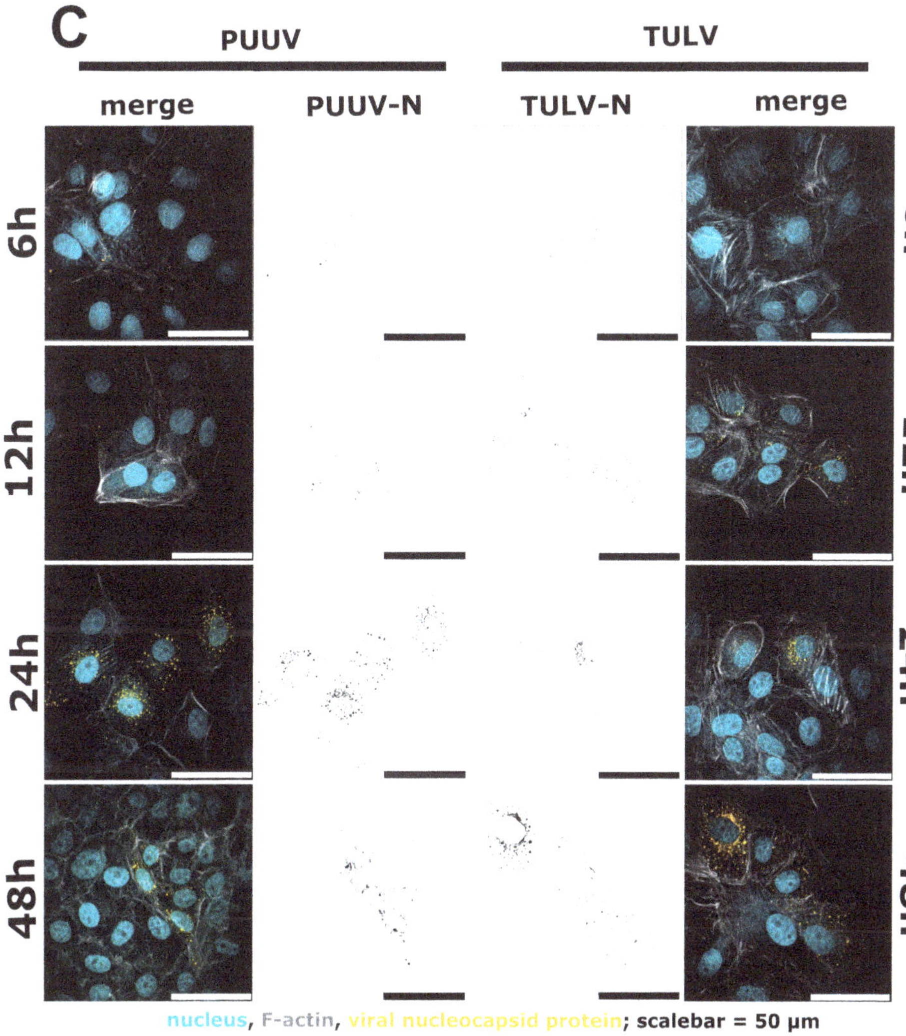

Figure 3. Quantification of infected Vero E6 cells 6 h, 12 h, 24 h and 48 h post-infection with either PUUV (**A**) or TULV (**B**). A comparison of the percentage of infected cells shows an increase in detected cells after 48 h compared to 6 h. We observed a correlation between the number of infected cells and the intensity per pixel. Shown are mean ± SD, n = 3. Exemplary images in (**C**). Merged image show nuclei in blue, F-actin in grey and the viral N protein in orange. Viral N is also shown in black as a separated image in the middle of the image panel.

3.3. Custom Analysis Software Tools Allow for the Quick Visualization of Image Data and the Calculation of Infection Rates

The image analysis pipeline discussed above produces signal intensity datasets in tabular format, which require further processing. To ease this step, we provide two versions of our open-source software equipped with a graphical user interface (Figure 4B–D) so that no programming skills are required. After downloading, the software to process CellProfiler or Nis Elements data is ready to use without further installation. The raw data can be visualized as a scatterplot (Figure 4C), whereby datasets are color-coded and can be further summarized as an average percentage of infected cells (Figure 4D). For more detailed information, please refer to the provided software manual packaged together with the Supplementary Software. Finally, to convert the fraction of infected cells into infectious particles per milliliter, three parameters are necessary: (1) the percentage of infected cells as given by our analysis tool (e.g., 50.9% converted as a fraction (0.509)); (2) the used cell numbers (e.g., 100,000 cells per well); and (3) the amount of virus suspension (e.g., 5 µL per well) used to include the used dilution (1:200) to finally calculate infectious particles per mL.

$$\text{Fraction of infected cells [decimal]} \times \text{cell number} \times \text{dilution [e.g., 5 µL = 1:200]}$$

$$= \text{infectious particles/Milliliter}$$

$$\text{Example: } 0.509 \times 100{,}000 \times 200 = 10{,}180{,}000 \text{ or } 10.18 \times 10^6 \text{ infectious particles per mL.}$$

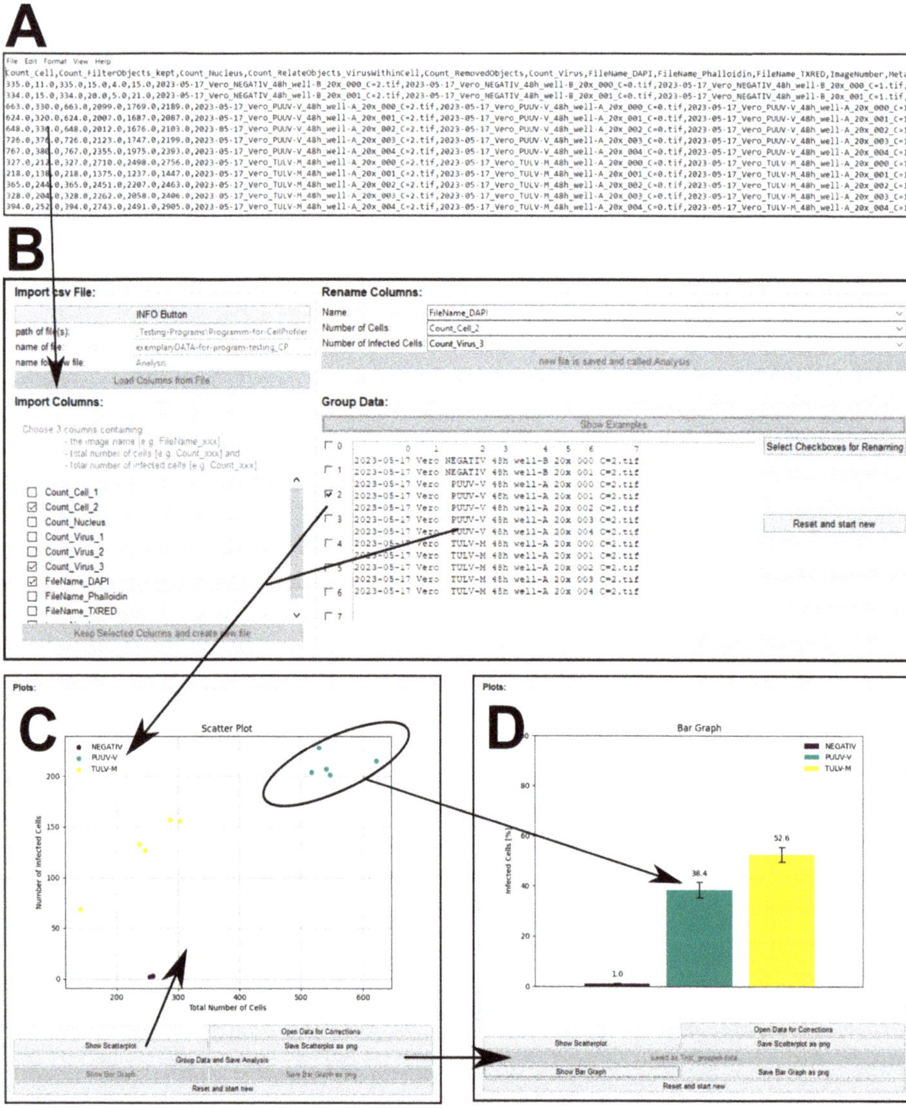

Figure 4. Custom software tool for processing CellProfiler output. (**A**) An exemplary csv file as a result of using CellProfiler to quantify PUUV and/or TULV infection after 48 h. (**B**) Graphical user

interface (GUI) of the software that allows for the processing of csv files. The software tool summarizes and groups' important data columns based on the image name to obtain visual output in the form of scatterplots (**C**) or bar graphs (**D**).

3.4. Automated Image Analysis Allows for the Extraction of Structural Parameters to Describe the Infection Status

During the analysis of our acquired images, we noticed that the size of the N protein structures per cell changes over the course of infection. This corresponds to different levels of N protein expression per cell, presumably corresponding to different stages of viral replication. We thus wanted to test if such changes could be visualized using our image analysis pipeline. Indeed, in addition to the percentage of infected cells, other parameters, such as the size of the N protein segments, can easily be extracted and displayed over time (Figure 5E and Supplementary Figure S9). Both PUUV-N and TULV-N particles are detected after 6 h as small and mostly round-shaped structures. While the number of particles increases over time, their shape also changes, which can be measured using the area of the segmented N protein signal. TULV-N, as well as forming bigger round particles, also form long fiber-like structures which can then be investigated in detail using higher magnification (Figure 5F,G).

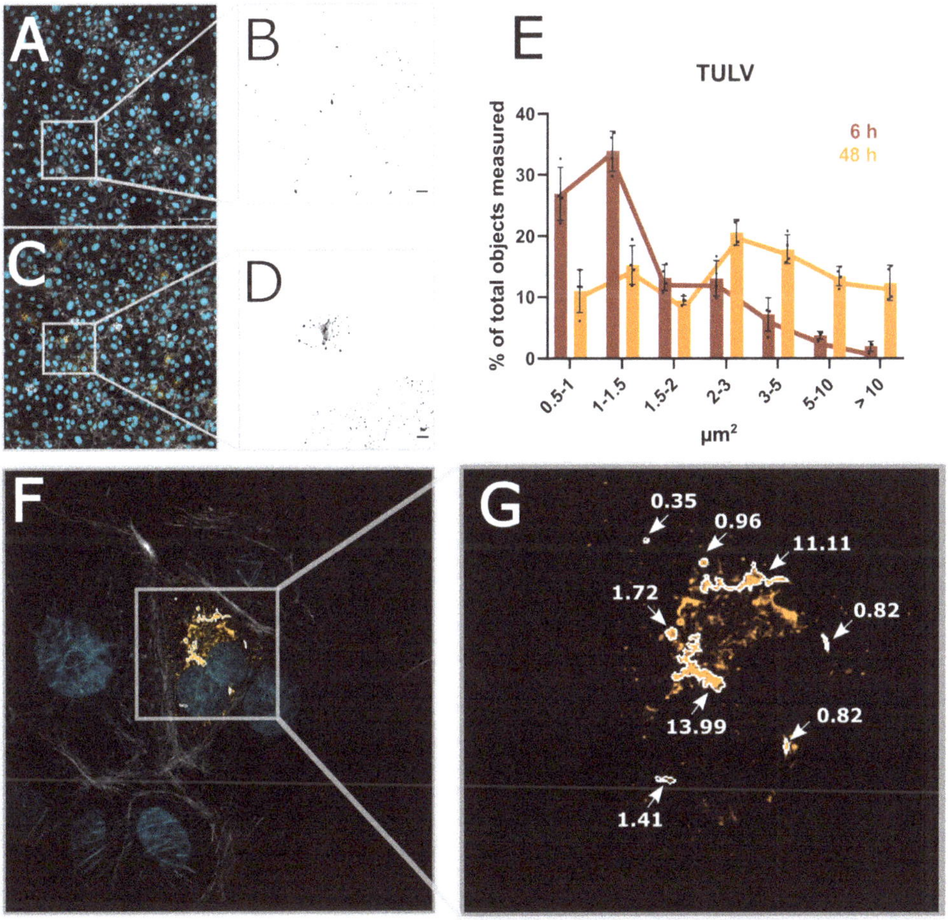

Figure 5. Automated image analysis allows for the extraction of structural parameters. (**A**) A micrograph taken with 20× magnification of Vero E6 cells infected with TULV 6 h post-infection.

(**B**) The viral N protein; scale bar = 10 μm. (**C,D**) The 48 h post-infection time point of TULV infection. (**E**) The size of TULV-N segmented particles at either 6 h or 48 h post-infection. (**F**) A 100× magnification of 48 hpi TULV-infected Vero E6 cells, (**G**) Exemplary TULV-N structures and the correlating size.

4. Conclusions

We developed an optimized workflow to determine the infectious titer of two orthohantavirus strains, PUUV and TULV, using flow cytometry and fluorescence imaging. Our improved workflow allows us to identify infected cells as early as 6 h post-infection, but we recommend using the 48 h time point for the robust recognition of infected cells. We want to point out that in our study, we typically used low MOIs to save material and avoid challenges in obtaining high titers. It would be interesting to determine the titer over a larger range of MOI and compare our methods with, for example, TCID50 curves. In this study, we provide a detailed protocol of the sample preparation procedure and include new user-friendly and open-source software tools to quickly process results from automated image analysis. In conclusion, our microscopic assay enables the fast identification of infected cells with low cell loss and minimized material requirements. Imaging-based quantification furthermore allows for the extraction of structural and positional features of the detected signal, which can later be used for the correlation of other phenotypes such as replication status, as we show, or cell cycle, for example.

Supplementary Materials: The following supporting information can be downloaded at: https://www.mdpi.com/article/10.3390/v16020269/s1, Figure S1: Comparison of the infectious titer of different enveloped RNA virus; Figure S2: Flow cytometry measurements of Vero E6 cells 6 h post infection with PUUV and TULV; Figure S3: Percentual decrease of Vero E6 cells due to consecutively washing steps in PBS during the flow cytometry staining protocol; Figure S4: Staining background created by different usage of antibodies (AB). Figure S5: Exemplary gating settings for flow cytometric analysis of infected cells; Figure S6: Software recognition of infected cells with a fibroblastic cell-type; Figure S7: Quantification of PUUV and TULV infectious titer using microscopy or flow cytometry; Figure S8: Release of infectious viruses after different time points; Figure S9: Size distribution of identified segments of PUUV-N identified in infected cells at different time points (6 h vs. 48 h) in Vero E6 cells.

Author Contributions: Conceptualization, C.S.; methodology, L.M.; software, L.M.; validation L.M.; formal analysis, L.M.; investigation, L.M.; resources, L.M. and C.S.; data curation, L.M.; writing—original draft preparation, L.M. and C.S.; writing—review and editing, all authors; visualization, L.M. and C.S.; supervision, C.S.; project administration, C.S.; funding acquisition, C.S. All authors have read and agreed to the published version of the manuscript.

Funding: C.S. acknowledges support from the Helmholtz Association (VH-NG-1526) within the Helmholtz Young Investigator Program and the Initiative and Networking Fund for Infection Research Greifswald (Project HANTadapt).

Institutional Review Board Statement: Not applicable.

Informed Consent Statement: Not applicable.

Data Availability Statement: Supplementary data are provided. Example raw data for testing the provided software are available at https://doi.org/10.5281/zenodo.10417810.

Acknowledgments: MGLU-2-R (CCLV-RIE 1304) cells were kindly provided by Rainer G. Ulrich, Friedrich-Loeffler-Institute, Greifswald—Insel Riems, Germany. Puumala virus (PUUV Vranica) was kindly provided by Sandra Essbauer, Bundeswehr Institute of Microbiology, Munich, Germany. Tula virus (TULV Moravia/Ma5302V) was kindly provided by Rainer G. Ulrich (Friedrich-Loeffler-Institute, Greifswald—Insel Riems, Germany) and Detlev H. Krüger (Institut für Virologie an der Medizinischen Fakultät (Charité) der Humboldt-Universität, Berlin, Germany). Illustrations were generated with BioRender.

Conflicts of Interest: The authors declare no conflict of interest.

References

1. Schmaljohn, C.S.; Hasty, S.E.; Harrison, S.A.; Dalrymple, J.M. Characterization of Hantaan Virions, the Prototype Virus of Hemorrhagic Fever with Renal Syndrome. *J. Infect. Dis.* **1983**, *148*, 1005–1012. [CrossRef]
2. Schmaljohn, C.S.; Dalrymple, J.M. Analysis of hantaan virus RNA: Evidence for a new genus of bunyaviridae. *Virology* **1983**, *131*, 482–491. [CrossRef] [PubMed]
3. Elliott, L.H.; Kiley, M.P.; McCormick, J.B. Hantaan Virus: Identification of Virion Proteins. *J. Gen. Virol.* **1984**, *65 Pt 8*, 1285–1293. [CrossRef] [PubMed]
4. Jonsson, C.B.; Figueiredo, L.T.M.; Vapalahti, O. A Global Perspective on Hantavirus Ecology, Epidemiology, and Disease. *Clin. Microbiol. Rev.* **2010**, *23*, 412–441. [CrossRef] [PubMed]
5. Vapalahti, O.; Kallio-Kokko, H.; Närvänen, A.; Julkunen, I.; Lundkvist, Å.; Plyusnin, A.; Lehvaslaiho, H.; Brummer-Korvenkontio, M.; Vaheri, A.; Lankinen, H. Human B-cell epitopes of Puumala virus nucleocapsid protein, the major antigen in early serological response. *J. Med. Virol.* **1995**, *46*, 293–303. [CrossRef] [PubMed]
6. Byström, J.W.; Näslund, J.; Trulsson, F.; Evander, M.; Lwande, O.W.; Ahlm, C.; Bucht, G. Quantification and kinetics of viral RNA transcripts produced in Orthohantavirus infected cells. *Virol. J.* **2018**, *15*, 18. [CrossRef] [PubMed]
7. Lee, H.W.; Lee, P.W.; Johnson, K.M. Isolation of the Etiologic Agent of Korean Hemorrhagic Fever. *J. Infect. Dis.* **1978**, *137*, 298–308. [CrossRef] [PubMed]
8. French, G.R.; Foulke, R.S.; Brand, O.A.; Eddy, G.A.; Lee, H.W.; Lee, P.W. Korean Hemorrhagic Fever: Propagation of the Etiologic Agent in a Cell Line of Human Origin. *Science* **1981**, *211*, 1046–1048. [CrossRef] [PubMed]
9. McCormick, J.; Palmer, E.; Sasso, D.; Kiley, M. Morphological Identification of the Agent of Korean Haemorrhagic Fever (Hantaan Virus) as A Member of the Bunyaviridae. *Lancet* **1982**, *319*, 765–768. [CrossRef] [PubMed]
10. Schmaljohn, C.S.; Hasty, S.E.; Dalrymple, J.M.; LeDuc, J.W.; Lee, H.W.; von Bonsdorff, C.-H.; Brummer-Korvenkontio, M.; Vaheri, A.; Tsai, T.F.; Regnery, H.L.; et al. Antigenic and Genetic Properties of Viruses Linked to Hemorrhagic Fever with Renal Syndrome. *Science* **1985**, *227*, 1041–1044. [CrossRef] [PubMed]
11. Tanishita, O.; Takahashi, Y.; Okuno, Y.; Yamanishi, K.; Takahashi, M. Evaluation of focus reduction neutralization test with peroxidase-antiperoxidase staining technique for hemorrhagic fever with renal syndrome virus. *J. Clin. Microbiol.* **1984**, *20*, 1213–1215. [CrossRef] [PubMed]
12. Bharadwaj, M.; Nofchissey, R.; Goade, D.; Koster, F.; Hjelle, B. Humoral Immune Responses in the Hantavirus Cardiopulmonary Syndrome. *J. Infect. Dis.* **2000**, *182*, 43–48. [CrossRef] [PubMed]
13. Heider, H.; Ziaja, B.; Priemer, C.; Lundkvist, Å.; Neyts, J.; Krüger, D.H.; Ulrich, R. A chemiluminescence detection method of hantaviral antigens in neutralisation assays and inhibitor studies. *J. Virol. Methods* **2001**, *96*, 17–23. [CrossRef] [PubMed]
14. Barriga, G.P.; Martínez-Valdebenito, C.; Galeno, H.; Ferrés, M.; Lozach, P.-Y.; Tischler, N.D. A rapid method for infectivity titration of Andes hantavirus using flow cytometry. *J. Virol. Methods* **2013**, *193*, 291–294. [CrossRef]
15. Kucinskaite-Kodze, I.; Petraityte-Burneikiene, R.; Zvirbliene, A.; Hjelle, B.; Medina, R.A.; Gedvilaite, A.; Razanskiene, A.; Schmidt-Chanasit, J.; Mertens, M.; Padula, P.; et al. Characterization of monoclonal antibodies against hantavirus nucleocapsid protein and their use for immunohistochemistry on rodent and human samples. *Arch. Virol.* **2011**, *156*, 443–456. [CrossRef] [PubMed]
16. Bates, M.; Jones, S.A.; Zhuang, X. Preparation of Photoswitchable Labeled Antibodies for STORM Imaging. *Cold Spring Harb. Protoc.* **2013**, *2013*, 540–541. [CrossRef] [PubMed]
17. Menke, L.; Sperber, H.S.; Aji, A.K.; Chiantia, S.; Schwarzer, R.; Sieben, C. Advances in fluorescence microscopy for orthohantavirus research. *Microscopy* **2023**, *72*, 191–203. [CrossRef] [PubMed]
18. Witte, R.; Andriasyan, V.; Georgi, F.; Yakimovich, A.; Greber, U.F. Concepts in Light Microscopy of Viruses. *Viruses* **2018**, *10*, 202. [CrossRef] [PubMed]
19. Carpenter, A.E.; Jones, T.R.; Lamprecht, M.R.; Clarke, C.; Kang, I.H.; Friman, O.; Guertin, D.A.; Chang, J.H.; Lindquist, R.A.; Moffat, J.; et al. CellProfiler: Image analysis software for identifying and quantifying cell phenotypes. *Genome Biol.* **2006**, *7*, R100. [CrossRef] [PubMed]

viruses

MDPI

Review

Bridging the Structural Gap: A Methodological Review of Cryo-Electron Microscopy for Underrepresented Viruses

Yoon Ho Park [1], Hyun Suk Jung [1], Sungjin Moon [2,*] and Chihong Song [3,*]

[1] Department of Biochemistry, College of Natural Sciences, Kangwon National University, Chuncheon 24341, Republic of Korea; yhpark99@kangwon.ac.kr (Y.H.P.)
[2] Department of Biological Science, Kangwon National University, Chuncheon 24341, Republic of Korea
[3] Department of Convergence Medicine, School of Medicine, Pusan National University, Yangsan 50612, Republic of Korea
* Correspondence: sungjin.moon@kangwon.ac.kr (S.M.); chsong@pusan.ac.kr (C.S.); Tel.: +82-33-250-8536 (S.M.); +82-51-510-8081 (C.S.)

Abstract

Cryo-electron microscopy (cryo-EM) has revolutionized structural virology, enabling routine structure determination at 2–4 Å resolution, with exceptional cases reaching 1.56 Å. The structural diversity of viruses across vertebrate, plant, and insect hosts provides fundamental insights into infection mechanisms, host–pathogen coevolution, and therapeutic target identification. However, analysis of Electron Microscopy Data Bank entries reveals notable disparities in structural coverage: among 11,717 eukaryotic virus structures (excluding bacteriophages), vertebrate viruses constitute 97.6% (n = 11,432) of deposited entries, while plant viruses (1.0%; n = 117) and insect viruses (1.4%; n = 168) remain significantly underrepresented. This bias stems from distinct technical barriers including size limitations for giant viruses exceeding 200 nm, the loss of asymmetric information during symmetry-imposed processing, and the morphological complexity of filamentous and pleomorphic viruses. Each barrier has driven the development of specialized methodological solutions: block-based local refinement overcomes through-focus variations in giant viruses, cryo-electron tomography (cryo-ET) validates and reveals asymmetric features lost in symmetrized reconstructions, and subtomogram averaging enables structural analysis of pleomorphic assemblies. This review synthesizes recent methodological advances, critically evaluates their capacity to address specific technical barriers, and proposes strategies for expanding structural investigations across underrepresented host systems to achieve comprehensive understanding of viral structural biology.

Keywords: Cryo-EM; vertebrate viruses; insect viruses; plant viruses; giant viruses

1. Introduction to Host-Specific Barriers in Viral Structure

Viruses represent one of the most diverse biological entities, infecting virtually every form of cellular life across vertebrate, plant, and insect kingdoms [1]. Understanding the structural basis of viral infection requires atomic-level resolution of viral components, from capsid proteins to envelope glycoproteins. Structural virology has undergone a transformative revolution in cryogenic electron microscopy (cryo-EM), evolving from a technique once termed "blobology" to a method capable of achieving near-atomic to atomic resolution structures [2,3].

The importance of characterizing virus structures across diverse host systems cannot be overstated. Animal viruses, particularly those causing human disease, have received disproportionate attention due to their direct medical relevance and associated research funding. Plant viruses, despite causing estimated annual crop losses exceeding $30 billion globally [4], remain comparatively understudied at the structural level. Insect viruses, including those infecting model organisms such as *Drosophila melanogaster*, offer unique opportunities to understand fundamental aspects of viral evolution and innate immunity, yet face substantial barriers limiting their structural characterization.

This review critically examines the current landscape of host-specific virus structure determination, analyzing the technical barriers that have created disparities in structural coverage across host systems. By identifying the specific challenges posed by viral size, symmetry, and morphology, and evaluating the methodological solutions developed to address each barrier, we aim to provide a framework for achieving more comprehensive structural coverage of the virosphere.

2. Host-Specific Bias in Structural Databases

Analysis of Electron Microscopy Data Bank (EMDB) entries (accessed December 2025) reveals notable disparities in structural coverage across virus host categories (Figure 1). Of 12,875 total virus-related entries, 1158 correspond to bacteriophages, leaving 11,717 eukaryotic virus structures. Among these, vertebrate viruses dominate at 97.6% (n = 11,432), while plant viruses account for merely 1.0% (n = 117) and insect viruses 1.4% (n = 168) (Figure 1). This reflects a complex interplay of funding priorities, research infrastructure, and intrinsic technical challenges associated with each virus category.

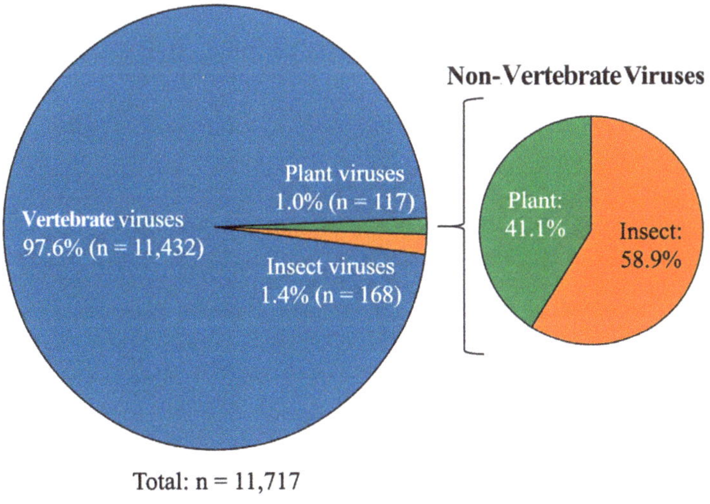

Figure 1. Host-specific distribution of eukaryotic virus structures in the Electron Microscopy Data Bank (EMDB). Host-specific distribution of eukaryotic virus structures (n = 11,717) in the Electron Microscopy Data Bank (EMDB). Analysis of EMDB entries (accessed December 2025) reveals notable disparities in structural coverage across virus host categories.

2.1. Vertebrate Viruses

Vertebrate viruses, particularly those causing human disease, represent the most extensively characterized category in structural virology. This dominance reflects decades of sustained research investment driven by public health priorities and the infrastructure

 https://doi.org/10.3390/v18020195

of medical research institutions. The extensive characterization of medically relevant families including Coronaviridae, Flaviviridae, Orthomyxoviridae, and Herpesviridae has established a rich structural foundation for understanding viral pathogenesis [5–7].

Vertebrate viruses exhibit structural diversity, ranging from small icosahedral particles (parvoviruses, ~25 nm) to large enveloped virions (herpesviruses, ~200 nm) and pleomorphic assemblies (orthomyxoviruses, influenza) [8]. Many medically important vertebrate viruses possess well-ordered icosahedral symmetry amenable to single-particle analysis, enabling routine high-resolution structure determination [9]. However, significant subsets including retroviruses (HIV), paramyxoviruses, and filoviruses exhibit pleomorphic enveloped morphologies that require specialized tomographic approaches [10].

Recent achievements demonstrate the capabilities of modern cryo-EM for vertebrate virus characterization. The determination of adeno-associated virus at 1.56 Å resolution represents a landmark where individual hydrogen atoms and ordered water molecules become visible [11,12]. Beyond static structures, cryo-EM has captured functional conformational dynamics, including the "rising" and "resting" conformations of norovirus P-domains and portal-like assemblies formed by calicivirus VP2 following receptor engagement [13,14].

2.2. Plant Viruses

Despite their substantial agricultural and economic impact, plant viruses constitute only 1.0% of EMDB eukaryotic virus structure entries, a mere 117 structures compared to over 11,000 vertebrate virus structures. This underrepresentation persists despite plant viruses causing estimated annual global crop losses exceeding $30 billion and threatening food security in developing regions. The reflects systematic biases in research funding that prioritize human and veterinary pathogens, as well as distinct technical challenges specific to plant virus structural biology [15].

The structural landscape of plant viruses has historically been dominated by well-behaved model systems. Tobacco mosaic virus (TMV), bromoviruses, tombusviruses, and geminiviruses have received moderate structural attention due to their amenability to high-yield purification and their rigid icosahedral or well-ordered helical symmetries [16–19]. However, these models do not represent the majority of agriculturally significant pathogens, particularly the flexuous filamentous viruses that cause the bulk of viral crop diseases.

Plant viruses display distinctive morphological features that distinguish them from most vertebrate viruses. A disproportionate number of economically important plant viruses, particularly members of Potyviridae (the largest family of plant-infecting RNA viruses), Closteroviridae, and Flexiviridae, exhibit flexuous filamentous morphologies rather than compact icosahedral shapes [20]. These elongated, flexible particles present significant challenges for high-resolution structure determination, particularly in image processing, due to their inherent conformational heterogeneity [21]. Furthermore, plant viruses encode movement proteins to facilitate cell-to-cell transport through plasmodesmata. Although these proteins represent unique structural targets not found among vertebrate viruses, their tight association with cellular membranes creates substantial hurdles for isolation and structural characterization [22].

Notably, the structure of sweet potato feathery mottle virus (Potyviridae) was determined at 2.9 Å resolution, revealing the conserved coat protein fold and its interaction with the single-stranded RNA genome [23]. Similarly, structures of sweet potato mild mottle virus and Pepino mosaic virus have been resolved at 2.6 Å and 3.9 Å, respectively, providing important insights into the assembly mechanisms of flexuous virions [23,24]. These breakthroughs demonstrate that the technical barriers limiting plant virus structural biology are becoming surmountable.

 https://doi.org/10.3390/v18020195

2.3. Insect Viruses

Insect viruses represent another underexplored category in structural virology, constituting only 1.4% (n = 168) of EMDB eukaryotic virus structure entries despite insects comprising the most species-rich animal class. Although insect viruses offer unique biological insights and hold practical value for biological pest control and recombinant protein expression systems, they remain significantly underrepresented in structural databases [25].

Insect viruses encompass structural diversity, including large occluded baculoviruses with complex biphasic life cycles, small RNA viruses (Dicistroviridae, Iflaviridae), and unique double-stranded RNA viruses (Birnaviridae, Reoviridae) [26,27]. Many insect-specific virus families lack close relatives among vertebrate viruses, potentially harboring novel protein architectures not represented in current structural databases. Baculoviruses are particularly notable for their occlusion bodies crystalline protein matrices embedding rod-shaped nucleocapsids representing a structural form unique among viruses [28].

In the context of host–pathogen interactions, *Drosophila melanogaster* serves as a powerful genetic model for dissecting innate immunity, particularly RNA interference pathways and virus–host coevolution [29]. The fruit fly lacks adaptive immunity, making it an ideal system for understanding fundamental antiviral mechanisms conserved across arthropods [30]. However, despite the extensive genetic characterization of the host, the structural biology of its native viruses remains relatively limited. While the structure of *Drosophila C virus* (DCV) has been well-characterized, providing insights into capsid stability and genome release, many other *Drosophila*-specific viruses identified through metagenomics remain structurally undefined [31]. Expanding structural studies to these neglected viruses is essential to fully exploit *Drosophila* as a comprehensive model for viral pathogenesis.

3. Technical Barriers Driving the Bias

The disparities in structural coverage across host systems arise not only from funding and priority differences but also from distinct technical barriers that differentially affect plant and insect viruses (Table 1). Understanding these barriers is essential for developing targeted solutions and achieving comprehensive structural coverage of the virosphere.

Table 1. Structural features and associated technical challenges in underrepresented viral lineages.

Target Category	Representative Models	Structural Feature	Technical Challenge	Methodological Solution
Giant Viruses	Mimivirus, Iridovirus, PBCV-1	Large diameter	Ewald sphere curvature effects	Block-Based Reconstruction (Local Refinement)
Plant Viruses	Potyviridae, Closteroviridae	Flexuous/Filamentous morphology	Conformational heterogeneity	Subtomogram Averaging (STA) or Helical Reconstruction
Insect Viruses	Caudovirales (phage-like), Podoviridae	Hidden Asymmetry (e.g., Portals)	Averaging artifacts	Cryo-ET Validation & Symmetry Relaxation

3.1. Size Barriers: Ewald Sphere Curvature and Depth of Field Effects in Giant Viruses

Giant viruses, including members of the Nucleocytoplasmic Large DNA Virus (NCLDV) superfamily, present fundamental physical challenges that standard cryo-EM approaches cannot overcome [32]. As viral dimensions increase beyond 200 nm, two related physical phenomena become limiting factors for high-resolution structure determination.

The electron microscope focuses on a specific plane within the sample. For giant viruses, the top and bottom of the capsid are significantly defocused relative to the central

https://doi.org/10.3390/v18020195

plane, introducing blur that degrades high-resolution information. This "depth of field" problem worsens proportionally with particle size, creating an effective resolution ceiling for large assemblies imaged using standard protocols [33].

In electron scattering, the assumption that the Ewald sphere can be approximated as a plane becomes invalid for large objects at high resolution. The curvature of the Ewald sphere introduces phase errors that manifest as resolution anisotropy and blurring in reconstructions [34]. For giant viruses, these effects become non-negligible at resolutions better than approximately 5–7 Å, historically limiting structures to moderate resolution insufficient for atomic model building.

While giant viruses are found across all host categories, certain plant-infecting NCLDVs (phycodnaviruses infecting algae) and insect-infecting viruses (iridoviruses, entomopoxviruses) fall into the giant virus category. The size barrier has historically prevented high-resolution structural analysis of these important virus families, contributing to the underrepresentation of plant and insect viruses in structural databases.

3.2. Symmetry Barriers: Obscuring of Functional Asymmetry by Icosahedral Averaging

Single-particle analysis (SPA) of icosahedral viruses routinely exploits 60-fold symmetry during reconstruction to significantly improve signal-to-noise ratios [35]. While this symmetry imposition enables high-resolution structure determination from modest dataset sizes, it creates a fundamental barrier: biologically relevant asymmetric features are averaged away during processing, resulting in loss of asymmetric information that may be functionally critical [36].

Recent studies have revealed that even viruses assumed to possess perfect icosahedral symmetry exhibit imperfect icosahedral symmetry when analyzed without symmetry imposition [37]. Alphaviruses, previously characterized exclusively by symmetry-imposed single-particle analysis, show asymmetric features that are functionally relevant but completely absent in symmetrized reconstructions. These include genome organization patterns, unique vertices associated with portal complexes, and localized structural heterogeneity.

Many plant viruses package their genomes in specific orientations or utilize unique vertices for genome delivery features invisible under imposed symmetry [38]. Insect viruses, particularly those with complex life cycles involving different host cell types, may exhibit structural variations between morphological forms that symmetry averaging obscures. The reliance on symmetry-imposed processing has potentially masked important biological insights for these underrepresented virus categories.

3.3. Morphology Barriers: Structural Heterogeneity and Flexibility in Filamentous and Pleomorphic Viruses

Standard single-particle analysis is optimized for compact, roughly spherical particles that can be accurately aligned and averaged. Viruses with non-spherical morphologies particularly filamentous and pleomorphic forms present distinct challenges that have historically limited their structural characterization.

Flexuous filamentous viruses, which include many of the most economically important plant pathogens (Potyviridae, Closteroviridae, Flexiviridae), exhibit continuous bending and twisting along their length. This conformational heterogeneity prevents straightforward helical reconstruction, as different segments of the same particle adopt different conformations. Additionally, filament length heterogeneity and end effects complicate particle selection and averaging strategies. Standard helical reconstruction assumes rigid, straight particles an assumption violated by flexuous viruses [39].

Enveloped viruses such as influenza, HIV, and baculoviruses incorporate host-derived lipid membranes and display surface glycoproteins in irregular patterns. These pleomor-

 https://doi.org/10.3390/v18020195

phic viruses vary in size, shape, and surface protein arrangement between individual particles, precluding the averaging strategies that enable high-resolution single-particle analysis. Each particle is essentially unique, eliminating the fundamental assumption of particle homogeneity underlying SPA.

The morphology barrier disproportionately affects plant viruses, where flexuous filamentous morphology predominates among economically important pathogens. For insect viruses, in the budded form of baculoviruses, the morphological form mediating systemic infection is pleomorphic and enveloped, explaining why baculovirus structural studies have focused predominantly on the occluded form despite the biological importance of budded virions. These morphological barriers have been primary drivers of the underrepresentation of plant and insect viruses in structural databases.

4. Methodological Solutions

Each technical barrier has driven the development of specialized methodological solutions. Understanding which method addresses which barrier enables strategic application of these approaches to underrepresented virus systems.

4.1. Block-Based Reconstruction: Mitigating Ewald Sphere and Focus Gradients via Local Refinement

The block-based reconstruction approach (also termed local refinement or divide-and-conquer strategy) computationally divides large viral particles into smaller blocks, each treated as an independent particle for local defocus refinement and reconstruction [34,40]. Within each small block, the assumptions of planar CTF and negligible Ewald sphere curvature hold true. The locally refined blocks are subsequently reassembled to generate the complete viral structure, effectively bypassing the physical limitations that arise at the whole-particle level.

Application of block-based reconstruction has enabled near-atomic resolution structures of multiple giant viruses. Singapore grouper iridovirus (SGIV), with a diameter of approximately 228 nm, was resolved at 3.5 Å, revealing the architecture of its major and minor capsid proteins [41]. Medusavirus capsid maturation was characterized at 7.3–9.9 Å resolution, providing insights into pentascale structural transitions [42]. Jyvaskylavirus, the first giant virus isolated from Finland, further demonstrates expanding structural coverage of NCLDVs [43]. The method has also been applied to jumbo bacteriophages, resolving the capsid structure of bacteriophage ΦKZ [44].

Block-based reconstruction introduces specific trade-offs. Computational costs scale with particle size, as each block requires independent processing. Block boundary effects can potentially introduce discontinuities at interfaces between locally refined regions, requiring sufficient overlap between blocks and careful validation. The method assumes blocks are sufficiently small to approximate planar CTF behavior, an assumption requiring verification for the largest viruses or highest target resolutions.

Block-based reconstruction directly enables structural analysis of giant plant-infecting viruses (phycodnaviruses) and giant insect-infecting viruses (iridoviruses, entomopoxviruses) that were previously intractable. Systematic application to these underrepresented giant virus families could increase structural coverage.

4.2. Cryo-Electron Tomography: Recovering Asymmetric Information via Symmetry-Free Reconstruction

Cryo-electron tomography (cryo-ET) collects a tilt series of images of the same specimen area at different angles, enabling reconstruction of three-dimensional tomograms

 https://doi.org/10.3390/v18020195

without imposing any symmetry [45]. Each viral particle is reconstructed individually, preserving asymmetric features that would be averaged away in single-particle analysis. By comparing symmetry-imposed SPA structures with cryo-ET reconstructions, researchers can identify structural features lost during symmetrized processing.

Studies employing cryo-ET validation have demonstrated that imperfect icosahedral symmetry is more common than previously appreciated. In alphaviruses, cryo-ET revealed asymmetric features that are functionally relevant but completely absent in symmetrized reconstructions [46]. Flaviviruses similarly exhibit asymmetric positioning of the nucleocapsid core relative to the glycoprotein shell [47]. Similar observations have been made for genome organization within viral capsids, where asymmetric nucleic acid packaging cannot be visualized under imposed symmetry [38]. These findings emphasize the importance of cryo-ET as a complementary validation technique.

Historically, cryo-ET achieved lower resolution than SPA for whole virions, but recent advances in STA have reached 3–4 Å resolution for symmetrized subunits, approaching the 2–4 Å routinely achieved by SPA [45,48,49]. Tilt-series collection increases radiation damage and data acquisition time. The "missing wedge" of data due to limited tilt range enhances misalignment and contributes to potential anisotropic resolution. Despite these limitations, cryo-ET provides essential validation of symmetry-imposed structures.

Cryo-ET validation is particularly valuable for plant viruses, where genome packaging orientation and portal complexes may be functionally critical yet invisible under imposed symmetry. For insect viruses with complex life cycles, asymmetric features distinguishing different morphological forms can now be characterized.

4.3. *Subtomogram Averaging: Resolving Structural Heterogeneity Through Local Unit Alignment*

Subtomogram averaging (STA) combines the three-dimensional imaging capability of cryo-ET with averaging of repeating structural units. After tomogram reconstruction, repetitive elements such as envelope glycoprotein spikes, capsid subunits, or helical segments are computationally extracted as "subtomograms" and aligned for averaging [50]. This approach does not require global particle homogeneity; instead, local structural units are averaged while preserving information about their spatial arrangement within heterogeneous assemblies.

STA has enabled structural analysis of previously intractable pleomorphic viruses. In situ visualization of influenza A virus assembly captured the spatial arrangement of hemagglutinin (HA) and neuraminidase (NA) on budding virions directly within host cellular context [51]. Herpesvirus capsid assembly intermediates were resolved within the cell nucleus, redefining canonical assembly pathways [52,53]. The membrane replication factories of hepatitis C virus (HCV) and Zika virus have been characterized, revealing potential targets for antiviral intervention [54,55].

For flexuous filamentous viruses, STA offers an alternative to conventional helical reconstruction. By extracting and averaging short segments rather than assuming global helical order, STA can accommodate the bending and conformational heterogeneity characteristic of plant flexiviruses. Recent applications to filamentous plant viruses have achieved sub-nanometer resolution while accommodating the flexibility that precluded earlier structural analysis [23].

Resolution achieved by STA has historically been 5–15 Å, with recent advances reaching 3–4 Å for symmetrized viral subunits [48–50,56]. Despite these improvements, computational demands for tomogram reconstruction and subtomogram alignment remain substantial, though hardware acceleration and improved algorithms continue to reduce these barriers [50]. Computational demands for tomogram reconstruction and subtomogram alignment remain substantial, though hardware acceleration and improved algorithms continue to reduce these barriers. The requirement for tilt-series collection limits throughput compared to single-particle data acquisition.

https://doi.org/10.3390/v18020195

STA directly addresses the morphology barrier that has significantly impacted plant virus structural coverage. The flexuous filamentous morphology characteristic of Potyviridae, Closteroviridae, and Flexiviridae families causing the majority of agricultural viral disease becomes tractable through STA approaches. For insect viruses, STA enables structural analysis of the pleomorphic budded form of baculoviruses, complementing existing knowledge of the occluded form.

Based on the technical characteristics discussed above, we propose a strategic decision-making workflow to guide the structural determination of underrepresented viruses (Figure 2).

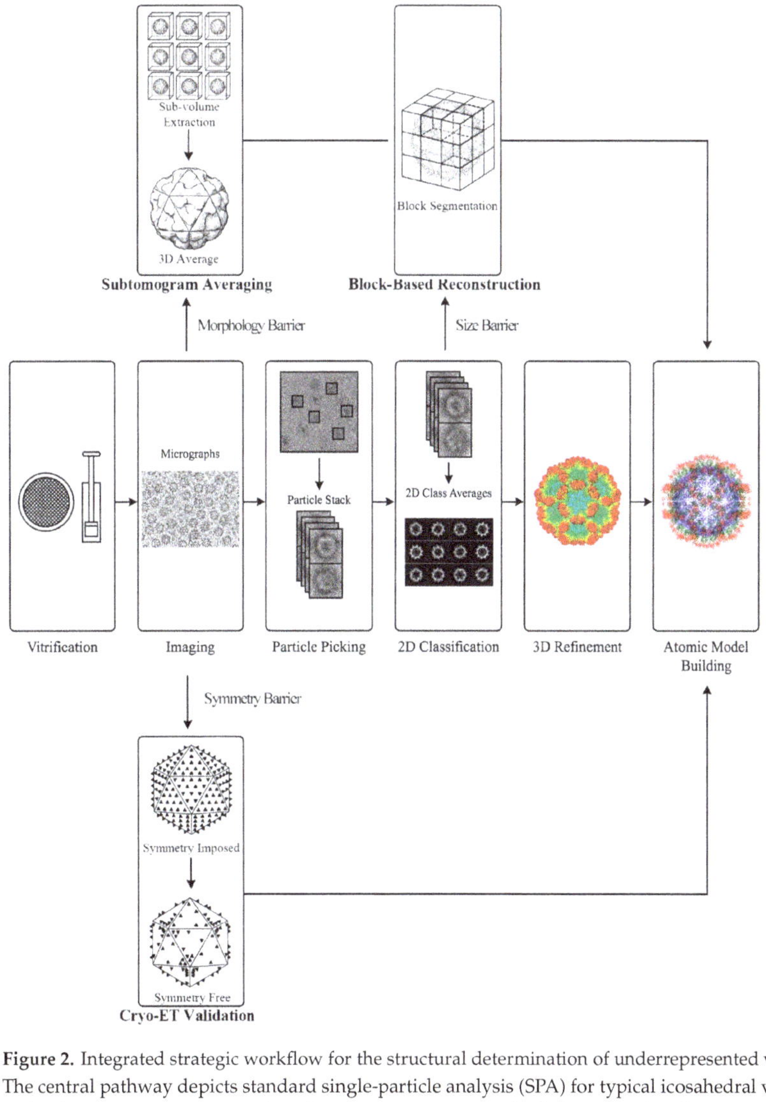

Figure 2. Integrated strategic workflow for the structural determination of underrepresented viruses. The central pathway depicts standard single-particle analysis (SPA) for typical icosahedral viruses, with schematic representations based on Sapovirus structure [57]. Specialized pathways overcome specific barriers: Subtomogram Averaging (STA) for morphological heterogeneity, Block-Based Reconstruction for size limitations, and Cryo-ET for symmetry validation. All strategies converge to a high-resolution atomic model.

 https://doi.org/10.3390/v18020195

5. Future Perspectives

The integration of in situ structural biology approaches represents the next frontier in viral structure determination. By combining cryo-focused ion beam (cryo-FIB) milling with cryo-ET, researchers can visualize viral infection processes within native cellular contexts [58]. This approach captures transient intermediates, host factor interactions, and membrane remodeling events invisible to purified particle analysis. For plant viruses, advances in cryo-FIB milling of plant tissues will enable visualization of plasmodesmatal transport and replication complex organization. For insect viruses, in situ approaches can reveal the cellular context of baculovirus assembly and the architecture of viral factories.

Recent demonstrations of high-resolution structure determination by cryo-EM at 100 keV accelerating voltage, together with high-resolution imaging on common 120 keV LaB6-source microscopes equipped with modern direct electron detectors, are lowering barriers to entry for the technique [59–61]. These developments will accelerate discovery of viral structures from diverse environmental niches and underrepresented host systems.

Integration of artificial intelligence (AI) and machine learning into cryo-EM workflows will revolutionize data processing. Automated particle picking, classification of heterogeneous assemblies, and structure prediction algorithms will accelerate structure determination while reducing operator bias [62,63]. For underrepresented virus families with limited prior structural information, AI-assisted approaches may prove particularly valuable by enabling structure determination from smaller datasets and facilitating model building for novel protein folds.

Emerging time-resolved cryo-EM approaches utilizing rapid mixing or light activation followed by rapid freezing promise to capture viral structural dynamics on millisecond timescales [64]. Visualization of receptor-triggered conformational changes, membrane fusion intermediates, and assembly pathways will provide mechanistic insights inaccessible to static structural snapshots [65].

6. Conclusions

This review has examined the host-specific bias in viral structural databases and the technical barriers driving this imbalance. The striking disparity in EMDB coverage vertebrate viruses (97.6%), plant viruses (1.0%), and insect viruses (1.4%) reflects not merely funding priorities but fundamental technical challenges that differentially affect virus categories.

Three distinct technical barriers have historically limited structural characterization of underrepresented virus systems. Size barriers, arising from depth of field limitations and Ewald sphere curvature, have prevented high-resolution analysis of giant viruses exceeding 200 nm, including plant-infecting phycodnaviruses and insect-infecting iridoviruses. Symmetry barriers have caused the loss of asymmetric information in icosahedral reconstructions, obscuring functionally important features such as portal complexes and genome packaging across all host categories. Morphology barriers have significantly impacted plant viruses, where the predominance of flexuous filamentous forms has rendered many economically important pathogens structurally intractable.

Each barrier has driven the development of targeted methodological solutions. Block-based local refinement overcomes size barriers by treating large particles as collections of smaller, independently refinable blocks, effectively bypassing physical limitations to achieve near-atomic resolution. cryo-ET overcomes symmetry barriers by reconstructing individual particles without symmetry imposition, revealing asymmetric features lost in conventional processing. STA overcomes morphology barriers by averaging locally

 https://doi.org/10.3390/v18020195

repetitive elements within globally heterogeneous assemblies, enabling the structural determination of flexible filamentous and pleomorphic viruses.

Strategic application of these methodological solutions to underrepresented virus systems offers a clear path toward a more balanced virosphere. Block-based reconstruction can now be systematically applied to giant plant and insect viruses, while subtomogram averaging should be prioritized for the flexuous filamentous plant viruses that cause the majority of agricultural disease. While implementing these strategies requires coordinated efforts linking virology expertise with cryo-EM facilities, the ultimate goal transcends administrative metrics. Bridging this structural gap is not merely an archival necessity but a fundamental step toward deciphering the universal principles of viral assembly and evolution. By illuminating the vast unexplored diversity of the virosphere from giant insect viruses to flexible plant pathogens we secure the knowledge base required to anticipate and mitigate emerging threats across biological kingdoms.

Funding: This research was supported by the Commercialization Promotion Agency for R&D Outcomes (COMPA) funded by the Ministry of Science and ICT (MSIT). (RS-2025-25436300 to HSJ), Basic Science Research Program through the National Research Foundation of Korea (NRF) funded by the Ministry of Education (RS-2024-00410337 to YHP, RS-2023-NR076408221821021300003 to S.M., RS-2024-00440289 to C.S.).

Data Availability Statement: No new data were created or analyzed in this study.

Conflicts of Interest: The authors declare no conflicts of interest.

List of Abbreviations

AI: Artificial Intelligence; Cryo-EM: Cryogenic Electron Microscopy; Cryo-ET: Cryo-Electron Tomography; Cryo-FIB: Cryo-Focused Ion Beam; CTF: Contrast Transfer Function; EMDB: Electron Microscopy Data Bank; HA: Hemagglutinin; HCV: Hepatitis C Virus; HIV: Human Immunodeficiency Virus; NA: Neuraminidase; NCLDV: Nucleocytoplasmic Large DNA Virus; SGIV: Singapore Grouper Iridovirus; SPA: Single-Particle Analysis; STA: Subtomogram Averaging; TMV: Tobacco Mosaic Virus.

References

1. Krupovic, M.; Koonin, E.V. Multiple origins of viral capsid proteins from cellular ancestors. *Proc. Natl. Acad. Sci. USA* **2017**, *114*, E2401–E2410. [CrossRef]
2. Cheng, Y. Single-particle cryo-EM—How did it get here and where will it go. *Science* **2018**, *361*, 876–880. [CrossRef]
3. Kühlbrandt, W. The resolution revolution. *Science* **2014**, *343*, 1443–1444. [CrossRef]
4. Jones, R.A.; Naidu, R.A. Global dimensions of plant virus diseases: Current status and future perspectives. *Annu. Rev. Virol.* **2019**, *6*, 387–409. [CrossRef]
5. Wrapp, D.; Wang, N.; Corbett, K.S.; Goldsmith, J.A.; Hsieh, C.-L.; Abiona, O.; Graham, B.S.; McLellan, J.S. Cryo-EM structure of the 2019-nCoV spike in the prefusion conformation. *Science* **2020**, *367*, 1260–1263. [CrossRef] [PubMed]
6. Sirohi, D.; Chen, Z.; Sun, L.; Klose, T.; Pierson, T.C.; Rossmann, M.G.; Kuhn, R.J. The 3.8 Å resolution cryo-EM structure of Zika virus. *Science* **2016**, *352*, 467–470. [CrossRef] [PubMed]
7. Harris, A.; Cardone, G.; Winkler, D.C.; Heymann, J.B.; Brecher, M.; White, J.M.; Steven, A.C. Influenza virus pleiomorphy characterized by cryoelectron tomography. *Proc. Natl. Acad. Sci. USA* **2006**, *103*, 19123–19127. [CrossRef]
8. Chapman, M.S.; Agbandje-McKenna, M. Atomic structure of viral particles. In *Parvoviruses*; Edward Arnold, Ltd.: London, UK, 2006; pp. 107–123.
9. Comas-Garcia, M. How structural biology has changed our understanding of icosahedral viruses. *J. Virol.* **2024**, *98*, e01111-23. [CrossRef] [PubMed]
10. Mattei, S.; Glass, B.; Hagen, W.J.; Kräusslich, H.-G.; Briggs, J.A. The structure and flexibility of conical HIV-1 capsids determined within intact virions. *Science* **2016**, *354*, 1434–1437. [CrossRef]
11. Xie, Q.; Yoshioka, C.K.; Chapman, M.S. Adeno-associated virus (AAV-DJ)—Cryo-EM structure at 1.56 Å resolution. *Viruses* **2020**, *12*, 1194. [CrossRef]

https://doi.org/10.3390/v18020195

12. Stagg, S.M.; Yoshioka, C.; Davulcu, O.; Chapman, M.S. Cryo-electron microscopy of adeno-associated virus. *Chem. Rev.* **2022**, *122*, 14018–14054. [CrossRef]
13. Song, C.; Takai-Todaka, R.; Miki, M.; Haga, K.; Fujimoto, A.; Ishiyama, R.; Oikawa, K.; Yokoyama, M.; Miyazaki, N.; Iwasaki, K. Dynamic rotation of the protruding domain enhances the infectivity of norovirus. *PLoS Pathog.* **2020**, *16*, e1008619. [CrossRef]
14. Conley, M.J.; McElwee, M.; Azmi, L.; Gabrielsen, M.; Byron, O.; Goodfellow, I.G.; Bhella, D. Calicivirus VP2 forms a portal-like assembly following receptor engagement. *Nature* **2019**, *565*, 377–381. [CrossRef] [PubMed]
15. Strange, R.N.; Scott, P.R. Plant disease: A threat to global food security. *Annu. Rev. Phytopathol.* **2005**, *43*, 83–116. [CrossRef]
16. Ge, P.; Zhou, Z.H. Hydrogen-bonding networks and RNA bases revealed by cryo electron microscopy suggest a triggering mechanism for calcium switches. *Proc. Natl. Acad. Sci. USA* **2011**, *108*, 9637–9642. [CrossRef]
17. Lucas, R.W.; Kuznetsov, Y.G.; Larson, S.B.; McPherson, A. Crystallization of brome mosaic virus and T = 1 brome mosaic virus particles following a structural transition. *Virology* **2001**, *286*, 290–303. [CrossRef]
18. Qu, C.; Liljas, L.; Opalka, N.; Brugidou, C.; Yeager, M.; Beachy, R.N.; Fauquet, C.M.; Johnson, J.E.; Lin, T. 3D domain swapping modulates the stability of members of an icosahedral virus group. *Structure* **2000**, *8*, 1095–1103. [CrossRef]
19. Zhang, W.; Olson, N.H.; Baker, T.S.; Faulkner, L.; Agbandje-McKenna, M.; Boulton, M.I.; Davies, J.W.; McKenna, R. Structure of the Maize streak virus geminate particle. *Virology* **2001**, *279*, 471–477. [CrossRef] [PubMed]
20. Zamora, M.; Méndez-López, E.; Agirrezabala, X.; Cuesta, R.; Lavín, J.L.; Sánchez-Pina, M.A.; Aranda, M.A.; Valle, M. Potyvirus virion structure shows conserved protein fold and RNA binding site in ssRNA viruses. *Sci. Adv.* **2017**, *3*, eaao2182. [CrossRef] [PubMed]
21. DiMaio, F.; Chen, C.-C.; Yu, X.; Frenz, B.; Hsu, Y.-H.; Lin, N.-S.; Egelman, E.H. The molecular basis for flexibility in the flexible filamentous plant viruses. *Nat. Struct. Mol. Biol.* **2015**, *22*, 642–644. [CrossRef]
22. Peiró, A.; Martínez-Gil, L.; Tamborero, S.; Pallás, V.; Sánchez-Navarro, J.A.; Mingarro, I. The Tobacco mosaic virus movement protein associates with but does not integrate into biological membranes. *J. Virol.* **2014**, *88*, 3016–3026. [CrossRef]
23. Chase, O.; Javed, A.; Byrne, M.J.; Thuenemann, E.C.; Lomonossoff, G.P.; Ranson, N.A.; López-Moya, J.J. CryoEM and stability analysis of virus-like particles of potyvirus and ipomovirus infecting a common host. *Commun. Biol.* **2023**, *6*, 433. [CrossRef]
24. Agirrezabala, X.; Mendez-Lopez, E.; Lasso, G.; Sanchez-Pina, M.A.; Aranda, M.; Valle, M. The near-atomic cryoEM structure of a flexible filamentous plant virus shows homology of its coat protein with nucleoproteins of animal viruses. *eLife* **2015**, *4*, e11795. [CrossRef]
25. Bonning, B.C.; Hammock, B.D. Development of recombinant baculoviruses for insect control. *Annu. Rev. Entomol.* **1996**, *41*, 191–210. [CrossRef]
26. Valles, S.; Chen, Y.; Firth, A.; Guérin, D.A.; Hashimoto, Y.; Herrero, S.; De Miranda, J.; Ryabov, E.; Consortium, I.R. ICTV virus taxonomy profile: Iflaviridae. *J. Gen. Virol.* **2017**, *98*, 527–528. [CrossRef]
27. Valles, S.; Chen, Y.; Firth, A.; Guérin, D.A.; Hashimoto, Y.; Herrero, S.; De Miranda, J.; Ryabov, E.; Consortium, I.R. ICTV virus taxonomy profile: Dicistroviridae. *J. Gen. Virol.* **2017**, *98*, 355–356. [CrossRef]
28. Ji, X.; Sutton, G.; Evans, G.; Axford, D.; Owen, R.; Stuart, D.I. How baculovirus polyhedra fit square pegs into round holes to robustly package viruses. *EMBO J.* **2009**, *29*, 505. [CrossRef]
29. Aliyari, R.; Ding, S.W. RNA-based viral immunity initiated by the Dicer family of host immune receptors. *Immunol. Rev.* **2009**, *227*, 176–188. [CrossRef] [PubMed]
30. Govind, S. Innate immunity in Drosophila: Pathogens and pathways. *Insect Sci.* **2008**, *15*, 29–43. [CrossRef] [PubMed]
31. Estrozi, L.; Agirre, J.; Imler, J.-L.; Santiago, E.; Navaza, J.; Schoehn, G.; Guérin, D.M. The cryo-EM Reconstruction of Drosophila C Virus (DCV) at 5.4 Å. *Biophys. J.* **2013**, *104*, 414a. [CrossRef]
32. Koonin, E.V.; Yutin, N. Evolution of the large nucleocytoplasmic DNA viruses of eukaryotes and convergent origins of viral gigantism. *Adv. Virus Res.* **2019**, *103*, 167–202. [PubMed]
33. Downing, K.H.; Glaeser, R.M. Estimating the effect of finite depth of field in single-particle cryo-EM. *Ultramicroscopy* **2018**, *184*, 94–99. [CrossRef] [PubMed]
34. Zhu, D.; Wang, X.; Fang, Q.; Van Etten, J.L.; Rossmann, M.G.; Rao, Z.; Zhang, X. Pushing the resolution limit by correcting the Ewald sphere effect in single-particle Cryo-EM reconstructions. *Nat. Commun.* **2018**, *9*, 1552. [CrossRef] [PubMed]
35. Scheres, S.H. RELION: Implementation of a Bayesian approach to cryo-EM structure determination. *J. Struct. Biol.* **2012**, *180*, 519–530. [CrossRef]
36. Baker, T.S.; Cheng, R.H. A model-based approach for determining orientations of biological macromolecules imaged by cryoelectron microscopy. *J. Struct. Biol.* **1996**, *116*, 120–130. [CrossRef]
37. Jose, J.; Hafenstein, S.L. Asymmetry in icosahedral viruses. *Curr. Opin. Virol.* **2022**, *54*, 101230. [CrossRef]
38. Geraets, J.A.; Dykeman, E.C.; Stockley, P.G.; Ranson, N.A.; Twarock, R. Asymmetric genome organization in an RNA virus revealed via graph-theoretical analysis of tomographic data. *PLoS Comput. Biol.* **2015**, *11*, e1004146. [CrossRef]

39. Egelman, E.H. The iterative helical real space reconstruction method: Surmounting the problems posed by real polymers. *J. Struct. Biol.* **2007**, *157*, 83–94. [CrossRef]
40. Rohou, A.; Grigorieff, N. CTFFIND4: Fast and accurate defocus estimation from electron micrographs. *J. Struct. Biol.* **2015**, *192*, 216–221. [CrossRef]
41. Zhao, Z.; Huang, Y.; Liu, C.; Zhu, D.; Gao, S.; Liu, S.; Peng, R.; Zhang, Y.; Huang, X.; Qi, J. Near-atomic architecture of Singapore grouper iridovirus and implications for giant virus assembly. *Nat. Commun.* **2023**, *14*, 2050. [CrossRef]
42. Watanabe, R.; Song, C.; Takemura, M.; Murata, K. Subnanometer structure of medusavirus capsid during maturation using cryo-electron microscopy. *J. Virol.* **2024**, *98*, e00436-24. [CrossRef]
43. de Freitas Almeida, G.M.; Arriaga, I.; de Azevedo, B.L.; Leppänen, M.; Abrahão, J.S.; Andreani, J.; Zabeo, D.; Ravantti, J.J.; Abrescia, N.G.; Sundberg, L.-R. Genomic and structural insights into Jyvaskylavirus, the first giant virus isolated from Finland. *eLife* **2025**, *13*, RP103492. [CrossRef]
44. Yang, Y.; Shao, Q.; Guo, M.; Han, L.; Zhao, X.; Wang, A.; Li, X.; Wang, B.; Pan, J.-A.; Chen, Z. Capsid structure of bacteriophage ΦKZ provides insights into assembly and stabilization of jumbo phages. *Nat. Commun.* **2024**, *15*, 6551. [CrossRef]
45. Turk, M.; Baumeister, W. The promise and the challenges of cryo-electron tomography. *FEBS Lett.* **2020**, *594*, 3243–3261. [CrossRef]
46. Chmielewski, D.; Su, G.-C.; Kaelber, J.T.; Pintilie, G.D.; Chen, M.; Jin, J.; Auguste, A.J.; Chiu, W. Cryogenic electron microscopy and tomography reveal imperfect icosahedral symmetry in alphaviruses. *PNAS Nexus* **2024**, *3*, pgae102. [CrossRef]
47. Therkelsen, M.D.; Klose, T.; Vago, F.; Jiang, W.; Rossmann, M.G.; Kuhn, R.J. Flaviviruses have imperfect icosahedral symmetry. *Proc. Natl. Acad. Sci. USA* **2018**, *115*, 11608–11612. [CrossRef]
48. Himes, B.A.; Zhang, P. emClarity: Software for high-resolution cryo-electron tomography and subtomogram averaging. *Nat. Methods* **2018**, *15*, 955–961. [CrossRef]
49. Tegunov, D.; Xue, L.; Dienemann, C.; Cramer, P.; Mahamid, J. Multi-particle cryo-EM refinement with M visualizes ribosome-antibiotic complex at 3.5 Å in cells. *Nat. Methods* **2021**, *18*, 186–193. [CrossRef] [PubMed]
50. Wan, W.; Briggs, J.A. Cryo-electron tomography and subtomogram averaging. *Methods Enzymol.* **2016**, *579*, 329–367. [PubMed]
51. Wachsmuth-Melm, M.; Peterl, S.; O'Riain, A.; Makroczyová, J.; Fischer, K.; Krischuns, T.; Vale-Costa, S.; Amorim, M.J.; Chlanda, P. Visualizing influenza A virus assembly by in situ cryo-electron tomography. *Nat. Commun.* **2025**, *16*, 9394. [CrossRef] [PubMed]
52. Oliver, S.L.; Chen, M.; Engel, L.; Hecksel, C.W.; Zhou, X.; Schmid, M.F.; Arvin, A.M.; Chiu, W. Cryogenic Electron Tomography Redefines Herpesvirus Capsid Assembly Intermediates Inside the Cell Nucleus. *bioRxiv* **2025**, bioRxiv:2025.06.27.661840. [CrossRef]
53. Yuan, S.; Wang, J.; Zhu, D.; Wang, N.; Gao, Q.; Chen, W.; Tang, H.; Wang, J.; Zhang, X.; Liu, H. Cryo-EM structure of a herpesvirus capsid at 3.1 Å. *Science* **2018**, *360*, eaao7283. [CrossRef]
54. Sykora, U.M.; O'Sullivan, T.J.; Halfon, Y.; Fontana, J.; Harris, M. The architecture of membrane structures involved in hepatitis C virus genome replication revealed under close-to-native conditions by cryo-electron tomography. *J. Gen. Virol.* **2025**, *106*, 002168. [CrossRef]
55. Ci, Y.; Liu, Z.-Y.; Zhang, N.-N.; Niu, Y.; Yang, Y.; Xu, C.; Yang, W.; Qin, C.-F.; Shi, L. Zika NS1–induced ER remodeling is essential for viral replication. *J. Cell Biol.* **2019**, *219*, e201903062. [CrossRef]
56. Mattei, S.; Tan, A.; Glass, B.; Müller, B.; Kräusslich, H.-G.; Briggs, J.A. High-resolution structures of HIV-1 Gag cleavage mutants determine structural switch for virus maturation. *Proc. Natl. Acad. Sci. USA* **2018**, *115*, E9401–E9410. [CrossRef]
57. Miyazaki, N.; Song, C.; Oka, T.; Miki, M.; Murakami, K.; Iwasaki, K.; Katayama, K.; Murata, K. Atomic structure of the human sapovirus capsid reveals a unique capsid protein conformation in caliciviruses. *J. Virol.* **2022**, *96*, e00298-22. [CrossRef] [PubMed]
58. Schaffer, M.; Mahamid, J.; Engel, B.D.; Laugks, T.; Baumeister, W.; Plitzko, J.M. Optimized cryo-focused ion beam sample preparation aimed at in situ structural studies of membrane proteins. *J. Struct. Biol.* **2017**, *197*, 73–82. [CrossRef] [PubMed]
59. Karia, D.; Koh, A.F.; Yang, W.; Cushing, V.I.; Basanta, B.; Mihaylov, D.B.; Khavnekar, S.; Vyroubal, O.; Malínský, M.; Sháněl, O. Sub-3 Å resolution protein structure determination by single-particle cryo-EM at 100 keV. *Structure* **2025**, *33*, 1717–1727.e1714. [CrossRef] [PubMed]
60. Venugopal, H.; Mobbs, J.; Taveneau, C.; Fox, D.R.; Vuckovic, Z.; Gulati, S.; Knott, G.; Grinter, R.; Thal, D.; Mick, S. High-resolution cryo-EM using a common LaB6 120-keV electron microscope equipped with a sub–200-keV direct electron detector. *Sci. Adv.* **2025**, *11*, eadr0438. [CrossRef]
61. McMullan, G.; Faruqi, A.; Henderson, R. Direct electron detectors. *Methods Enzymol.* **2016**, *579*, 1–17.
62. Bepler, T.; Morin, A.; Rapp, M.; Brasch, J.; Shapiro, L.; Noble, A.J.; Berger, B. Positive-unlabeled convolutional neural networks for particle picking in cryo-electron micrographs. *Nat. Methods* **2019**, *16*, 1153–1160. [CrossRef] [PubMed]
63. Jamali, K.; Kimanius, D.; Scheres, S.H. A graph neural network approach to automated model building in cryo-EM maps. *arXiv* **2022**, arXiv:2210.00006.

https://doi.org/10.3390/v18020195

64. Harder, O.F.; Barrass, S.V.; Drabbels, M.; Lorenz, U.J. Fast viral dynamics revealed by microsecond time-resolved cryo-EM. *Nat. Commun.* **2023**, *14*, 5649. [CrossRef] [PubMed]
65. Kephart, S.M.; Hom, N.; Lee, K.K. Visualizing intermediate stages of viral membrane fusion by cryo-electron tomography. *Trends Biochem. Sci.* **2024**, *49*, 916–931. [CrossRef]

 https://doi.org/10.3390/v18020195

viruses

MDPI

Review

Zooming in and out: Exploring RNA Viral Infections with Multiscale Microscopic Methods

Cheng-An Lyu [1,2], Yao Shen [1,*] and Peijun Zhang [1,2,3,*]

[1] Division of Structural Biology, Wellcome Centre for Human Genetics, University of Oxford, Oxford OX3 7BN, UK; cheng-an.lyu@ndm.ox.ac.uk
[2] Chinese Academy of Medical Sciences Oxford Institute, University of Oxford, Oxford OX3 7BN, UK
[3] Diamond Light Source, Harwell Science and Innovation Campus, Didcot OX11 0DE, UK
* Correspondence: yao.shen@strubi.ox.ac.uk (Y.S.); peijun.zhang@strubi.ox.ac.uk (P.Z.)

Abstract: RNA viruses, being submicroscopic organisms, have intriguing biological makeups and substantially impact human health. Microscopic methods have been utilized for studying RNA viruses at a variety of scales. In order of observation scale from large to small, fluorescence microscopy, cryo-soft X-ray tomography (cryo-SXT), serial cryo-focused ion beam/scanning electron microscopy (cryo-FIB/SEM) volume imaging, cryo-electron tomography (cryo-ET), and cryo-electron microscopy (cryo-EM) single-particle analysis (SPA) have been employed, enabling researchers to explore the intricate world of RNA viruses, their ultrastructure, dynamics, and interactions with host cells. These methods evolve to be combined to achieve a wide resolution range from atomic to sub-nano resolutions, making correlative microscopy an emerging trend. The developments in microscopic methods provide multi-fold and spatial information, advancing our understanding of viral infections and providing critical tools for developing novel antiviral strategies and rapid responses to emerging viral threats.

Keywords: RNA virus; viral infection; virus–host interactions; multiscale microscopic methods; cryo-soft X-ray tomography (cryo-SXT); serial cryo-focused ion beam/scanning electron microscopy cryo-FIB/SEM volume imaging; cellular cryo-tomography (cryo-ET); cryo-EM single-particle analysis (SPA)

1. Introduction

RNA viruses represent a diverse and highly adaptive group of pathogens that play a central role in shaping the landscape of biology and human health [1]. RNA viruses use ribonucleic acid (RNA) as their genetic material and can be classified into single-stranded RNA (ssRNA) viruses, double-stranded RNA (dsRNA) viruses, and retroviruses, depending on their genome type and replication strategy. Single/double-stranded RNA viruses, such as influenza, SARS-CoV-2, and Ebola, rely on RNA-dependent RNA polymerases for genome replication. In contrast, retroviruses such as the human immunodeficiency virus (HIV) employ reverse transcriptase to produce viral DNA, which is then integrated into the host genome by integrase. RNA viruses are characterized by high mutation rates and rapid evolution, allowing them to quickly escape host immune responses and develop resistance to antiviral treatments. This adaptability underscores the significant public health threats posed by RNA viruses, as evidenced by recent outbreaks and pandemics caused by the Ebola virus and SARS-CoV-2 [2,3].

Continued research into RNA viruses remains crucial for understanding their pathology and has facilitated breakthroughs in vaccine development, antiviral treatments, and our comprehension of the immune system [4–6]. The advancement of various imaging techniques, notably electron microscopy (EM), has proven robust in delineating complex structures of RNA viruses, viral life cycles, exploring virus–host interactions, and unravelling host immune responses [7,8]. This review aims to provide a comprehensive overview of cryo-imaging techniques that offer multiscale insights into RNA virus biology that allow for a holistic view of virus-infected cells in a native, hydrated state. These cryo-techniques, organized by observation scale from large to small, include cryo-soft X-ray tomography (cryo-SXT), serial cryo-focused ion beam/scanning electron microscopy cryo-FIB/SEM volume imaging, cellular cryo-tomography (cryoET), and cryo-electron microscopy (cryo-EM) single-particle analysis (SPA) (Figure 1). When used together, they can provide unprecedented details of ultrastructural features of virus-infected cells, and contextual information that bridges the gap between molecular mechanisms and cellular responses. In this paper, we emphasize the importance of microscopy techniques in RNA virus research and showcase how these diverse imaging modalities can be used synergistically for comprehensive studies of RNA viral infection. We summarize the achievable resolutions of these techniques in Figure 1, followed by their respective workflows in Figure 2, and applications in Figures 3 and 4.

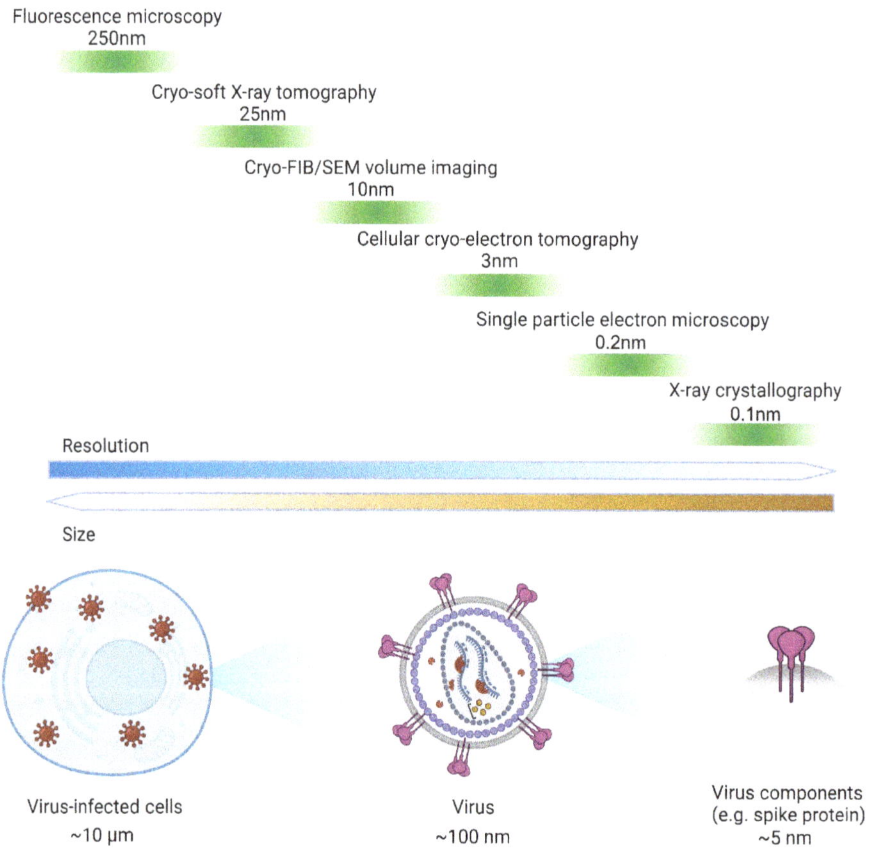

Figure 1. Schematic diagram showing the resolution capabilities of various imaging techniques in the research of RNA virus. Upper: resolution ranges of various imaging and structural methods. Lower: Schematic representations of the size scales of virus-infected cells, virus particles, and virus components (e.g., spike proteins). Different approaches can be utilized depending on the scale of the research object.

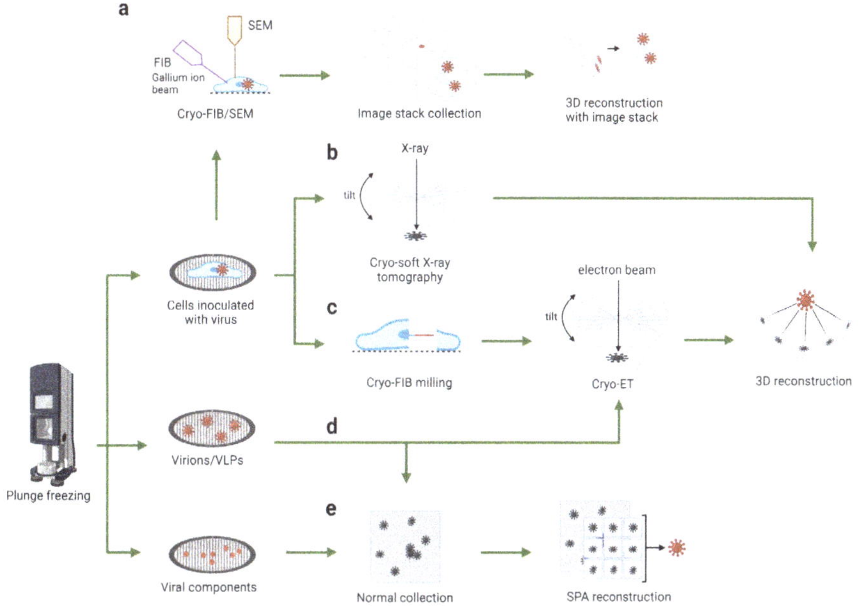

Figure 2. General workflow for microscopical research on RNA viruses. Samples are applied to grids and undergo a vitrification process (the most well-known procedure being plunge freezing), with subsequent imaging procedures depending on the sample. (**a**,**b**) For samples containing whole mammalian cells, serial cryo-focused ion beam/scanning electron microscopy (cryo-FIB/SEM) (**a**) and cryo-soft X-ray tomography (cryo-SXT) (**b**) can be used for volume imaging. (**c**) To achieve high-resolution imaging of regions of interest (ROI) within the cell (e.g., virus particles), a thin lamella of 80–250 nm can be prepared using FIB milling and imaged using cryo-ET. (**d**) Virions and VLPs can be imaged and structurally characterized using both cryo-ET or SPA. (**e**) For purified viral components (e.g., spikes), SPA is often the preferred method for structural analysis.

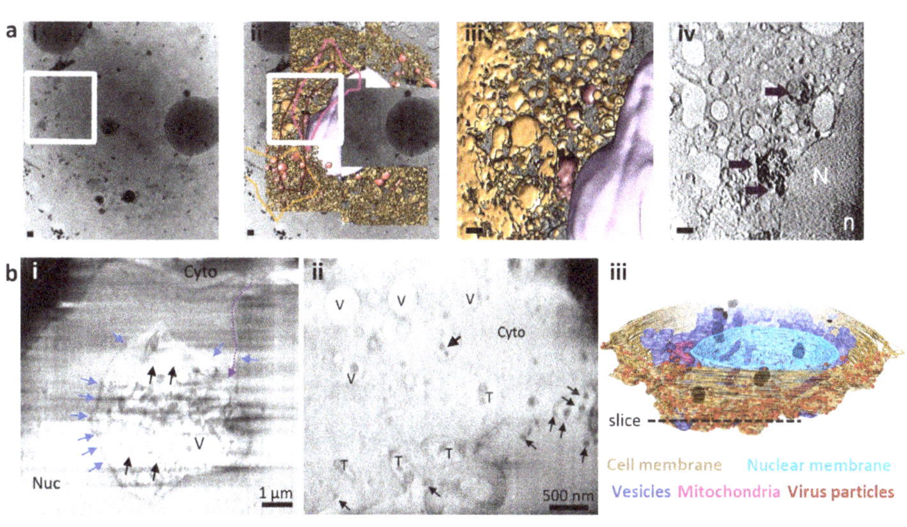

Figure 3. Cryo-SXT and cryo-FIB/SEM techniques are suitable for imaging and reconstructing intact virus-infected cells. (**a**) Three-dimensional reconstruction of whole-cell volumes of HCV replicon-bearing

cells by cryo-SXT (scale bars, 1 μm) [9]. (**i**,**ii**) Tile-scanned projection (**i**) and overlaid with segmentation (**ii**) showing the area selected for SXT (white square); The enlarged view of segmented volume (**iii**) and tomographic slice (**iv**) from the boxed areas are also shown. Threshold-based isosurface segmentation of the surface boundaries identifies the different organelles present in the cells: abnormal mitochondria in red and purple, ER in yellow, modified ER in brown and nuclear envelope in pink. Purple arrows point to the abnormal mitochondria. (**b**) Characterization of SARS-CoV-2-infected Vero cells using cryo-FIB/SEM volume imaging [10]. Black arrows, SARS-CoV-2 virus particles; blue arrows, nuclear pores; dashed purple arrow, invagination path; Nuc nucleus; Cyto cytoplasm; V vesicles; T tunnels. (**i**) A representative cryo-FIB/SEM slice showing the invagination of cytoplasm into the nuclear space (scale bar, 1 μm). (**ii**) A representative cryo-FIB/SEM slice showing the distribution of SARS-CoV-2 particles (scale bar, 500 nm). (**iii**) Surface rendering of the segmented volume of a SARS-CoV-2-infected cell as shown in (**ii**). Segmented organelles and virus particles are labeled with indicated colors. The dashed lines indicate the position of the slice shown in (**ii**).

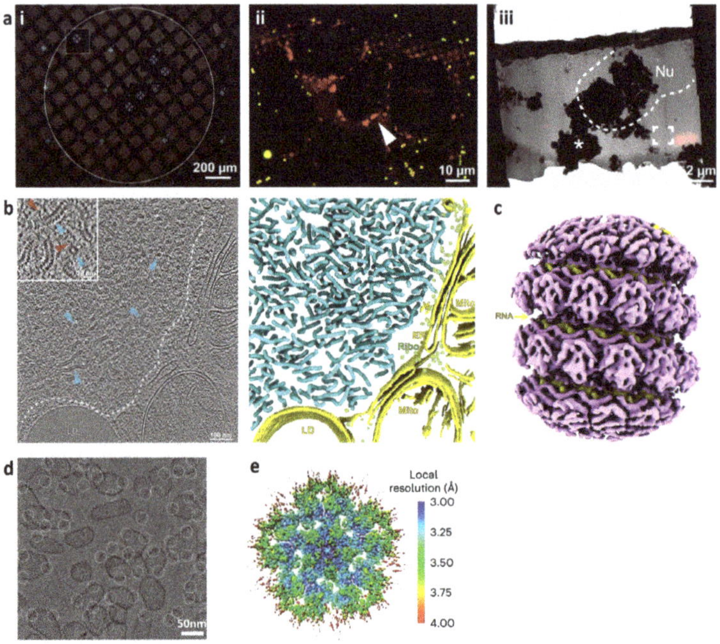

Figure 4. Cryo-ET and cryo-EM for high-resolution characterization of viral structure in situ. (**a**) A cryo-correlative light and electron microscopy (CLEM) workflow targeting unlabelled MuV near fluorescently labelled mCherry-G3BP1 stress granules (SGs) [11]. (**i**) MuV-infected HeLa cells were seeded on an EM grid and imaged by phase contrast light microscope (scale bar, 200 μm). (**ii**) Infected cells with red fluorescent SGs (white arrowhead) are targeted for FIB-milling (scale bar, 10 μm). (**iii**) Lamella TEM map overlaid with cryo-confocal images confirming the presence of SGs post-milling and guiding cryo-ET data collection (scale bar, 2 μm) (Nu: nucleus). (**b**) Representative tomographic slice (left) and corresponding 3D segmentation (right) for in-cell MuV characterization (scale bar, 100 nm). Side and top views of MuV nucleocapsids are marked with cyan arrowheads and colored accordingly. The inset shows nucleocapsids have extended flexible densities (red arrowheads). Both the tomographic slice and segmentation are labelled with lipid droplets (LD), mitochondria (Mito), endoplasmic reticulum (ER), ribosomes (Ribo), and vesicles (V) [11]. (**c**) Subtomogram average of in-cell MuV nucleocapsids resolved to 6.5 Å (EMD-13137), with RNA densities in yellow. (**d**) Representative micrograph of in vitro assembled HIV-1 capsids with picked pentamer particles (white circles) (scale bar, 50 nm) [12]. (**e**) High-resolution reconstruction of HIV-1 capsid pentamer structure colored according to local resolution (EMD-28186) [12].

2. Seeing through the Whole Cell: Cryo-Soft X-ray Tomography

Cryo-SXT has emerged as a robust tool to image whole intact cells in a high throughput manner, ideally for characterizing cells upon viral infection that can provide unique insights into the ultrastructural details of virus-infected cells. Compared with conventional X-ray tomography, it uses low-energy X-rays, typically in the "water window" range of 284 to 543 eV, to image hydrated, intact cells, without the need for chemical fixing, staining or sectioning [13–15]. Under this condition, carbon-rich biological samples produce high contrast while oxygen-rich structures such as the cytosol remain virtually transparent, enabling the visualization of cellular compartments. Despite its slightly lower achievable resolution compared to electron microscopy, about 25 to 50 nanometers depending on the zone plate, cryo-SXT excels at imaging relatively thick samples (up to 15 μm) rapidly. It can capture an entire three-dimensional 3D volume of $15 \times 15 \times 15\ \mu m^3$ in about 5 min and allow the examination of large, complex cellular structures and significant ultrastructural alterations caused by infection [14,16]. Moreover, when integrated with fluorescence microscopy, viral particles can be localized for targeted imaging, enabling the characterization of viral trafficking and viral life cycle in cells [17]. These distinctive features of cryo-SXT make it a robust approach for investigating RNA virus infections, offering details for cellular response to viral infection, viral life cycle, and providing details that are not accessible with other imaging modalities.

The cryo-SXT workflow is a systematic process, in which cells are seeded or deposited on sterilized microscopy grids or housed in thin-walled glass capillary tubes. For virus research, cells are inoculated with the virus and incubated under appropriate biosafety conditions. Vitrification is performed by high-pressure freezing (HPF) or plunge freezing, with gold particles or fluorescent stains added for tomogram alignment and data analysis. In the transmission X-ray microscope (TXM), these grids are first imaged in visible light to select regions of interest (ROI) for X-ray imaging. Tomograms of ROIs are acquired by capturing a series of transmission images at various tilt angles of the sample to capture structural information at different orientations (Figure 2). Depending on the samples, the angular range can be up to 140 degrees for EM grids and 360 degrees for capillary tubes through the entire cell with a thickness up to 16 μm [18]. Image stacks are normalized and aligned using the established processing workflow such as AREC3D [19–22], and the reconstructed volumes are produced for detailed analysis.

Leveraging cryo-SXT, there has been an increasing amount of research in various RNA virus infections [10,18,23–28]. We herein take the research on hepatitis C virus (HCV) and reovirus as examples to illustrate the contribution of cryo-SXT to the research of RNA viruses. HCV is an enveloped positive-strand RNA (+ssRNA) virus notorious for its ability to induce serious liver damage [29]. To characterize the cell response to HCV infection and antiviral treatment, Pérez-Berná et al. [9,27] obtained 3D maps of the whole cells harboring HCV replicons at different stages of infection, as well as after the treatment of antiviral drugs using cryo-SXT. In Figure 3a, the HCV replicon-bearing cell slices are shown, with projections (Figure 3a(i,ii)) and volume slices (Figure 3a(iii,iv)) providing detailed views of the selected ROI and identified organelles. In their research, significant changes in the structure of the endoplasmic reticulum (ER) and mitochondria were seen in nearly native, whole, hydrated cells during HCV infection and antiviral treatments. HCV replication relies on special networks in HCV-infected cells formed by vesicular and tubular structures derived from the ER. Meanwhile, abnormal mitochondria were observed, which indicates the mitochondrial dysfunction induced by the virus. Notably, these morphological changes were reverted by clinically approved antivirals sofosbuvir and daclatasvir. These results suggest that the cellular morphology observed by cryo-SXT can be used as a criterion for evaluating the effectiveness of other potential antiviral drugs. Their study highlights the promise of this method to aid in drug development during the preclinical stage, which allows for the assessment of the possible influence of candidate compounds on the cellular ultrastructure.

In parallel, reovirus is an icosahedral, non-enveloped dsRNA virus that can be potentially applied to oncolytic therapy [30]. Non-enveloped viruses do not have the membrane-fusion ability, making their cell-entry process largely unknown. To track the entry process of reovirus and modification of endosomes from early to late stages post-infection, Kounatidis et al. developed a correlative microscopy system combining cryo-SXT and other techniques and achieved direct data correlation in 3D [24]. By combining cryo-SXT with 3D-structure illumination microscopy (3D-SIM), the study provided high-resolution 3D views of the cellular compartments involved in viral trafficking, allowing for the visualization of vesicles carrying viral particles and their transformations over time. This approach enabled the identification of multi-vesicular bodies and other vesicle structures that form during infection, providing insights into how reoviruses escape from endosomes into the cytoplasm without disrupting the overall vesicle integrity. These findings underscore the utility of cryo-SXT in capturing the dynamic interactions between reoviruses and host cell organelles, contributing to a more comprehensive understanding of viral pathogenesis.

These studies demonstrate that cryo-SXT is effective in the characterization of virus–host interactions, from imaging viral particles within infected cells to investigating changes to cell ultrastructure and organelles post-infection to facilitate the development of future treatments.

3. Peeling Back the Layers: Serial Cryo-FIB/SEM Volume Imaging

Another rapidly evolving imaging technique allowing whole-cell imaging is serial cryo-focused ion beam/scanning electron microscopy (cryo-FIB/SEM) volume imaging. This technique allows the visualization of relatively thick frozen-hydrated biological specimens (e.g., virus-infected cells) at resolutions ~10 nm. It employs a dual-beam, enabling the serial sectioning of thick samples (above 300 nm up to hundreds of microns) using a focused ion beam coupled with subsequent imaging with SEM. Briefly, biological samples (e.g., cells and tissues) that have been either plunge-frozen or high-pressure-frozen undergo serial sectioning using a focused ion beam, typically composed of gallium ions or plasma ions to mill away a thin slice of the sample. The milling process is regulated in a way to allow for precise removal of thin slices of material ranging from 3 to 15 nm. Immediately following the milling, the sample surface is imaged by collecting backscattered and secondary electrons, using an integrated scanning electron beam to capture high-resolution details about the sample. The sequential cryo-focused ion beam (cryo-FIB) milling, and SEM imaging are repeated iteratively (up to hundreds to thousands of times) until a series of images covering the entire volume of interest is obtained. Due to these repeated steps, imaging of an entire mammalian cell can take up to days. The collected 2D images are then computationally reconstructed into a 3D volume for the whole cell, providing details into the spatial organization of cellular compartments, the 3D architecture of cellular structures, such as organelles, membranes, and other subcellular components in their native, hydrated state. Moreover, this technique can also be combined with florescent microscopy [31] for cryo-FIB lamella preparation to target a region of interest, allowing for penetration of electron beam imaging using cryo-ET to obtain high-resolution information of the targeted region, which will be discussed in the following section.

Given its straightforward sample preparation process, serial cryo-FIB/SEM volume imaging has been successfully applied to various biological samples ranging from cells to multicellular organisms and tissues [32–38]. Its capability to reveal the spatial organization of cellular compartments has made it increasingly popular for studying viral infections, with a notable surge in research on SARS-CoV-2 in recent years [10,39].

SARS-CoV-2 coronaviruses are small, enveloped viruses containing a single-stranded, positive RNA genome. While the viral components have been extensively characterized to an atomic resolution [40,41], understanding of the SARS-CoV-2 coronavirus infection at the cellular level is limited. Using a multi-scale imaging technique including volume imaging using cryo-SXT, serial cryo-FIB/SEM volume imaging, and cellular tomography imaging (explained in the next section), Mendonca et al. [10] provided a comprehensive characterization of SARS-CoV-2 infection within frozen hydrated Vero cells (Figure 3b). Volume imaging reveals drastic cytopathic alterations of SARS-CoV-2-infected cells, including both mitochondrial defects in morphology and number, as well as nucleus defects resulting from cytoplasmic invagination (Figure 3b(i)). In addition to the visualization of the cellular compartment, virus particles represented by electron-dense particles can be unambiguously identified at the cell periphery, inside the cytoplasm and within vesicles (Figure 3b(i–iii)), highlighting the robustness of volume imaging in virus research. Combined with cryo-ET imaging on cell lamella (explained in the next section), this work shed light on the SARS-CoV-2 life cycle from viral RNA replication and transport to virus assembly and egress within a native cellular context, providing a holistic view of the SARS-CoV-2 infection within cells that could not have been possible without the combination of different imaging techniques.

4. Zooming in on a Single Layer: Cellular Cryo-Electron Tomography

Cryo-ET represents one of the most important imaging techniques in structural biology that bridges the resolution gap between volumetric imaging and high-resolution structural determination methods such as SPA and X-ray crystallography. Unlike the later methods, which excel in revealing atomic or near-atomic details of isolated macromolecules, cryo-ET provides a unique window to characterize macromolecular complexes within their natural cellular environments. Cryo-ET is most effective when imaging samples thinner than 300 nm, thus can be directly used for biological samples, including purified virions and viruses located within thin regions such as the cell periphery [10,42]. For thicker samples, such as whole cells, where electron penetration is insufficient, cryo-FIB can be used to prepare thin lamellae of 80–250 nm thickness and subsequently imaged using cryo-ET. This combination allows imaging capabilities to extend deep into the cell, enabling the characterization of complete viral life cycles, their interactions with host cells, and the cellular immune response during viral infection.

Cryo-ET requires capturing a series of two-dimensional projections of a specimen using a microscope stage that is tilted at different angles in relation to the incident beam [43–45]. The concatenate projections are widely referred to as the 'tilt-series', spanning an angular range of $\pm 60^{\circ}$ with increments of a few degrees per image of the target region. Due to the mechanical properties of the stage and the limited tilt angle used, cryo-ET imaging often leads to incomplete sampling of Fourier space, leading to the so called "missing wedge" problem [46]. To minimize radiation damage while retaining high-resolution information, several data collection strategies can be applied, including continuous, bidirectional, and dose-symmetric tilt schemes. The dose-symmetric tilt scheme, which starts at a low angle and alternates between progressively positive and negative angles, is the most used cryo-ET data collection strategy due to its optimized, near-symmetric information transfer [47]. Post-collection, the tilt series undergo motion correction and alignment using various software packages (e.g., emClarity [48,49], RELION [50], IMOD [51]) to generate tomograms and offer a three-dimensional depiction of the target region on the biological specimens [52–55].

Collecting a group of tomograms, or sometimes even single tomograms, can offer a thorough examination of biological processes. Moreover, when tomograms contain multiple copies of a macromolecule of interest, subtomogram averaging (STA) [43,48–50] can achieve near-atomic resolution structures by aligning, extracting, and merging multiple segments from the tomograms, enhancing the signal-to-noise ratio for specific features such as viral proteins or ribonucleoprotein complexes (RNP) [56,57]. Together, cryo-ET combined with STA is powerful for studying diverse viral structures with a high level of detail [48,58].

Advancements in cryo-ET methods over the past two decades have provided valuable insights into viral assemblies and viral protein structures from in vitro isolated virions, including those of the Ebola virus [59,60], Marburg virus [60], HIV-1 [61], and SARS-CoV-2 [57,62,63], facilitating the rapid response to viral outbreaks and vaccine development. Recently, developments in cellular cryo-ET have enabled the characterization of virus particles and vaccine candidates within cells to a comparable resolution, as well as their interactions with host cells [10,11,41,64–66]. This is exemplified by studies on bluetongue virus (BTV) and mumps virus (MuV) infection, respectively.

Bluetongue virus (BTV) is a dsRNA virus in the family of Reoviridae that can cause diarrhea in humans and death to livestock [67]. The BTV virion consists of three layers of capsid protein encapsulating ten segments of dsRNA genome. Despite the well-known structures of the BTV particles, the genome packaging and viral assembly process remain unclear. Leveraging in-cell cryo-ET advancement in studies of reovirus and rotavirus infection processes [68,69], Xia et al. [64] captured six assembly intermediates of BTV to intermediate resolutions (up to 7.2 Å) in infected baby hamster kidney (BHK21) cells. By combining these structures with high-resolution SPA of isolated viral particles, they demonstrated a comprehensive process of virus assembly. This process begins with the co-assembly of newly synthesized ssRNA segments with the first layer of capsid (viral protein VP6 and VP3) in the virus inclusion body, followed by the sequential assembly of the outer two capsid layers (second layer VP7 and third layer VP2), and continues with the egress of immature virus particles mediated by a virus-encoded transmembrane protein and the outermost capsid (VP2).

The integration of cryo-correlative light and electron microscopy (cryo-CLEM) with cryo-ET has proven to be highly effective for characterizing viral life cycles in cells. Using a cryo-correlative imaging workflow, Zhang et al. [11] investigated the reactivation mechanism of MuV replication under oxidative stress conditions in long-term infected HeLa cells. They prepared thin cell lamellae containing MuV factories by targeting the closely located and fluorescently labeled G3BP stress granule assembly factor 1 (G3BP1) (Figure 4a). Cryo-ET imaging revealed altered MuV nucleocapsid morphologies under oxidative stress conditions, with an increased abundance of straight filaments (Figure 4b). Using in-cell subtomogram averaging, they achieved a reconstruction of the MuV capsid at 6.5 Å resolution, which adopted a genome-accessible conformation (Figure 4c). This conformation, along with additional densities in the viral capsid, elucidates how partner proteins phosphoprotein P and polymerase L access the viral RNA to activate replication. This work sheds light on how MuV–host interplay disrupts the metastable balance during long-term infection under host stress that allows the activation of viral replication and exemplifies a robust cryo-correlative workflow to study other RNA infections.

5. Spotlight on Key Components: Single-Particle Analysis

SPA is the most established cryo-EM technique for determining the structures of macromolecules in their near-native state at near-atomic resolution [70]. SPA conventionally does not provide information on viral particles within cells, but it does provide high resolutions that are not attainable with other microscopic imaging techniques. Nevertheless, a recent trend has emerged where SPA technique is being utilized to analyze protein particles in thin lamellae [71].

SPA workflow begins with preparing and rapidly freezing in vitro purified or assembled macromolecules to preserve particles in a thin aqueous layer (<150 nm). Subsequently, images containing projections of randomly oriented particles are recorded using an electron microscope and direct detection cameras are then processed computationally to generate 3D reconstructions of complexes. Specifically, individual particles are detected, aligned, and classified into groups based on their projections, enhancing signal and reducing noise. A three-dimensional model is reconstructed from these 2D images consisting of varied angles of particle orientation. This initial model undergoes iterative 3D classification, refinement, and validation to ensure accuracy in the angular assignment and resolution. SPA is particularly valuable for studying flexible or dynamic large protein assemblies with a high degree of heterogeneity (compositional and conformational). Routine and automatic data collection enables efficient data collection (5000 micrographs per day). These large amounts of data available can provide near-atomic resolution of RNA virus components such as the viral capsid [72,73], viral envelope, spikes, and viral replication machinery [74] and facilitate vaccine and small molecule therapeutics development [75]. Notably, viral components like viral capsid which were traditionally imaged using cryo-ET have recently been characterized using SPA, resulting in higher resolution and novel mechanistic insights [12,76,77].

The work of Schirra et al. on the HIV-1 capsid [12] illustrates the power of SPA in revealing novel molecular mechanisms. In Figure 4d, in vitro-assembled VLPs with declinations encircled are shown, while Figure 4e depicts the map after further local refinement with imposed five-fold symmetry, colored according to local resolution. Their study identified a Thr-Val-Gly-Gly motif in the viral capsid protein (CA) that acts as a molecular switch, controlling the formation of hexamers and pentamers in the fullerene capsid structure. This finding not only shed light on the assembly process of the HIV-1 capsid but also highlighted its role in modulating post-assembly functions and interactions with antiviral agents like Lenacapavir. Similarly, Stacey et al. [76] further dissected the structural complexities of the HIV-1 capsid. By using SPA, they revealed two structural switches in the CA that regulate capsid curvature and host factor binding, providing insights into the evolutionary advantage of the capsid conical morphology and its implications for cell entry mechanisms. The potential of SPA in viral structural analysis extends beyond HIV-1. Sugita et al.'s study on the Ebola virus nucleocapsid [77] offers a prime example. By achieving a near-atomic resolution structure of the recombinant Ebola virus nucleoprotein–RNA complex, their research provided a detailed view of the virus's genome encapsulation and the dynamic transitions between RNA-bound and RNA-free states. This study elucidated the nucleocapsid assembly mechanism and highlighted key structural features for potential antiviral drug development.

Moreover, Lazic et al.'s exploration of integrated differential phase contrast scanning transmission electron microscopy (iDPC-STEM) marked a technological advancement in SPA [78]. Other approaches, such as cryo-electron ptychography, also explored the possibility of the STEM being used in the field of SPA [79]. Demonstrating the efficacy of iDPC-STEM, they successfully obtained high-resolution structures of tobacco mosaic virus (TMV) at 3.5 Å. Despite not exceeding the resolution of that of commonly used CTEM, iDPC-STEM holds the potential to significantly reduce radiation damage during data collection. The scanning mode in STEM deposits electrons sequentially one pixel at a time, which allows energy to spread and dissipate towards nonilluminated areas, potentially weakening the impact and damage at exposed spots. It has been shown that spot scanning of 100 nm beams in CTEM mitigates beam-induced motion and improves contrast. In addition, alternative scan patterns can further reduce beam damage. This advancement opens new possibilities for achieving even higher resolution structures with cryo-EM, further pushing the boundaries of our understanding of viral architectures.

6. Conclusions and Outlook

Advancements in multiscale imaging techniques have significantly enhanced our understanding of RNA virus biology, shedding light on the highly dynamic interactions between viruses and their hosts, as well as the complex cellular responses. Whole-cell imaging techniques offer valuable insights into the ultrastructural alterations induced by viral invasion, while high-resolution imaging techniques provide detailed delineation of molecular mechanisms. On the other hand, RNA virus research and the development of multi-modal imaging methods are highly intertwined, as virus research has also contributed to the advancement of imaging techniques. For instance, the study of HIV-1 capsids has led to the development of a set of tools with extensive implications for in situ structural biology by bringing cryo-ET to subnanometer and near-atomic resolution [59,80–82].

Looking ahead, continued advancements and emerging trends hold great promise for characterizing viral infections from a holistic perspective. Ongoing and future efforts are essential to bridge the gap between "in vitro" and "in cellular" RNA virus research. This includes integrating various imaging modalities, incorporating fluorescent microscopy methods for correlative imaging approaches to enhance precision in targeted imaging, and refining reconstruction algorithms by leveraging artificial intelligence (AI) and machine learning. Enhancing accessibility, automation, and a diverse array of approaches adaptable to various scales will offer crucial tools for developing novel antiviral strategies and mounting rapid responses to emerging viral threats.

Author Contributions: P.Z. conceived the work. C.-A.L., Y.S. and P.Z. wrote the manuscript. All authors have read and agreed to the published version of the manuscript.

Funding: This work was supported by the UK Wellcome Trust Investigator Award (206422/Z/17/Z), the ERC Advanced Grant (101021133) and partially supported by Chinese Academy of Medical Sciences Oxford Institute core funding grant (CNR00050). Dr. Yao Shen was supported by a CIHR fellowship from the Canadian Institutes of Health Research (funding reference number: 194032). Cheng-An Lyu was supported by a Chinese Academy of Medical Sciences Oxford Institute DPhil Studentship.

Conflicts of Interest: The authors declare that there are no competing interests associated with the manuscript.

References

1. Wasik, B.R.; Turner, P.E. On the biological success of viruses. *Annu. Rev. Microbiol.* **2013**, *67*, 519–541. [CrossRef] [PubMed]
2. Msemburi, W.; Karlinsky, A.; Knutson, V.; Aleshin-Guendel, S.; Chatterji, S.; Wakefield, J. The WHO estimates of excess mortality associated with the COVID-19 pandemic. *Nature* **2023**, *613*, 130–137. [CrossRef]
3. Jacob, S.T.; Crozier, I.; Fischer, W.A., 2nd; Hewlett, A.; Kraft, C.S.; Vega, M.A.; Soka, M.J.; Wahl, V.; Griffiths, A.; Bollinger, L.; et al. Ebola virus disease. *Nat. Rev. Dis. Primers* **2020**, *6*, 13. [CrossRef] [PubMed]
4. Finlay, B.B.; McFadden, G. Anti-immunology: Evasion of the host immune system by bacterial and viral pathogens. *Cell* **2006**, *124*, 767–782. [CrossRef]
5. De Armas-Rillo, L.; Valera, M.S.; Marrero-Hernandez, S.; Valenzuela-Fernandez, A. Membrane dynamics associated with viral infection. *Rev. Med. Virol.* **2016**, *26*, 146–160. [CrossRef]
6. Hernandez-Gonzalez, M.; Larocque, G.; Way, M. Viral use and subversion of membrane organization and trafficking. *J. Cell Sci.* **2021**, *134*, jcs252676. [CrossRef]
7. Kondylis, P.; Schlicksup, C.J.; Zlotnick, A.; Jacobson, S.C. Analytical Techniques to Characterize the Structure, Properties, and Assembly of Virus Capsids. *Anal. Chem.* **2019**, *91*, 622–636. [CrossRef] [PubMed]
8. Adrian, M.; Dubochet, J.; Lepault, J.; McDowall, A.W. Cryo-electron microscopy of viruses. *Nature* **1984**, *308*, 32–36. [CrossRef]
9. Perez-Berna, A.J.; Benseny-Cases, N.; Rodriguez, M.J.; Valcarcel, R.; Carrascosa, J.L.; Gastaminza, P.; Pereiro, E. Monitoring reversion of hepatitis C virus-induced cellular alterations by direct-acting antivirals using cryo soft X-ray tomography and infrared microscopy. *Acta Crystallogr. D Struct. Biol.* **2021**, *77*, 1365–1377. [CrossRef]
10. Mendonca, L.; Howe, A.; Gilchrist, J.B.; Sheng, Y.; Sun, D.; Knight, M.L.; Zanetti-Domingues, L.C.; Bateman, B.; Krebs, A.S.; Chen, L.; et al. Correlative multi-scale cryo-imaging unveils SARS-CoV-2 assembly and egress. *Nat. Commun.* **2021**, *12*, 4629. [CrossRef]
11. Zhang, X.; Sridharan, S.; Zagoriy, I.; Eugster Oegema, C.; Ching, C.; Pflaesterer, T.; Fung, H.K.H.; Becher, I.; Poser, I.; Muller, C.W.; et al. Molecular mechanisms of stress-induced reactivation in mumps virus condensates. *Cell* **2023**, *186*, 1877–1894.e27. [CrossRef] [PubMed]
12. Schirra, R.T.; Dos Santos, N.F.B.; Zadrozny, K.K.; Kucharska, I.; Ganser-Pornillos, B.K.; Pornillos, O. A molecular switch modulates assembly and host factor binding of the HIV-1 capsid. *Nat. Struct. Mol. Biol.* **2023**, *30*, 383–390. [CrossRef] [PubMed]
13. Groen, J.; Sorrentino, A.; Aballe, L.; Oliete, R.; Valcarcel, R.; Okolo, C.; Kounatidis, I.; Harkiolaki, M.; Perez-Berna, A.J.; Pereiro, E. A 3D Cartographic Description of the Cell by Cryo Soft X-ray Tomography. *J. Vis. Exp. JoVE* **2021**, 169. [CrossRef]
14. Garriga, D.; Chichon, F.J.; Calisto, B.M.; Ferrero, D.S.; Gastaminza, P.; Pereiro, E.; Perez-Berna, A.J. Imaging of Virus-Infected Cells with Soft X-ray Tomography. *Viruses* **2021**, *13*, 2109. [CrossRef]
15. Harkiolaki, M.; Darrow, M.C.; Spink, M.C.; Kosior, E.; Dent, K.; Duke, E. Cryo-soft X-ray tomography: Using soft X-rays to explore the ultrastructure of whole cells. *Emerg. Top. Life Sci.* **2018**, *2*, 81–92. [CrossRef]
16. Okolo, C.A. A guide into the world of high-resolution 3D imaging: The case of soft X-ray tomography for the life sciences. *Biochem. Soc. Trans.* **2022**, *50*, 649–663. [CrossRef]
17. Carzaniga, R.; Domart, M.C.; Duke, E.; Collinson, L.M. Correlative cryo-fluorescence and cryo-soft X-ray tomography of adherent cells at European synchrotrons. *Methods Cell Biol.* **2014**, *124*, 151–178. [CrossRef]
18. Loconte, V.; Chen, J.H.; Cortese, M.; Ekman, A.; Le Gros, M.A.; Larabell, C.; Bartenschlager, R.; Weinhardt, V. Using soft X-ray tomography for rapid whole-cell quantitative imaging of SARS-CoV-2-infected cells. *Cell Rep. Methods* **2021**, *1*, 100117. [CrossRef]
19. Ekman, A.; Weinhardt, V.; Chen, J.H.; McDermott, G.; Le Gros, M.A.; Larabell, C. PSF correction in soft X-ray tomography. *J. Struct. Biol.* **2018**, *204*, 9–18. [CrossRef]

20. Parkinson, D.Y.; Knoechel, C.; Yang, C.; Larabell, C.A.; Le Gros, M.A. Automatic alignment and reconstruction of images for soft X-ray tomography. *J. Struct. Biol.* **2012**, *177*, 259–266. [CrossRef]
21. Liu, J.; Li, F.; Chen, L.; Guan, Y.; Tian, L.; Xiong, Y.; Liu, G.; Tian, Y. Quantitative imaging of Candida utilis and its organelles by soft X-ray Nano-CT. *J. Microsc.* **2018**, *270*, 64–70. [CrossRef] [PubMed]
22. Vyas, N.; Kunne, S.; Fish, T.M.; Dobbie, I.M.; Harkiolaki, M.; Paul-Gilloteaux, P. Protocol for image registration of correlative soft X-ray tomography and super-resolution structured illumination microscopy images. *STAR Protoc.* **2021**, *2*, 100529. [CrossRef] [PubMed]
23. Cortese, M.; Goellner, S.; Acosta, E.G.; Neufeldt, C.J.; Oleksiuk, O.; Lampe, M.; Haselmann, U.; Funaya, C.; Schieber, N.; Ronchi, P.; et al. Ultrastructural Characterization of Zika Virus Replication Factories. *Cell Rep.* **2017**, *18*, 2113–2123. [CrossRef]
24. Kounatidis, I.; Stanifer, M.L.; Phillips, M.A.; Paul-Gilloteaux, P.; Heiligenstein, X.; Wang, H.; Okolo, C.A.; Fish, T.M.; Spink, M.C.; Stuart, D.I.; et al. 3D Correlative Cryo-Structured Illumination Fluorescence and Soft X-ray Microscopy Elucidates Reovirus Intracellular Release Pathway. *Cell* **2020**, *182*, 515–530.e517. [CrossRef]
25. Nahas, K.L.; Connor, V.; Scherer, K.M.; Kaminski, C.F.; Harkiolaki, M.; Crump, C.M.; Graham, S.C. Near-native state imaging by cryo-soft-X-ray tomography reveals remodelling of multiple cellular organelles during HSV-1 infection. *PLoS Pathog.* **2022**, *18*, e1010629. [CrossRef]
26. Okolo, C.; Jadhav, A.; Fish, T.; Nahas, K.; Watts, A.; Harkiolaki, M. Applications of Soft X-ray Tomography for the Direct Observation of Native Cellular Events. *Microsc. Microanal.* **2023**, *29*, 1147–1148. [CrossRef] [PubMed]
27. Perez-Berna, A.J.; Rodriguez, M.J.; Chichon, F.J.; Friesland, M.F.; Sorrentino, A.; Carrascosa, J.L.; Pereiro, E.; Gastaminza, P. Structural Changes In Cells Imaged by Soft X-ray Cryo-Tomography During Hepatitis C Virus Infection. *ACS Nano* **2016**, *10*, 6597–6611. [CrossRef]
28. Castro, V.; Perez-Berna, A.J.; Calvo, G.; Pereiro, E.; Gastaminza, P. Three-Dimensional Remodeling of SARS-CoV2-Infected Cells Revealed by Cryogenic Soft X-ray Tomography. *ACS Nano* **2023**, *17*, 22708–22721. [CrossRef]
29. Biggins, S.W.; Bambha, K.M.; Terrault, N.A.; Inadomi, J.; Shiboski, S.; Dodge, J.L.; Gralla, J.; Rosen, H.R.; Roberts, J.P. Projected future increase in aging hepatitis C virus-infected liver transplant candidates: A potential effect of hepatocellular carcinoma. *Liver Transplant.* **2012**, *18*, 1471–1478. [CrossRef]
30. Thirukkumaran, C.; Morris, D.G. Oncolytic viral therapy using reovirus. *Methods Mol. Biol.* **2009**, *542*, 607–634. [CrossRef]
31. Scher, N.; Rechav, K.; Paul-Gilloteaux, P.; Avinoam, O. In situ fiducial markers for 3D correlative cryo-fluorescence and FIB-SEM imaging. *iScience* **2021**, *24*, 102714. [CrossRef] [PubMed]
32. Schertel, A.; Snaidero, N.; Han, H.M.; Ruhwedel, T.; Laue, M.; Grabenbauer, M.; Mobius, W. Cryo FIB-SEM: Volume imaging of cellular ultrastructure in native frozen specimens. *J. Struct. Biol.* **2013**, *184*, 355–360. [CrossRef]
33. Zhu, Y.; Sun, D.; Schertel, A.; Ning, J.; Fu, X.; Gwo, P.P.; Watson, A.M.; Zanetti-Domingues, L.C.; Martin-Fernandez, M.L.; Freyberg, Z.; et al. Serial cryoFIB/SEM Reveals Cytoarchitectural Disruptions in Leigh Syndrome Patient Cells. *Structure* **2021**, *29*, 82–87.e3. [CrossRef] [PubMed]
34. Wu, G.H.; Mitchell, P.G.; Galaz-Montoya, J.G.; Hecksel, C.W.; Sontag, E.M.; Gangadharan, V.; Marshman, J.; Mankus, D.; Bisher, M.E.; Lytton-Jean, A.K.R.; et al. Multi-scale 3D Cryo-Correlative Microscopy for Vitrified Cells. *Structure* **2020**, *28*, 1231–1237.e3. [CrossRef]
35. Sviben, S.; Gal, A.; Hood, M.A.; Bertinetti, L.; Politi, Y.; Bennet, M.; Krishnamoorthy, P.; Schertel, A.; Wirth, R.; Sorrentino, A.; et al. A vacuole-like compartment concentrates a disordered calcium phase in a key coccolithophorid alga. *Nat. Commun.* **2016**, *7*, 11228. [CrossRef] [PubMed]
36. Akiva, A.; Nelkenbaum, O.; Schertel, A.; Yaniv, K.; Weiner, S.; Addadi, L. Intercellular pathways from the vasculature to the forming bone in the zebrafish larval caudal fin: Possible role in bone formation. *J. Struct. Biol.* **2019**, *206*, 139–148. [CrossRef]
37. Vidavsky, N.; Masic, A.; Schertel, A.; Weiner, S.; Addadi, L. Mineral-bearing vesicle transport in sea urchin embryos. *J. Struct. Biol.* **2015**, *192*, 358–365. [CrossRef]
38. Dumoux, M.; Glen, T.; Smith, J.L.R.; Ho, E.M.L.; Perdigao, L.M.A.; Pennington, A.; Klumpe, S.; Yee, N.B.Y.; Farmer, D.A.; Lai, P.Y.A.; et al. Cryo-plasma FIB/SEM volume imaging of biological specimens. *eLife* **2023**, *12*, e83623. [CrossRef]
39. Baena, V.; Conrad, R.; Friday, P.; Fitzgerald, E.; Kim, T.; Bernbaum, J.; Berensmann, H.; Harned, A.; Nagashima, K.; Narayan, K. FIB-SEM as a Volume Electron Microscopy Approach to Study Cellular Architectures in SARS-CoV-2 and Other Viral Infections: A Practical Primer for a Virologist. *Viruses* **2021**, *13*, 611. [CrossRef]
40. Wolff, G.; Limpens, R.; Zevenhoven-Dobbe, J.C.; Laugks, U.; Zheng, S.; de Jong, A.W.M.; Koning, R.I.; Agard, D.A.; Grunewald, K.; Koster, A.J.; et al. A molecular pore spans the double membrane of the coronavirus replication organelle. *Science* **2020**, *369*, 1395–1398. [CrossRef]
41. Klein, S.; Cortese, M.; Winter, S.L.; Wachsmuth-Melm, M.; Neufeldt, C.J.; Cerikan, B.; Stanifer, M.L.; Boulant, S.; Bartenschlager, R.; Chlanda, P. SARS-CoV-2 structure and replication characterized by in situ cryo-electron tomography. *Nat. Commun.* **2020**, *11*, 5885. [CrossRef] [PubMed]
42. Chmielewski, D.; Schmid, M.F.; Simmons, G.; Jin, J.; Chiu, W. Chikungunya virus assembly and budding visualized in situ using cryogenic electron tomography. *Nat. Microbiol.* **2022**, *7*, 1270–1279. [CrossRef] [PubMed]
43. Wan, W.; Briggs, J.A. Cryo-Electron Tomography and Subtomogram Averaging. *Methods Enzymol.* **2016**, *579*, 329–367. [CrossRef] [PubMed]

44. Lucic, V.; Rigort, A.; Baumeister, W. Cryo-electron tomography: The challenge of doing structural biology in situ. *J. Cell Biol.* **2013**, *202*, 407–419. [CrossRef]
45. Schaffer, M.; Mahamid, J.; Engel, B.D.; Laugks, T.; Baumeister, W.; Plitzko, J.M. Optimized cryo-focused ion beam sample preparation aimed at in situ structural studies of membrane proteins. *J. Struct. Biol.* **2017**, *197*, 73–82. [CrossRef] [PubMed]
46. Ding, G.; Liu, Y.; Zhang, R.; Xin, H.L. A joint deep learning model to recover information and reduce artifacts in missing-wedge sinograms for electron tomography and beyond. *Sci. Rep.* **2019**, *9*, 12803. [CrossRef]
47. Hagen, W.J.H.; Wan, W.; Briggs, J.A.G. Implementation of a cryo-electron tomography tilt-scheme optimized for high resolution subtomogram averaging. *J. Struct. Biol.* **2017**, *197*, 191–198. [CrossRef]
48. Himes, B.A.; Zhang, P. emClarity: Software for high-resolution cryo-electron tomography and subtomogram averaging. *Nat. Methods* **2018**, *15*, 955–961. [CrossRef]
49. Ni, T.; Frosio, T.; Mendonca, L.; Sheng, Y.; Clare, D.; Himes, B.A.; Zhang, P. High-resolution in situ structure determination by cryo-electron tomography and subtomogram averaging using emClarity. *Nat. Protoc.* **2022**, *17*, 421–444. [CrossRef]
50. Bharat, T.A.; Scheres, S.H. Resolving macromolecular structures from electron cryo-tomography data using subtomogram averaging in RELION. *Nat. Protoc.* **2016**, *11*, 2054–2065. [CrossRef]
51. Henderson, L.D.; Beeby, M. High-Throughput Electron Cryo-tomography of Protein Complexes and Their Assembly. *Methods Mol. Biol.* **2018**, *1764*, 29–44. [CrossRef] [PubMed]
52. Kremer, J.R.; Mastronarde, D.N.; McIntosh, J.R. Computer visualization of three-dimensional image data using IMOD. *J. Struct. Biol.* **1996**, *116*, 71–76. [CrossRef]
53. Zheng, S.Q.; Palovcak, E.; Armache, J.P.; Verba, K.A.; Cheng, Y.; Agard, D.A. MotionCor2: Anisotropic correction of beam-induced motion for improved cryo-electron microscopy. *Nat. Methods* **2017**, *14*, 331–332. [CrossRef]
54. Zheng, S.; Wolff, G.; Greenan, G.; Chen, Z.; Faas, F.G.A.; Barcena, M.; Koster, A.J.; Cheng, Y.; Agard, D.A. AreTomo: An integrated software package for automated marker-free, motion-corrected cryo-electron tomographic alignment and reconstruction. *J. Struct. Biol. X* **2022**, *6*, 100068. [CrossRef] [PubMed]
55. Noble, A.J.; Stagg, S.M. Automated batch fiducial-less tilt-series alignment in Appion using Protomo. *J. Struct. Biol.* **2015**, *192*, 270–278. [CrossRef]
56. Huang, Y.; Wang, T.; Zhong, L.; Zhang, W.; Zhang, Y.; Yu, X.; Yuan, S.; Ni, T. Molecular architecture of coronavirus double-membrane vesicle pore complex. *Nature* **2024**, *633*, 224–231. [CrossRef]
57. Yao, H.; Song, Y.; Chen, Y.; Wu, N.; Xu, J.; Sun, C.; Zhang, J.; Weng, T.; Zhang, Z.; Wu, Z.; et al. Molecular Architecture of the SARS-CoV-2 Virus. *Cell* **2020**, *183*, 730–738.e13. [CrossRef] [PubMed]
58. Bharat, T.A.M.; Russo, C.J.; Lowe, J.; Passmore, L.A.; Scheres, S.H.W. Advances in Single-Particle Electron Cryomicroscopy Structure Determination applied to Sub-tomogram Averaging. *Structure* **2015**, *23*, 1743–1753. [CrossRef]
59. Wan, W.; Kolesnikova, L.; Clarke, M.; Koehler, A.; Noda, T.; Becker, S.; Briggs, J.A.G. Structure and assembly of the Ebola virus nucleocapsid. *Nature* **2017**, *551*, 394–397. [CrossRef]
60. Wan, W.; Clarke, M.; Norris, M.J.; Kolesnikova, L.; Koehler, A.; Bornholdt, Z.A.; Becker, S.; Saphire, E.O.; Briggs, J.A. Ebola and Marburg virus matrix layers are locally ordered assemblies of VP40 dimers. *eLife* **2020**, *9*, e59225. [CrossRef]
61. Schur, F.K.; Obr, M.; Hagen, W.J.; Wan, W.; Jakobi, A.J.; Kirkpatrick, J.M.; Sachse, C.; Krausslich, H.G.; Briggs, J.A. An atomic model of HIV-1 capsid-SP1 reveals structures regulating assembly and maturation. *Science* **2016**, *353*, 506–508. [CrossRef] [PubMed]
62. Ke, Z.; Oton, J.; Qu, K.; Cortese, M.; Zila, V.; McKeane, L.; Nakane, T.; Zivanov, J.; Neufeldt, C.J.; Cerikan, B.; et al. Structures and distributions of SARS-CoV-2 spike proteins on intact virions. *Nature* **2020**, *588*, 498–502. [CrossRef]
63. Liu, C.; Mendonca, L.; Yang, Y.; Gao, Y.; Shen, C.; Liu, J.; Ni, T.; Ju, B.; Liu, C.; Tang, X.; et al. The Architecture of Inactivated SARS-CoV-2 with Postfusion Spikes Revealed by Cryo-EM and Cryo-ET. *Structure* **2020**, *28*, 1218–1224.e4. [CrossRef] [PubMed]
64. Xia, X.; Sung, P.Y.; Martynowycz, M.W.; Gonen, T.; Roy, P.; Zhou, Z.H. RNA genome packaging and capsid assembly of bluetongue virus visualized in host cells. *Cell* **2024**, *187*, 2236–2249.e17. [CrossRef] [PubMed]
65. Ni, T.; Mendonca, L.; Zhu, Y.; Howe, A.; Radecke, J.; Shah, P.M.; Sheng, Y.; Krebs, A.S.; Duyvesteyn, H.M.E.; Allen, E.; et al. ChAdOx1 COVID vaccines express RBD open prefusion SARS-CoV-2 spikes on the cell surface. *iScience* **2023**, *26*, 107882. [CrossRef]
66. Watanabe, Y.; Mendonca, L.; Allen, E.R.; Howe, A.; Lee, M.; Allen, J.D.; Chawla, H.; Pulido, D.; Donnellan, F.; Davies, H.; et al. Native-like SARS-CoV-2 Spike Glycoprotein Expressed by ChAdOx1 nCoV-19/AZD1222 Vaccine. *ACS Cent. Sci.* **2021**, *7*, 594–602. [CrossRef]
67. Baldwin, C.L.; Yirsaw, A.; Gillespie, A.; Le Page, L.; Zhang, F.; Damani-Yokota, P.; Telfer, J.C. gammadelta T cells in livestock: Responses to pathogens and vaccine potential. *Transbound. Emerg. Dis.* **2020**, *67* (Suppl. S2), 119–128. [CrossRef] [PubMed]
68. Sutton, G.; Sun, D.; Fu, X.; Kotecha, A.; Hecksel, C.W.; Clare, D.K.; Zhang, P.; Stuart, D.I.; Boyce, M. Assembly intermediates of orthoreovirus captured in the cell. *Nat. Commun.* **2020**, *11*, 4445. [CrossRef]
69. Shah, P.N.M.; Gilchrist, J.B.; Forsberg, B.O.; Burt, A.; Howe, A.; Mosalaganti, S.; Wan, W.; Radecke, J.; Chaban, Y.; Sutton, G.; et al. Characterization of the rotavirus assembly pathway in situ using cryoelectron tomography. *Cell Host Microbe* **2023**, *31*, 604–615.e4. [CrossRef]
70. Elmlund, D.; Elmlund, H. Cryogenic electron microscopy and single-particle analysis. *Annu. Rev. Biochem.* **2015**, *84*, 499–517. [CrossRef]

71. Cheng, J.; Liu, T.; You, X.; Zhang, F.; Sui, S.F.; Wan, X.; Zhang, X. Determining protein structures in cellular lamella at pseudo-atomic resolution by GisSPA. *Nat. Commun.* **2023**, *14*, 1282. [CrossRef] [PubMed]
72. Highland, C.M.; Tan, A.; Ricana, C.L.; Briggs, J.A.G.; Dick, R.A. Structural insights into HIV-1 polyanion-dependent capsid lattice formation revealed by single particle cryo-EM. *Proc. Natl. Acad. Sci. USA* **2023**, *120*, e2220545120. [CrossRef] [PubMed]
73. Bester, S.M.; Wei, G.; Zhao, H.; Adu-Ampratwum, D.; Iqbal, N.; Courouble, V.V.; Francis, A.C.; Annamalai, A.S.; Singh, P.K.; Shkriabai, N.; et al. Structural and mechanistic bases for a potent HIV-1 capsid inhibitor. *Science* **2020**, *370*, 360–364. [CrossRef] [PubMed]
74. Yan, L.; Ge, J.; Zheng, L.; Zhang, Y.; Gao, Y.; Wang, T.; Huang, Y.; Yang, Y.; Gao, S.; Li, M.; et al. Cryo-EM Structure of an Extended SARS-CoV-2 Replication and Transcription Complex Reveals an Intermediate State in Cap Synthesis. *Cell* **2021**, *184*, 184–193.e110. [CrossRef] [PubMed]
75. Rapp, M.; Shapiro, L.; Frank, J. Contributions of single-particle cryoelectron microscopy toward fighting COVID-19. *Trends Biochem. Sci.* **2022**, *47*, 117–123. [CrossRef]
76. Stacey, J.C.V.; Tan, A.; Lu, J.M.; James, L.C.; Dick, R.A.; Briggs, J.A.G. Two structural switches in HIV-1 capsid regulate capsid curvature and host factor binding. *Proc. Natl. Acad. Sci. USA* **2023**, *120*, e2220557120. [CrossRef]
77. Sugita, Y.; Matsunami, H.; Kawaoka, Y.; Noda, T.; Wolf, M. Cryo-EM structure of the Ebola virus nucleoprotein-RNA complex at 3.6 A resolution. *Nature* **2018**, *563*, 137–140. [CrossRef]
78. Lazic, I.; Wirix, M.; Leidl, M.L.; de Haas, F.; Mann, D.; Beckers, M.; Pechnikova, E.V.; Muller-Caspary, K.; Egoavil, R.; Bosch, E.G.T.; et al. Single-particle cryo-EM structures from iDPC-STEM at near-atomic resolution. *Nat. Methods* **2022**, *19*, 1126–1136. [CrossRef]
79. Pei, X.; Zhou, L.; Huang, C.; Boyce, M.; Kim, J.S.; Liberti, E.; Hu, Y.; Sasaki, T.; Nellist, P.D.; Zhang, P.; et al. Cryogenic electron ptychographic single particle analysis with wide bandwidth information transfer. *Nat. Commun.* **2023**, *14*, 3027. [CrossRef]
80. Ni, T.; Zhu, Y.; Yang, Z.; Xu, C.; Chaban, Y.; Nesterova, T.; Ning, J.; Bocking, T.; Parker, M.W.; Monnie, C.; et al. Structure of native HIV-1 cores and their interactions with IP6 and CypA. *Sci. Adv.* **2021**, *7*, eabj5715. [CrossRef]
81. Mendonca, L.; Sun, D.; Ning, J.; Liu, J.; Kotecha, A.; Olek, M.; Frosio, T.; Fu, X.; Himes, B.A.; Kleinpeter, A.B.; et al. CryoET structures of immature HIV Gag reveal six-helix bundle. *Commun. Biol.* **2021**, *4*, 481. [CrossRef] [PubMed]
82. Schur, F.K.; Hagen, W.J.; Rumlova, M.; Ruml, T.; Muller, B.; Krausslich, H.G.; Briggs, J.A. Structure of the immature HIV-1 capsid in intact virus particles at 8.8 A resolution. *Nature* **2015**, *517*, 505–508. [CrossRef] [PubMed]

viruses

MDPI

Data Descriptor

CryoVirusDB: An Annotated Dataset for AI-Based Virus Particle Identification in Cryo-EM Micrographs

Rajan Gyawali [1,2,†], Ashwin Dhakal [1,2,†], Liguo Wang [3] and Jianlin Cheng [1,2,*]

1 Department of Electrical Engineering and Computer Science, University of Missouri, Columbia, MO 65211, USA; rgkg2@missouri.edu (R.G.); ad256@umsystem.edu (A.D.)
2 NextGen Precision Health, University of Missouri, Columbia, MO 65211, USA
3 Laboratory for BioMolecular Structure (LBMS), Brookhaven National Laboratory, Upton, NY 11973, USA; lwang1@bnl.gov
* Correspondence: author: chengji@missouri.edu
† These authors contributed equally to this work.

Abstract

With the advancements in instrumentation, image processing algorithms, and computational capabilities, single-particle cryo-electron microscopy (cryo-EM) has achieved atomic resolution in determining the 3D structures of viruses. The virus structures play a crucial role in studying their biological function and advancing the development of antiviral vaccines and treatments. Despite the effectiveness of artificial intelligence (AI) in general image processing, its development for identifying and extracting virus particles from cryo-EM micrographs has been hindered by the lack of manually labeled high-quality datasets. To fill the gap, we introduce CryoVirusDB, a labeled dataset containing the coordinates of expert-picked virus particles in cryo-EM micrographs. CryoVirusDB comprises 9941 micrographs from nine datasets representing seven distinct non-enveloped viruses exhibiting icosahedral or pseudo-icosahedral symmetry, along with coordinates of 339,398 labeled virus particles. It can be used to train and test AI and machine learning (e.g., deep learning) methods to accurately identify virus particles in cryo-EM micrographs for building atomic 3D structural models for viruses.

Keywords: cryo-electron microscopy; AI-based virus particle picking; structural biology; viral structure; labeled dataset

1. Introduction

Cryo-EM is a method of capturing 2D images of biological molecules and assemblies at cryogenic temperatures. Advancements in both instrumentation and computational methodologies have established cryo-EM as an essential tool for interrogating the structures and dynamics of biological macromolecular complexes including large virus particles [1–4]. Single particle experiment and data analysis in cryo-EM involve vitrifying biological specimens and collecting micrographs of individual particles in an electron microscope (Figure 1A), followed by picking and extracting particle images (Figure 1B), applying image processing for correction and alignment (Figure 1C), and performing three-dimensional (3D) reconstruction of macromolecular complexes [4,5] (Figure 1D).

In the realm of virology, cryo-EM has been instrumental in studying and determining the 3D structures and morphology of various viruses such as poliovirus, Ebola, HIV, and coronaviruses [6,7]. Particularly during the COVID-19 pandemic, cryo-EM played a pivotal role in understanding the intricate structure of the SARS-CoV-2 spike protein [8–10]. This knowledge has facilitated the development of highly effective vaccines. For instance, scientists have been able to design immunogens that mimic the spike protein's shape, eliciting targeted immune responses [11–13]. Moreover, cryo-EM has revolutionized epitope mapping, enabling the identification of specific binding sites [14] and facilitating the exploration of antibody mutations for the rapid discovery and development of precise vaccines and antiviral treatments [15,16].

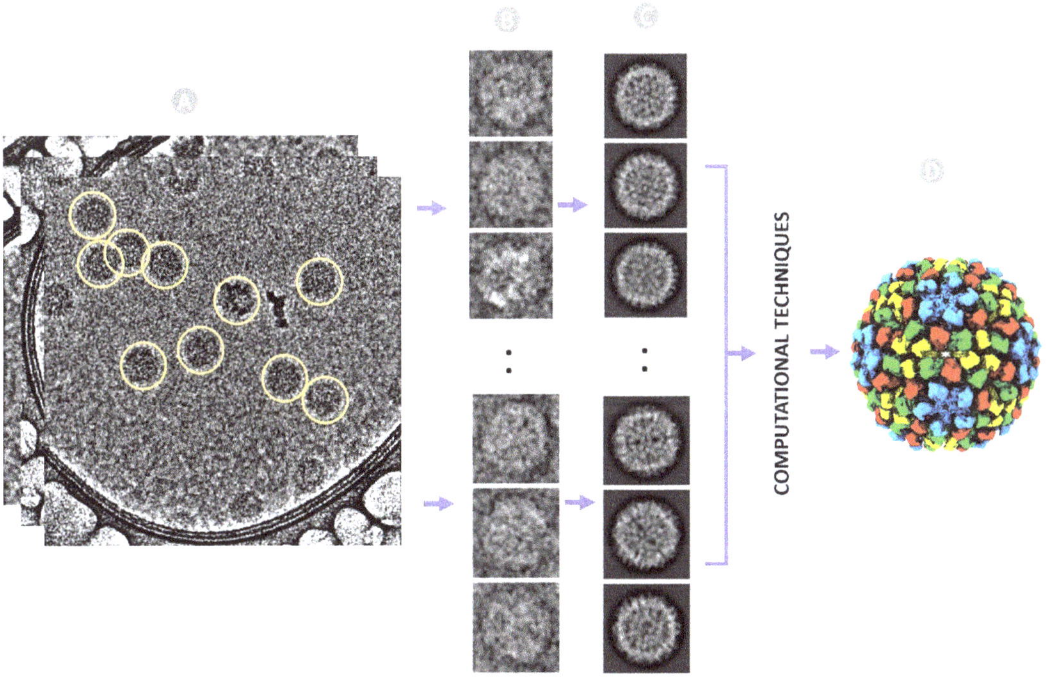

Figure 1. An overview of cryo-EM single particle analysis from particle selection to 3D reconstruction of virus (EMPIAR 11060). (**A**) Stack of ideal micrographs where the true virus particles are picked (encircled yellow). (**B**) Extracted virus particles from micrographs with fixed box size. (**C**) Multiple 2D classes to facilitate stack cleaning and the removal of false particles. (**D**) Reconstructed 3D structure of the virus from 2D images using a series of computational techniques. Subunits/domain regions are displayed in different colors.

To achieve high-resolution 3D reconstructions of virus structures, the initial step of accurately recognizing and extracting virus particles from 2D image projections (micrographs) is crucial. Currently, three virus particle-picking approaches are employed: manual virus particle picking, template-based picking [17,18], and AI-based picking [19]. Manual picking is laborious and time-consuming, requiring specialized expertise for precise identification, which cannot be used by beginners. Challenges in the manual picking arise from low signal-to-noise ratios, low particle contrast, and the unpredictability of individual particle appearances due to orientation variations. Template-based virus particle picking requires users to pick some initial particles as templates for software tools to search for more particles, which suffers from the presence of ice contamination, carbon areas, and overlapping aggregated particles in micrographs. AI-based particle picking [20–24] has the best potential to automate the process and overcome the problems of the manual pick-

https://doi.org/10.3390/fluids10110297

ing and template-based matching, but the development of sophisticated AI-based virus particle-picking methods is largely hindered by the lack of high-quality labeled training and test data of virus particles.

To harness the power of cutting-edge AI technologies in automatic virus particle recognition and picking, we created a comprehensive and expert-labeled dataset—CryoVirusDB [25] in this work. This open-access dataset aims to expedite the development of automated virus particle-picking workflows, and ultimately advance the viral research and the design of therapeutic interventions. CryoVirusDB includes 9941 micrographs from nine datasets representing seven distinct non-enveloped viruses with icosahedral or pseudo-icosahedral symmetry, along with coordinates of 339,398 labeled virus particles. The particle count distribution per micrograph varies considerably across datasets, as illustrated in Figure 2. The detailed statistics of CryoVirusDB are reported in Tables 1 and 2.

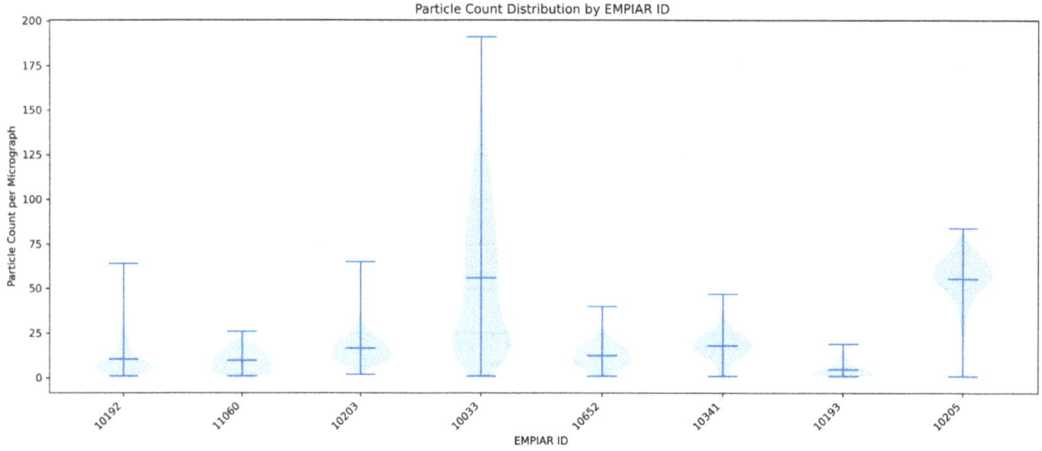

Figure 2. Particle count distribution across CryoVirusDB datasets. Violin plots show the distribution of particle counts per micrograph for each EMPIAR dataset. Horizontal lines indicate mean count (center), maximum count (top) and minimum count (bottom) of virus particles.

Table 1. The statistics of micrographs and particles of nine viruses in CryoVirusDB.

SN	EMPIAR ID	Virus Type	Number of Micrographs	Micrograph Size	Particle Diameter (px)	Defocus Range (μm)	Number of Virus Particles	Average Particles per Micrograph
1	10192 [26]	Feline calicivirus	1000	(4096, 4096)	470	−1.2 to −3.5	9660	9.66
2	11060 [27]	Nudaurelia capensis omega virus	1276	(4096, 4096)	516	−0.70 to −2.2	11,916	9.34
3	10203 [28]	Macrobrachium rosenbergii nodavirus	1000	(3838, 3710)	377	−1.0 to −2.5	16,601	16.60
4	10033 [29]	Human parechovirus 3	1000	(4096, 4096)	350	−0.42 to −2.34	55,732	55.73
5	10652 [30]	Coxsackievirus B4	1127	(3838, 3710)	374	−0.60 to −3.0	11,144	9.89
6	10341 [31]	Bovine enterovirus	1274	(4096, 4096)	376	−0.75 to −3.5	22,694	17.81
7	10193 [26]	Feline calicivirus	1000	(4096, 4096)	516	−1.2 to −3.5	96,126	96.13
8	10205 [32]	Cowpea mosaic virus	1000	(4096, 4096)	310	−0.50 to −3.5	81,037	81.04
9	10555 [33]	Nudaurelia capensis omega virus	1264	(4096, 4096)	564	−0.70 to −2.7	34,488	27.28
		Total	9941				339,398	34.14

 https://doi.org/10.3390/fluids10110297

Table 2. Metrics of EM data acquisition and grid preparation used in importing micrographs for virus particle picking.

SN	EMPIAR ID	Pixel Size (Å)	Electron Dose (e/Å^2)	Detector
1	10192	1.065	63	FEI FALCON III (4k × 4k)
2	11060	1.065	46	FEI FALCON III (4k × 4k)
3	10203	1.06	36	GATAN K2 QUANTUM (4k × 4k)
4	10033	1.14	36	FEI FALCON II (4k × 4k)
5	10652	1.06	47	GATAN K2 SUMMIT (4k × 4k)
6	10341	1.065	49.5	FEI FALCON III (4k × 4k)
7	10193	1.065	63	FEI FALCON III (4k × 4k)
8	10205	1.065	67.5	FEI FALCON III (4k × 4k)
9	10555	1.0651	72	FEI FALCON III (4k × 4k)

Note: All datasets were acquired using an FEI TITAN KRIOS microscope (Thermo Fisher Scientific, Hillsboro, OR, USA) at 300 kV acceleration voltage with 2.7 mm spherical aberration. Micrographs are in mrc format.

2. Materials and Methods

2.1. Raw Data Acquisition and Preprocessing

The cryo-EM virus micrographs and their associated metadata were downloaded from the EMPIAR web portal [34] using Python API and FTP scripts. The comprehensive metadata include the EMPIAR ID for each cryo-EM dataset of a virus along with the corresponding identifiers such as Electron Microscopy Data Bank (EMDB) ID and Protein Data Bank (PDB) ID. Additionally, the dataset size, resolution, total number of micrographs, image specifications (size and type), pixel size, micrograph file extension, gain/motion correction file extension (if any), FTP and Globus paths for micrograph/gain files, and relevant publication information were recorded.

2.2. Software and Computation Tools

CryoSPARC [35] was employed for multiple stages of the particle-picking pipeline, including motion correction, patch-based contrast transfer function (CTF) estimation, interactive manual particle selection, template-based automated particle picking, two-dimensional (2D) classification, ab initio three-dimensional (3D) reconstruction, and homogeneous refinement. The software's graphical user interface facilitated visual inspection and quality control of picked particles throughout the curation process.

For comparative validation of particle-picking quality, Topaz [22] was used as a representative artificial intelligence (AI)-based particle-picking method. Topaz was executed with default parameters to provide an unbiased comparison with our manually curated approach.

Visualization and inspection of particle stacks and 3D density maps were conducted using UCSF ChimeraX [36]. Custom Python scripts (Python version 3.8) were developed for automated data acquisition from the EMPIAR database, metadata management, and file format conversions. The Python function download_micrographs() can be used to download micrographs directly from EMPIAR website, and data format conversion functionalities can be found in convert_cs_file_to_star.py and convert_star_to_csv_file.py. Details on the usage of these files and functions can be found at https://github.com/BioinfoMachineLearning/CryoVirusDB (accessed on 1 December 2025).

3. Results

CryoVirusDB, a labeled dataset for AI-based training for detection of icosahedral virus particles from cryo-EM micrographs was developed. The dataset contains the cryo-EM

 https://doi.org/10.3390/fluids10110297

micrographs and the associated coordinate information for training and validation of the AI models. To ensure the generalization capability of these models, we have maintained the dataset diversity by selecting 9 representative EMPIAR virus datasets with 7 unique viruses that encompass a broad range of particle sizes, shapes, density distributions, noise levels, and variations in ice thickness and carbon areas. The viruses in CryoVirusDB represent diverse taxonomic groups. Specifically, the database includes viruses from three families within the Order Picornavirales: Caliciviridae (Feline calicivirus), Picornaviridae (Human parechovirus 3, Coxsackievirus B4, and Bovine enterovirus), and Secoviridae (Cowpea mosaic virus). Additionally, it includes Nudaurelia capensis omega virus from the family Alphatetraviridae (Order Hepelivirales), and Macrobrachium rosenbergii nodavirus from the family Sarthroviridae (unassigned order) [37]. The database encompasses viruses that infect diverse hosts across multiple kingdoms, including humans (Human parechovirus 3, Coxsackievirus B4), animals (Bovine enterovirus, Feline calicivirus), arthropods (Nudaurelia capensis omega virus, Macrobrachium rosenbergii nodavirus), and plants (Cowpea mosaic virus), demonstrating the broad ecological and biological relevance of CryoVirusDB.

3.1. Particle-Picking Workflow

3.1.1. Micrograph Import

For each EMPIAR virus dataset, cryo-EM micrographs were imported into CryoSPARC using the corresponding microscopy parameters obtained from the EMPIAR metadata. Essential acquisition parameters including pixel size (Å), accelerating voltage (kV), spherical aberration coefficient (mm), and total electron dose (e/Å^2) were specified during the import process to ensure accurate downstream processing. The complete set of imaging parameters for all nine datasets is provided in Table 2.

3.1.2. Motion Correction and Patch-Based CTF Estimation of Micrographs

In our study, we used motion-corrected micrographs in CryoSPARC [35] for patch-based Contrast Transfer Function (CTF) estimation. Since CTF functions can vary substantially among micrographs and cannot be precisely predefined, accurately identifying CTF parameters for each micrograph is crucial. This precision is necessary for proper corrections and achieving high-resolution 3D reconstructions. Two stages (estimating the CTF and correcting it) were applied to the CTF analysis.

We employed the patch-based CTF tool in CryoSPARC to generate output micrographs containing information about their average defocus and the defocus landscape. Upon particle extraction, this information was automatically used to allocate a local defocus value to each particle based on its position in the landscape. The one-dimensional search across defocus values for a micrograph is shown in Figure 3A and Thon rings in the power spectra of images are shown in Figure 3B.

The plot in Figure 3C serves primarily as a verification for the successful execution of background subtraction and envelope function fitting. The X-axis represents frequency in inverse angstroms. The radially averaged power spectrum is depicted in black, where high values correspond to the bright portions of the Thon rings and low values to the dark regions. The orange curve represents the envelope function, aiming to model the expected falloff of Thon rings up to the Nyquist resolution, accounting for aberrations. Lastly, the fitted Contrast Transfer Function (CTF), scaled by the envelope function, is presented in blue. This oscillating plot is crucial for confirming the proper execution of background subtraction and envelope-fitting procedures. In the plot in Figure 3D, we assessed the background strength (depicted by the black line) within the area where thicker ice leads to an augmented background referred to as relative ice thickness.

https://doi.org/10.3390/fluids10110297

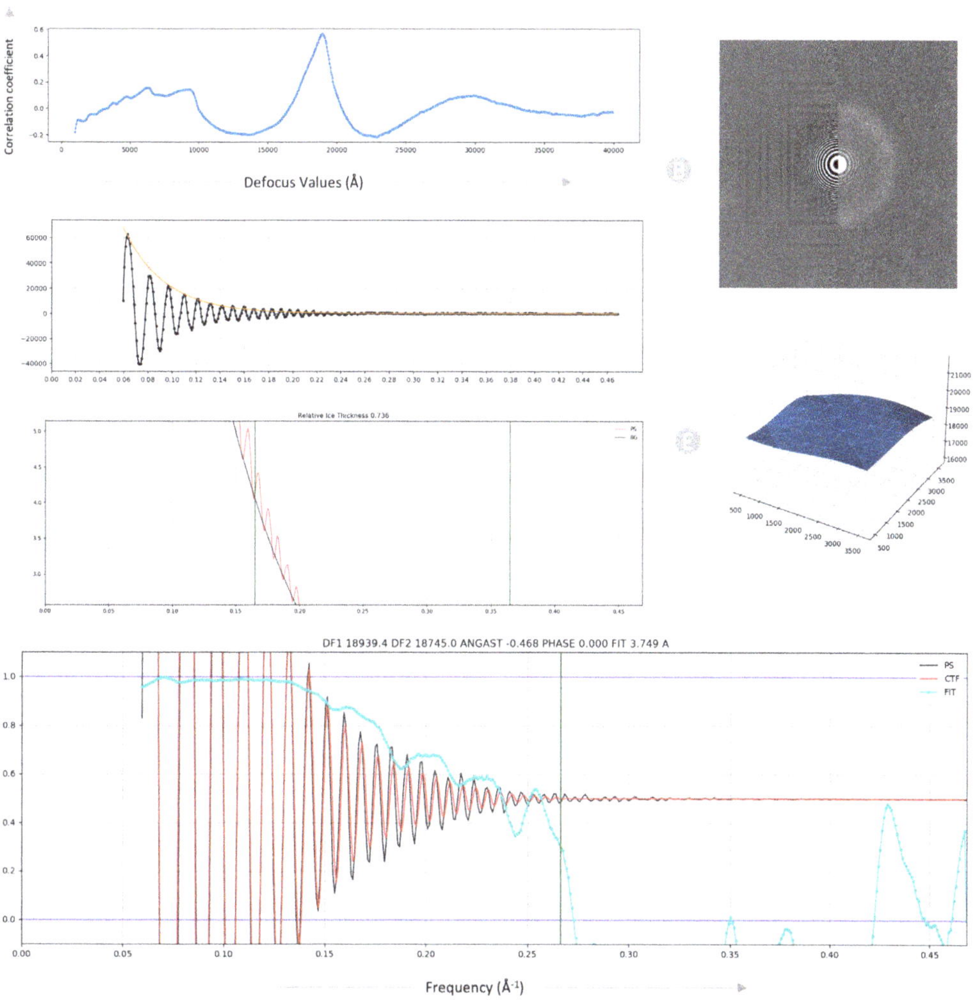

Figure 3. Diagnostic plots of CTF for EMPIAR 11060. (**A**) 1D search over varying defocus values. (**B**) Thon rings visible in the Fourier transform. (**C**) Fitted envelope function diagnostic plot. (**D**) Power spectrum showing the relative ice thickness. (**E**) 2D Patch result. (**F**) CTF fit plot. The frequency, measured in inverse angstroms ($Å^{-1}$), is represented on the *X*-axis, while the correlation metric between the power spectrum (PS) and CTF value is shown on the *Y*-axis. The black line corresponds to the observed experimental power spectrum, the red line represents the calculated CTF, and the cyan line indicates the cross-correlation (FIT).

The 3D surface plot (Figure 3E) shows the local defocus estimated throughout the micrograph. The surface plot in blue illustrates the defocus that has been fitted for each position along the micrograph. The x- and y-coordinates align with the micrograph's coordinates, while the z-coordinate represents the defocus values. The X, Y, and Z axes are all expressed in Angstrom. The CTF fit plot, illustrated in Figure 3F, depicts the alignment between the simulated and observed Thon rings in the micrograph, accounting for variations in defocus and astigmatism. The cyan curve indicates the cross-correlation fit level. The CTF fit resolution (3.749 angstroms) is the resolution at which this value drops

 https://doi.org/10.3390/fluids10110297

below a threshold. The vertical green line in the plot signifies the frequency at which the fit deviates from the cross-correlation threshold of 0.3, indicating a successful fit.

3.1.3. Manual Particle Picking and 2D Class Formation

Following the CTF estimation, we manually identified and selected true virus particles interactively from aligned and motion-corrected micrographs with the aim of generating some particle templates. We specify the particle diameter based on the virus particles' size and shape. Picking particles directly from raw noisy micrographs is challenging (Figure 4A). So, we adjusted the 'Contrast Intensity Override' using a low-pass filter while inspecting micrographs to achieve the most distinct view for particle selection (Figure 4B). Figure 4C shows the picked particles for subsequent 2D alignment steps.

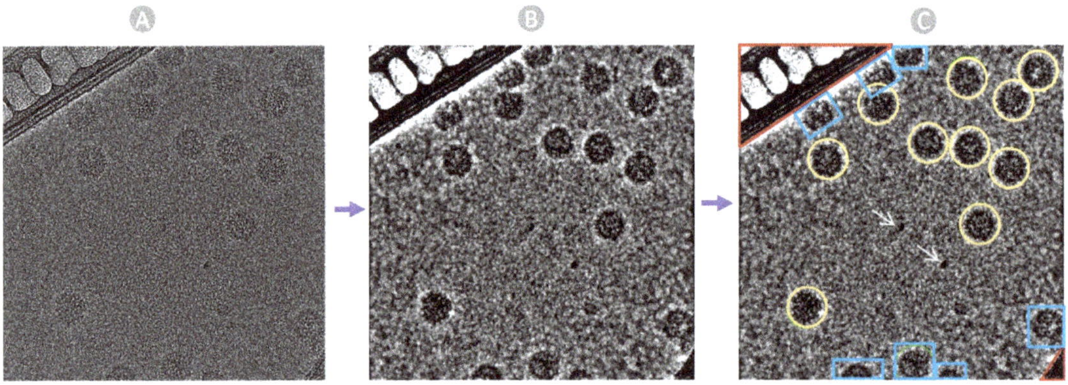

Figure 4. The manual picking process. (**A**) Raw micrograph obtained from EMPIAR. (**B**) Preprocessed micrograph with low-pass filter: 28 A to ease particle recognition and picking. (**C**) Manually picked true virus particle encircled in yellow, carbon region colored in red, ice patches and artifacts noted by white arrows, and cut particles in edges colored in blue.

Manually selecting particles from raw micrographs with smaller defocus values proved to be quite challenging. To create a comprehensive set of ground-truth templates covering a broad range of defocus values, we manually picked particles from numerous micrographs exhibiting diverse defocus and CTF fit values. Given the time-intensive nature of manual picking, we chose a small subset of micrographs (around 15% of the micrographs) specifically for generating templates. The detailed information about the manually picked particles and the micrographs considered for the manual picking can be found in Supplementary Table S1.

After manually picking the virus particles, the coordinates of particle centers and the designated box size were used to extract particles from the original micrographs. During this process, the box size was defined to provide ample padding, usually ranging from 25% to 50% extra space around the particles. The manually selected particles underwent a 2D classification step, where we categorized and chose the most favorable classes. This classification step organized particles into distinct 2D classes, streamlining the cleaning of the particle stack and removal of undesirable particles. Finally, we assessed the quality of the particles and eliminated classes containing unwanted particles. The remaining particle classes were used by the template-based picking for the identification of high-quality particle classes.

https://doi.org/10.3390/fluids10110297

3.1.4. Template-Based Picking

After exporting the optimal particle classes, we employed templates created in the '2D Class Formation' step in CryoSPARC. We followed an iterative approach, wherein the output from 'template-based picking and inspection' was once again utilized in the '2D Class Formation' step to select only the high-quality 2D particles, discarding the false positives. This cycle was repeated until we obtained high-resolution particles that encompassed all possible viewing directions of the virus particle.

Using CryoSPARC's Template Picker job, we employed the high-resolution templates to precisely pick virus particles that aligned with the geometry of the target structure. We set specific constraints, such as the particle diameter in angstrom and a minimum distance between the particles for generating templates based on the SK97 sampling algorithm [38].

3.1.5. Manual Particle Inspection and Extraction

The acquired particles above underwent manual inspection, in which we scrutinized and refined the picked particles using different thresholds. In CryoSPARC, we fine-tuned parameters such as the low-pass filter, normalized cross-correlation (NCC), and power threshold (Figure 5A) to eliminate false positives. The 2D colored histogram plots were employed to carefully analyze the median pick scores of micrographs against defocus, aiding in the extraction of coordinates for high-quality virus particles as depicted in Figure 5B.

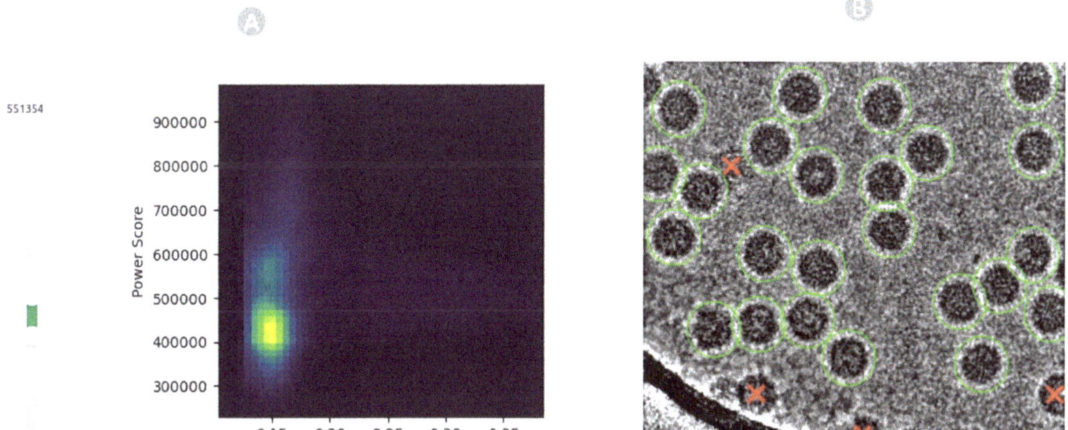

Figure 5. Particle quality inspection. (**A**) Particle filtration achieved by manipulating the NCC score on the *X*-axis and local power score on the *Y*-axis for EMPIAR 11060. (**B**) High-quality true virus particles (depicted by green circles) chosen through the template-based picking process and eliminated radiation-damaged, cut, and false-positive particles, represented by red cross markers.

Ultimately, we applied a 2D classification step to perform a final examination of the selected particles. The Select 2D job categorized particles into various 2D classes (usually 50 in our case), aiding in stack cleaning and the elimination of undesired particles. This process is valuable not only for assessing particle quality before entering the 3D reconstruction phase but also for qualitatively exploring the distribution of views within the dataset. Following 2D classification, certain classes are identified as "junk" classes, representing non-particle images, ice crystals, or instances of two particles being conjoined.

 https://doi.org/10.3390/fluids10110297

Consequently, we filtered out the particles associated with these "junk" classes from the picked particles. More information about the overall intermediate metadata and the final set of true virus particles can be found in Supplementary Table S1.

These final true particles were exported in the form of particle stacks and star files, which include a lot of information about the particles in micrographs such as the X-coordinate, Y-coordinate, Angle-Psi, Origin X (Ang), Origin Y (Ang), Defocus U, Defocus V, Defocus Angle, Phase Shift, CTF B Factor, Optics Group, and Class Number. The star files are the widely accepted file format in cryo-EM processing software. These star files are also converted to csv files which are easy to use in python scripts and contain the necessary coordinates for generating ground truth labels while training and validating the machine learning models.

3.2. Data Organization of CryoVirusDB

CryoVirusDB includes 9 virus subsets (each including approximately 1100 cryo-EM micrographs) along with the labeled coordinates of the virus particles in the micrographs. The total size of the CryoVirusDB database is 634 GB. The organizational structure of the directories of CryoVirusDB is depicted in Figure 6.

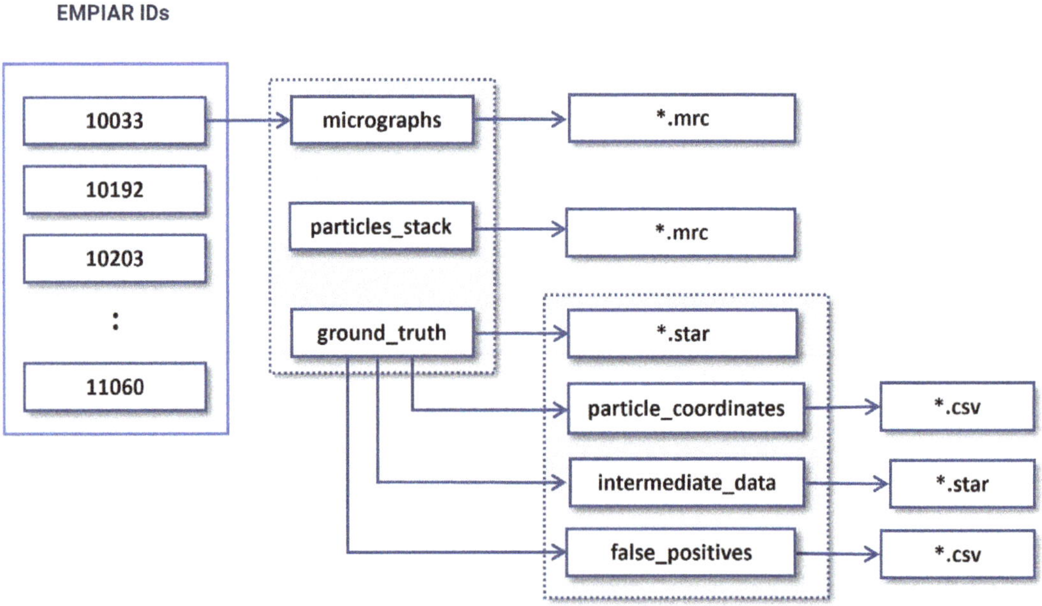

Figure 6. The directory structure of CryoVirusDB. The numbers in the blocks on the left side are the respective EMPIAR IDs.

3.3. Data Validation

For training machine learning models, high-quality curated data is essential. Following the data validation practices established in CryoPPP [4], we validated the quality of picked particles through both 2D classification analysis and 3D reconstruction methods. This comprehensive validation approach ensures that the particle coordinates in CryoVirusDB meet the stringent quality standards required for developing robust AI-based particle-picking algorithms.

 https://doi.org/10.3390/fluids10110297

3.3.1. 2D Particle Class Validation

We compared our picked virus particles with a popular AI-based particle-picking method, Topaz [22], considering factors such as the total number of classes, number of picked particles, 2D resolution, and visual orientation. Our manually picked particles have a better 2D class resolution than Topaz. It is noteworthy that a higher particle count alone does not ensure higher resolution. Instead, selecting a substantial number of high-quality particles across a broad angular distribution is crucial for achieving both high 2D and 3D resolution. The 2D class comparisons for two databases, EMPIAR 10205 and EMPIAR 10193 (each containing 1000 micrographs), are shown in Table 3 and Figure 7. In both cases, Topaz picked many more particles than CryoVirusDB but it had a worse weighted average resolution of 2D classes. The weighted average resolution of 2D classes is calculated by multiplying each class's resolution by its particle count, summing these products, and dividing by the total number of particles across all classes.

Table 3. 2D classification result comparison for EMPIAR 10205 and EMPIAR 10193.

EMPIAR 10205		
	2D Particle Class Statistics (Topaz)	**2D Particle Class Statistics (CryoVirusDB)**
Number of Picked Particles	155,953	81,037
Weighted Average Resolution of 2D classes ($N = 50$)	9.41 Å	6.59 Å
Weighted Average Resolution of 2D classes ($N = 10$)	13.42 Å	10.96 Å
EMPIAR 10193		
	2D Particle Class Statistics (Topaz)	**2D Particle Class Statistics (CryoVirusDB)**
Number of Picked Particles	239,852	96,126
Weighted Average Resolution of 2D classes ($N = 50$)	18.52 Å	15.02 Å
Weighted Average Resolution of 2D classes ($N = 10$)	23.68 Å	21.72 Å

We also assessed the density projections derived from the intermediate output during the ab initio reconstruction phase, as depicted at the bottom of each block in Figure 7. The plot illustrates the integrated density values along the perpendicular direction to that plane. The heatmap's color scheme represents scalar density values at each pixel, with the intensity of color indicating the magnitude of density. This indicates the high quality of the virus particles in CryoVirusDB.

3.3.2. 3D Density Map Validation

We reconstructed 3D density maps from the particles in CryoVirusDB and from those picked by Topaz for two datasets, EMPIAR 10205 and EMPIAR 10193, each comprising 1000 micrographs. The ab initio density map reconstruction and homogenous refinement were carried out in CryoSPARC using the generated star files that included the selected particles. To ensure an unbiased evaluation, we repeated the ab initio 3D reconstruction experiment with three distinct random seeds for each method.

Figure 8 presents a comparison of the resolution and distribution direction of the reconstructed 3D density maps. The Fourier Shell Correlation (FSC) plots include a 'loose

 https://doi.org/10.3390/fluids10110297

mask' curve that utilizes an automatically generated mask with a 15 Å falloff, and a 'tight mask' curve that employs an auto-generated mask with a falloff of 6 Å for all FSC plots.

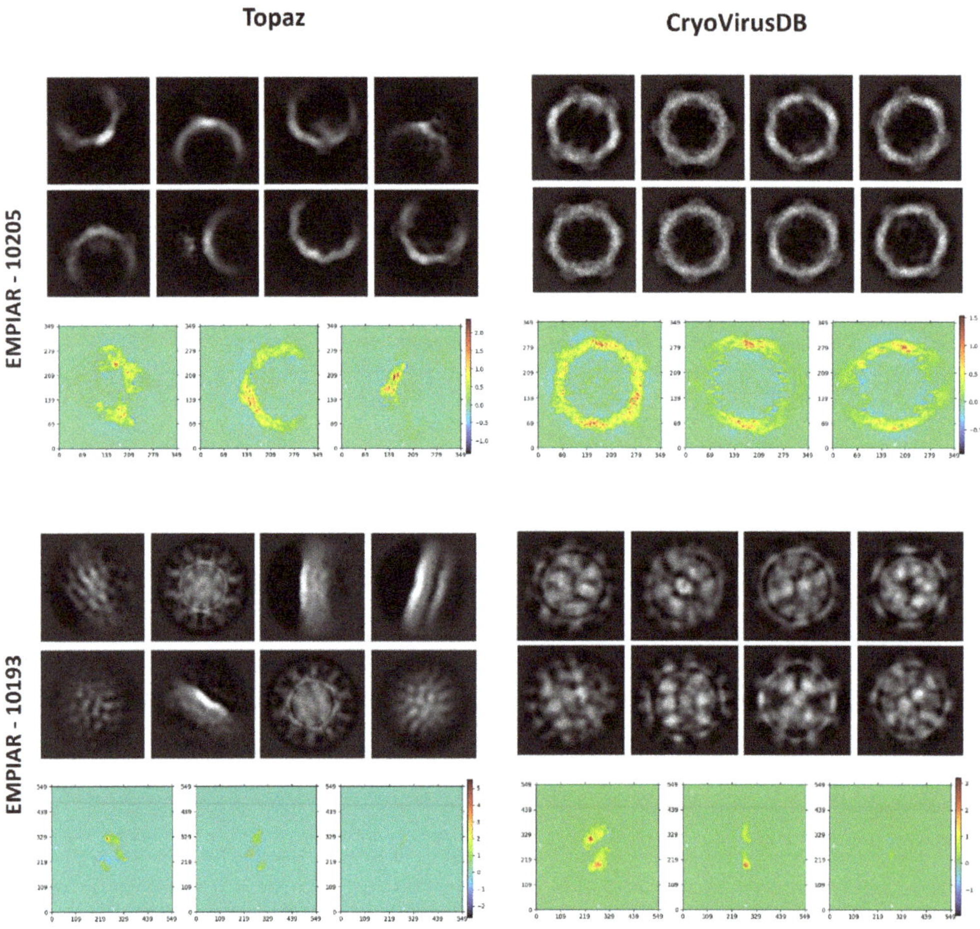

Figure 7. The comparison of 2D particle classification and density projections from the intermediate output of the ab initio reconstruction phase for EMPIAR 10205 and EMPIAR 10193. For each EMPIAR ID, the 2D classes are visualized at the top and the density projects are visualized at the bottom. The color scheme in the heatmap corresponds to the scalar density values at each pixel.

In the case of EMPIAR 10205, Topaz picked around 75,000 more particles compared to our manual picking. Despite this, the density map reconstructed from particles selected by CryoVirusDB achieved a resolution of 4.34 Å, substantially better than Topaz's resolution of 6.48 Å. This indicates the high quality of the picked particles in CryoVirusDB. For EMPIAR 10193, the resolution of CryoVirusDB is 5.16 Å, also better than 5.74 Å of Topaz.

The heightened intensity of the red color in the direction distribution shown in the lower section of each block in Figure 8 corresponds to an increased number of particles in the elevation vs. azimuth plots. CryoVirusDB demonstrated superior particle picking by capturing a substantial number of particles with a wide angular distribution, evident in the red coloration on the heatmap for both validation cases.

 https://doi.org/10.3390/fluids10110297

The detailed comparison of the 3D density map reconstruction of three trials for CryoVirusDB and Topaz is provided in Table 4. The density maps constructed from the labeled particles in CryoVirusDB consistently exhibit a higher quality than Topaz in terms of multiple resolution metrics, even though the number of particles in CryoVirusDB is much smaller than the number of particles picked by Topaz, indicating that Topaz may pick some false positives and/or miss some true positives representing different views.

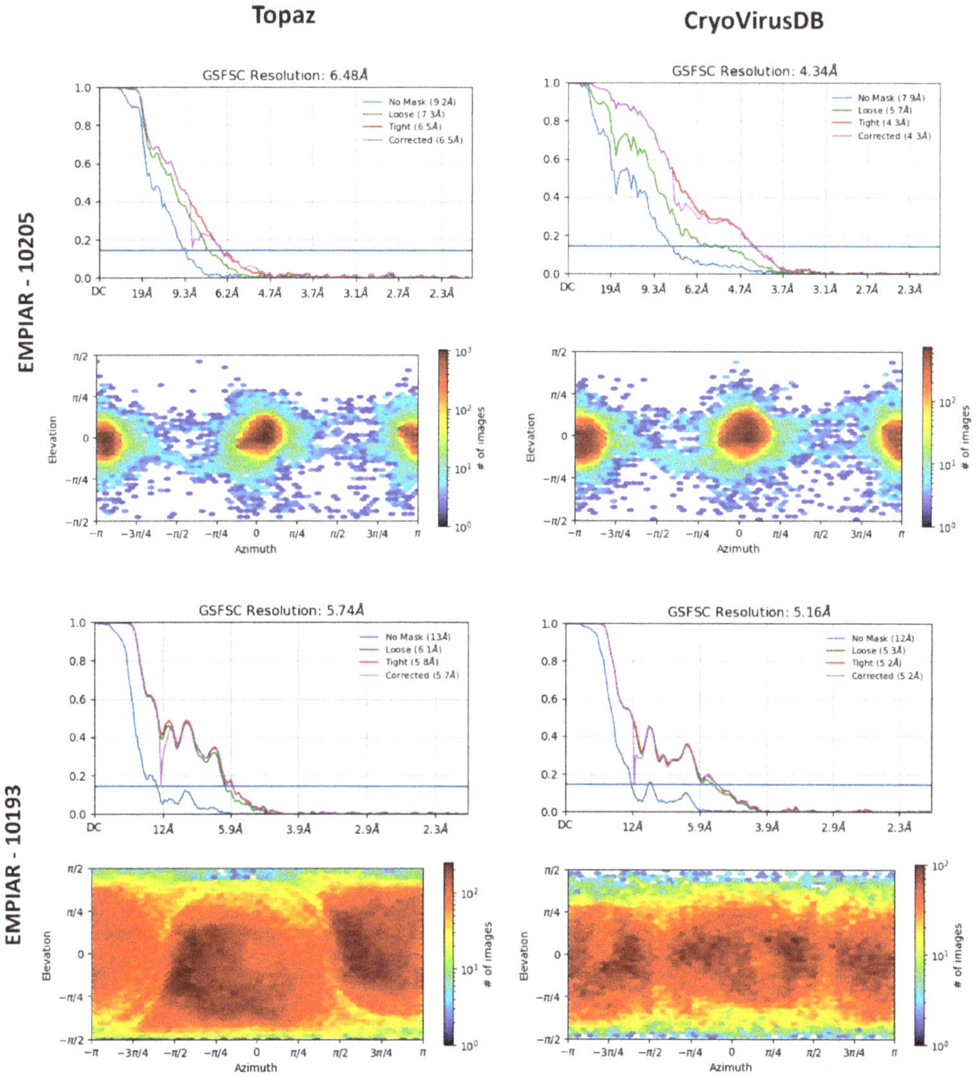

Figure 8. The comparison of reconstructed 3D density map resolution and direction distribution obtained by Topaz and CryoVirusDB on EMPIAR 10205 and EMPIAR 10193.

https://doi.org/10.3390/fluids10110297

3.4. Machine Learning Workflows with CryoVirusDB

3.4.1. Denoising and Preprocessing Cryo-EM Micrographs

Micrographs in mrc format (micrographs directory) serve as input for AI models. To enhance signal-to-noise ratio, we recommend applying Gaussian filters for noise reduction. Standard normalization procedures should be applied to ensure consistent intensity distributions across micrographs using the formula [pixel = (pixel − μ)/σ], where pixel intensity is normalized relative to the mean (μ) and standard deviation (σ).

Table 4. 3D density map result comparison for EMPIAR 10205 and EMPIAR 10193. Bold fonts highlight the resolution of the best of the three trials for each method.

EMPIAR 10205						
	3D Density Map Statistics (Topaz)			**3D Density Map Statistics (CryoVirusDB)**		
Number of Picked Particles	155,953			81,037		
GSFSC Resolution (Å)	Trial 1	Trial 2	Trial 3	Trial 1	Trial 2	Trial 3
	6.97	6.59	**6.48**	**4.34**	4.40	4.47
No Mask Resolution (Å)	9.3	9.5	**9.2**	**7.9**	8.8	8.1
Loose Mask Resolution (Å)	7.7	7.2	**7.3**	**5.7**	7.1	6.3
Tight Mask Resolution (Å)	6.8	6.6	**6.5**	**4.3**	4.5	4.4
Corrected Mask Resolution (Å)	7	6.6	**6.5**	**4.3**	4.4	4.5
EMPIAR 10193						
	3D Density Map Statistics (Topaz)			**3D Density Map Statistics (CryoVirusDB)**		
Number of Picked Particles	239,852			96,126		
GSFSC Resolution (Å)	Trial 1	Trial 2	Trial 3	Trial 1	Trial 2	Trial 3
	5.86	**5.74**	5.82	**5.16**	5.22	5.18
No Mask Resolution (Å)	12	**13**	11	**12**	9.4	9.1
Loose Mask Resolution (Å)	5.9	**6.1**	5.8	**5.3**	5.8	5.5
Tight Mask Resolution (Å)	5.8	**5.8**	5.7	**5.2**	5.2	5.2
Corrected Mask Resolution (Å)	5.9	**5.7**	5.6	**5.2**	5.2	5.3

Subsequently, normalized micrographs should be converted to 8-bit grayscale, collapsing multi-channel intensity information into a single channel. This standardizes pixel values to a 0–255 range and reduces input size, minimizing computational load during model training. Various filtering and contrast enhancement techniques can be employed to further improve image quality. The preprocessing pipeline for denoising micrographs can be adapted from established resources [20,24,39].

3.4.2. Generating Labels for Input Micrographs

The dataset includes expert-annotated particle coordinates within cryo-EM micrographs (ground_truth/particle_coordinates.csv). After preprocessing, corresponding labels must be generated for supervised learning using one of two widely adopted methods. First, coordinates can be stored in Common Objects in Context (COCO) format [40], a standard for object detection that structures labels and metadata in JSON format, enabling annotation of multiple particles within a single micrograph with category labels, bounding boxes, areas, micrograph references, and unique particle IDs (see Section 2.1.3 of CryoTransformer [20]).

https://doi.org/10.3390/fluids10110297

Alternatively, ground-truth segmentation masks can be generated from CSV coordinate data containing particle centers and diameters, serving as target labels during training for pixel-wise comparison between predicted and ground-truth masks for loss calculation, which is particularly effective for semantic segmentation tasks (refer to CryoSegNet [24] for detailed methodology).

3.4.3. Machine Learning Model Development: Inputs and Expected Outputs

Supervised machine learning models—including convolutional neural networks (CNNs), transformers, artificial neural networks (ANNs), or segmentation models—can be designed to accurately predict bounding boxes of virus particles within cryo-EM micrographs while minimizing false positives. The input layer receives preprocessed micrographs, while the output layer predicts particle coordinates (X, Y).

These predicted coordinates are then used to generate .star files, which can be imported into widely used cryo-EM processing software packages such as CryoSPARC [35] or RELION [17] for further analysis and refinement.

4. Discussion and Conclusions

Here, we present a comprehensive dataset, CryoVirusDB, which offers researchers a rich dataset for training and testing AI and machine learning algorithms for virus particle identification from cryo-EM micrographs. With nearly 10,000 high-resolution cryo-EM micrographs and over 330,000 labeled virus particles spanning 9 datasets from 7 distinct non-enveloped icosahedral viruses, this database facilitates 3D reconstruction of virus structures from 2D particles, which is crucial for understanding viral structural mechanisms and developing targeted therapeutics.

CryoVirusDB uses curated subsets (around 1000 micrographs per EMPIAR ID) rather than complete datasets to balance computational feasibility with high-quality ground truth generation. While this reduces final map resolution compared to full-dataset reconstructions (4.3 Å vs. 2.7 Å for EMPIAR-10205 and 5.2 Å vs. 3.8 Å for EMPIAR-10193), our validation shows that manually curated particles consistently outperform automated picks (Topaz) at equivalent particle counts. This demonstrates that CryoVirusDB's primary goal—providing superior training data for AI models—is achieved despite lower absolute resolution. Researchers can leverage models trained on our high-quality subsets to analyze complete datasets for maximum resolution.

While CryoVirusDB focuses on non-enveloped icosahedral viruses, the deep learning models trained on this dataset can potentially be extended to enveloped viruses through transfer learning approaches. Enveloped viruses such as influenza and SARS-CoV-2 present distinct challenges including pleomorphic morphology, ambiguous membrane boundaries, and flexible spike proteins. However, models pre-trained on CryoVirusDB capture fundamental cryo-EM image features—noise patterns, defocus variations, edge characteristics, and contamination signatures—that are independent of specific particle morphologies. These learned representations can be leveraged through fine-tuning on smaller annotated datasets of enveloped viruses. This transfer learning strategy would require substantially less manual annotation effort compared to training models from scratch, while enabling AI-based pickers to generalize across diverse viral architectures and advancing automated particle identification throughout structural virology.

Supplementary Materials: The following supporting information can be downloaded at https://www.mdpi.com/article/10.3390/fluids10110297/s1. Table S1: Intermediate metadata associated with various steps of data curation.

 https://doi.org/10.3390/fluids10110297

Author Contributions: J.C. conceived the research. R.G., A.D. and J.C. designed the methodology and experiment. R.G. and A.D. wrote the scripts and codes for dataset preprocessing. R.G. and A.D. curated the data. J.C. and L.W. conceptualized data validation. R.G. and A.D. drafted the manuscript. J.C. and L.W. revised manuscript. All authors participated in result discussions and data analysis, and made contributions to the final manuscript. All authors have read and agreed to the published version of the manuscript.

Funding: This work was supported by a National Institute of Health grant (grant #: R01GM146340).

Data Availability Statement: The GitHub repository https://github.com/BioinfoMachineLearning/CryoVirusDB (accessed on 1 December 2025) provides the instructions on how to download and use the data. It also contains all of the scripts used in different stages of data curation. Particle coordinate annotations for CryoVirusDB are publicly available on Zenodo [25] at https://zenodo.org/records/10397742 (accessed on 1 December 2025).

Conflicts of Interest: The authors declare no conflicts of interest.

Abbreviations

The following abbreviations are used in this manuscript:

AI	Artificial Intelligence
ML	Machine Learning
CTF	Contrast Transfer Function
Cryo-EM	Cryogenic Electron Microscopy
CSV	Comma-Separated Values
EMDB	Electron Microscopy Data Bank
EMPIAR	Electron Microscopy Public Image Archive
FSC	Fourier Shell Correlation
GSFSC	Gold Standard Fourier Shell Correlation
NCC	Normalized Cross-Correlation
PDB	Protein Data Bank
PS	Power Spectrum
Å	Angstrom
kV	Kilovolt
e/Å^2	Electron Dose per Square Angstrom
µm	Micrometer

References

1. Renaud, J.-P.; Chari, A.; Ciferri, C.; Liu, W.-T.; Rémigy, H.-W.; Stark, H.; Wiesmann, C. Cryo-EM in drug discovery: Achievements, limitations and prospects. *Nat. Rev. Drug Discov.* **2018**, *17*, 471–492. [CrossRef] [PubMed]
2. Chua, E.Y.; Mendez, J.H.; Rapp, M.; Ilca, S.L.; Tan, Y.Z.; Maruthi, K.; Kuang, H.; Zimanyi, C.M.; Cheng, A.; Eng, E.T.; et al. Better, Faster, Cheaper: Recent Advances in Cryo—Electron Microscopy. *Annu. Rev. Biochem.* **2022**, *91*, 1–32. [CrossRef] [PubMed]
3. Wu, M.; Lander, G.C. Present and Emerging Methodologies in Cryo-EM Single-Particle Analysis. *Biophys. J.* **2020**, *119*, 1281–1289. [CrossRef] [PubMed]
4. Dhakal, A.; Gyawali, R.; Wang, L.; Cheng, J. A large expert-curated cryo-EM image dataset for machine learning protein particle picking. *Sci. Data* **2023**, *10*, 392. [CrossRef]
5. Dhakal, A.; Gyawali, R.; Wang, L.; Cheng, J. CryoPPP: A Large Expert-Labelled Cryo-EM Image Dataset for Machine Learning Protein Particle Picking. *bioRxiv* **2023**. [CrossRef]
6. Hryc, C.F.; Chen, D.H.; Chiu, W. Near-atomic resolution cryo-EM for molecular virology. *Curr. Opin. Virol.* **2011**, *1*, 110–117. [CrossRef]
7. Jiang, W.; Tang, L. Atomic cryo-EM structures of viruses. *Curr. Opin. Struct. Biol.* **2017**, *46*, 122–129. [CrossRef]
8. Wrapp, D.; Wang, N.; Corbett, K.S.; Goldsmith, J.A.; Hsieh, C.-L.; Abiona, O.; Graham, B.S.; McLellan, J.S. Cryo-EM structure of the 2019-nCoV spike in the prefusion conformation. *Science* **2020**, *367*, 1260–1263. [CrossRef]
9. Walls, A.C.; Park, Y.-J.; Tortorici, M.A.; Wall, A.; McGuire, A.T.; Veesler, D. Structure, Function, and Antigenicity of the SARS-CoV-2 Spike Glycoprotein. *Cell* **2020**, *181*, 281–292.e6. [CrossRef]

 https://doi.org/10.3390/fluids10110297

10. Yan, R.; Zhang, Y.; Li, Y.; Xia, L.; Guo, Y.; Zhou, Q. Structural basis for the recognition of SARS-CoV-2 by full-length human ACE2. *Science* **2020**, *367*, 1444–1448. [CrossRef]
11. Hauser, B.M.; Sangesland, M.; Denis, K.J.S.; Lam, E.C.; Case, J.B.; Windsor, I.W.; Feldman, J.; Caradonna, T.M.; Kannegieter, T.; Diamond, M.S.; et al. Rationally designed immunogens enable immune focusing following SARS-CoV-2 spike imprinting. *Cell Rep.* **2022**, *38*, 110561. [CrossRef]
12. Ong, E.; Huang, X.; Pearce, R.; Zhang, Y.; He, Y. Computational design of SARS-CoV-2 spike glycoproteins to increase immunogenicity by T cell epitope engineering. *Comput. Struct. Biotechnol. J.* **2021**, *19*, 518–529. [CrossRef] [PubMed]
13. Castro, K.M.; Scheck, A.; Xiao, S.; Correia, B.E. Computational design of vaccine immunogens. *Curr. Opin. Biotechnol.* **2022**, *78*, 102821. [CrossRef] [PubMed]
14. Dhakal, A.; McKay, C.; Tanner, J.J.; Cheng, J. Artificial intelligence in the prediction of protein-ligand interactions: Recent advances and future directions. *Brief. Bioinform.* **2022**, *23*, bbab476. [CrossRef] [PubMed]
15. Earl, L.A.; Subramaniam, S. Cryo-EM of viruses and vaccine design. *Proc. Natl. Acad. Sci. USA* **2016**, *113*, 8903–8905. [CrossRef]
16. Dhakal, A.; Gyawali, R.; Cheng, J. Predicting Protein-Ligand Binding Structure Using E(n) Equivariant Graph Neural Networks. *bioRxiv* **2023**. bioRxiv:2023.02.26.23286462. [CrossRef]
17. Scheres, S.H.W. RELION: Implementation of a Bayesian approach to cryo-EM structure determination. *J. Struct. Biol.* **2012**, *180*, 519–530. [CrossRef]
18. Tang, G.; Peng, L.; Baldwin, P.R.; Mann, D.S.; Jiang, W.; Rees, I.; Ludtke, S.J. EMAN2: An extensible image processing suite for electron microscopy. *J. Struct. Biol.* **2007**, *157*, 38–46. [CrossRef]
19. Dhakal, A.; Gyawali, R.; Wang, L.; Cheng, J. Artificial intelligence in cryo-EM protein particle picking: Recent advances and remaining challenges. *Brief. Bioinform.* **2025**, *26*, bbaf011. [CrossRef]
20. Dhakal, A.; Gyawali, R.; Wang, L.; Cheng, J. CryoTransformer: A transformer model for picking protein particles from cryo-EM micrographs. *Bioinformatics* **2024**, *40*, btae109. [CrossRef]
21. He, F.; Yang, Z.; Gao, M.; Poudel, B.; Dhas, N.S.E.S.; Gyawali, R.; Dhakal, A.; Cheng, J.; Xu, D. Adapting Segment Anything Model (SAM) through Prompt-based Learning for Enhanced Protein Identification in Cryo-EM Micrographs. *arXiv* **2023**, arXiv:2311.16140. [CrossRef]
22. Bepler, T.; Morin, A.; Rapp, M.; Brasch, J.; Shapiro, L.; Noble, A.J.; Berger, B. Positive-unlabeled convolutional neural networks for particle picking in cryo-electron micrographs. *Nat. Methods* **2019**, *16*, 1153–1160. [CrossRef] [PubMed]
23. Wagner, T.; Merino, F.; Stabrin, M.; Moriya, T.; Antoni, C.; Apelbaum, A.; Hagel, P.; Sitsel, O.; Raisch, T.; Prumbaum, D.; et al. SPHIRE-crYOLO is a fast and accurate fully automated particle picker for cryo-EM. *Commun. Biol.* **2019**, *2*, 1–13. [CrossRef] [PubMed]
24. Gyawali, R.; Dhakal, A.; Wang, L.; Cheng, J. CryoSegNet: Accurate cryo-EM protein particle picking by integrating the foundational AI image segmentation model and attention-gated U-Net. *Brief. Bioinform.* **2024**, *25*, bbae282. [CrossRef] [PubMed]
25. Gyawali, R.; Dhakal, A.; Wang, L.; Jianlin, C. CryoVirusDB: An Annotated Dataset for AI-Based Virus Particle Identification in Cryo-EM Micrographs. *Zenodo* **2023**. [CrossRef]
26. Conley, M.J.; McElwee, M.; Azmi, L.; Gabrielsen, M.; Byron, O.; Goodfellow, I.G.; Bhella, D. Calicivirus VP2 forms a portal-like assembly following receptor engagement. *Nature* **2019**, *565*, 377–381. [CrossRef]
27. Castells-Graells, R.; Hesketh, E.L.; Johnson, J.E.; Ranson, N.A.; Lawson, D.M.; Lomonossoff, G.P. Decoding Virus Maturation with Cryo-EM Structures of Intermediates. EMPIAR. Available online: https://www.ebi.ac.uk/empiar/EMPIAR-11060/ (accessed on 1 December 2025).
28. Ho, K.L.; Gabrielsen, M.; Beh, P.L.; Kueh, C.L.; Thong, Q.X.; Streetley, J.; Tan, W.S.; Bhella, D. Structure of the Macrobrachium rosenbergii nodavirus: A new genus within the Nodaviridae? *PLoS Biol.* **2018**, *16*, e3000038. [CrossRef]
29. Shakeel, S.; Westerhuis, B.M.; Domanska, A.; Koning, R.I.; Matadeen, R.; Koster, A.J.; Bakker, A.Q.; Beaumont, T.; Wolthers, K.C.; Butcher, S.J. Multiple capsid-stabilizing interactions revealed in a high-resolution structure of an emerging picornavirus causing neonatal sepsis. *Nat. Commun.* **2016**, *7*, 11387. [CrossRef]
30. Flatt, J.W.; Domanska, A.; Seppälä, A.L.; Butcher, S.J. Identification of a conserved virion-stabilizing network inside the interprotomer pocket of enteroviruses. *Commun. Biol.* **2021**, *4*, 1–8. [CrossRef]
31. Chandler-Bostock, R.; Mata, C.P.; Bingham, R.J.; Dykeman, E.C.; Meng, B.; Tuthill, T.J.; Rowlands, D.J.; Ranson, N.A.; Twarock, R.; Stockley, P.G. Assembly of infectious enteroviruses depends on multiple, conserved genomic RNA-coat protein contacts. *PLoS Pathog.* **2020**, *16*, e1009146. [CrossRef]
32. Thompson, R.F.; Iadanza, M.G.; Hesketh, E.L.; Rawson, S.; Ranson, N.A. Collection, pre-processing and on-the-fly analysis of data for high-resolution, single-particle cryo-electron microscopy. *Nat. Protoc.* **2019**, *14*, 100–118. [CrossRef]

 https://doi.org/10.3390/fluids10110297

33. Castells-Graells, R.; Ribeiro, J.R.S.; Domitrovic, T.; Hesketh, E.L.; Scarff, C.A.; Johnson, J.E.; Ranson, N.A.; Lawson, D.M.; Lomonossoff, G.P. Plant-expressed virus-like particles reveal the intricate maturation process of a eukaryotic virus. *Commun. Biol.* **2021**, *4*, 619. [CrossRef] [PubMed]
34. Iudin, A.; Korir, P.K.; Somasundharam, S.; Weyand, S.; Cattavitello, C.; Fonseca, N.; Salih, O.; Kleywegt, G.J.; Patwardhan, A. EMPIAR: The Electron Microscopy Public Image Archive. *Nucleic Acids Res.* **2023**, *51*, D1503–D1511. [CrossRef] [PubMed]
35. Punjani, A.; Rubinstein, J.L.; Fleet, D.J.; Brubaker, M.A. cryoSPARC: Algorithms for rapid unsupervised cryo-EM structure determination. *Nat. Methods* **2017**, *14*, 290–296. [CrossRef] [PubMed]
36. Pettersen, E.F.; Goddard, T.D.; Huang, C.C.; Meng, E.C.; Couch, G.S.; Croll, T.I.; Morris, J.H.; Ferrin, T.E. UCSF ChimeraX: Structure visualization for researchers, educators, and developers. *Protein Sci.* **2021**, *30*, 70–82. [CrossRef]
37. De Castro, E.; Hulo, C.; Masson, P.; Auchincloss, A.; Bridge, A.; Le Mercier, P. ViralZone 2024 provides higher-resolution images and advanced virus-specific resources. *Nucleic Acids Res.* **2024**, *52*, D817–D821. [CrossRef]
38. Baldwin, P.R.; Penczek, P.A. The Transform Class in SPARX and EMAN2. *J. Struct. Biol.* **2007**, *157*, 250–261. [CrossRef]
39. Bepler, T.; Kelley, K.; Noble, A.J.; Berger, B. Topaz-Denoise: General deep denoising models for cryoEM and cryoET. *Nat. Commun.* **2020**, *11*, 5208. [CrossRef]
40. Lin, T.Y.; Maire, M.; Belongie, S.; Hays, J.; Perona, P.; Ramanan, D.; Dollár, P.; Zitnick, C.L. Microsoft COCO: Common Objects in Context. In *Computer Vision—ECCV 2014*; Fleet, D., Pajdla, T., Schiele, B., Tuytelaars, T., Eds.; Springer International Publishing: Berlin/Heidelberg, Germany, 2014; pp. 740–755.

 https://doi.org/10.3390/fluids10110297

MDPI AG
Grosspeteranlage 5
4052 Basel
Switzerland
Tel.: +41 61 683 77 34

Viruses Editorial Office
E-mail: viruses@mdpi.com
www.mdpi.com/journal/viruses

www.ingramcontent.com/pod-product-compliance
Lightning Source LLC
LaVergne TN
LVHW070449061126
851858LV00011B/194